Fortschritte der Chemie organischer Naturstoffe

Progress in the Chemistry of Organic Natural Products

47

Founded by L. Zechmeister
Edited by W. Herz, H. Grisebach, G. W. Kirby,
and Ch. Tamm

Authors:
S. Elson, I. Howe, M. Jarman,
P. G. McDougal, Judith Polonsky,
N. R. Schmuff, R. Southgate

Springer-Verlag Wien GmbH

Dr. W. Herz, Professor of Chemistry, Department of Chemistry,
The Florida State University, Tallahassee, Florida, U.S.A.

Prof. Dr. H. Grisebach, Biologisches Institut II, Lehrstuhl für Biochemie der Pflanzen,
Albert-Ludwigs-Universität, Freiburg i. Br., Federal Republic of Germany

G. W. Kirby, Sc. D., Regius Professor of Chemistry, Chemistry Department,
The University, Glasgow, Scotland

Prof. Dr. Ch. Tamm, Institut für Organische Chemie der Universität Basel,
Basel, Switzerland

With 16 Figures

© 1985 by Springer-Verlag Wien
Softcover reprint of the hardcover 1st edition 1985

ISBN 978-3-7091-8792-0   ISBN 978-3-7091-8790-6 (eBook)
DOI 10.1007/978-3-7091-8790-6

Contents

List of Contributors

ELSON, Dr. S., Beecham Pharmaceuticals, Research Division, Brockham Park, Betchworth, Surrey, U.K.

HOWE, Dr. I., Drug Metabolism Team, Section of Drug Development, Cancer Research Campaign Laboratory, Institute of Cancer Research, Clifton Avenue, Sutton, Surrey, SM2 5PX, U.K.

JARMAN, Dr. M., Drug Metabolism Team, Section of Drug Development, Cancer Research Campaign Laboratory, Institute of Cancer Research, Clifton Avenue, Sutton, Surrey, SM2 5PX, U.K.

McDOUGAL, Professor P. G., Department of Chemistry, Georgia Institute of Technology, Atlanta, GA 30332, U.S.A.

POLONSKY, Dr. JUDITH, Institut de Chimie des Substances Naturelles, C.N.R.S., F-91190 Gif-sur-Yvette, France.

SCHMUFF, Dr. N. R., Questel, Inc., 1625 Eye Street, N.W., Washington, DC 20006, U.S.A.

SOUTHGATE, Dr. R., Beecham Pharmaceuticals, Research Division, Brockham Park, Betchworth, Surrey, U.K.

Naturally Occurring β-Lactams

By R. Southgate and S. Elson, Beecham Pharmaceuticals, Research Division, Brockham Park, Betchworth, Surrey, U.K.

With 8 Figures

Contents

I. Introduction

Since the discovery of penicillin, the vast majority of new natural products possessing the β-lactam ring structure have fallen into that family of compounds known as β-lactam antibiotics. At present the only natural products possessing a β-lactam and not normally associated with this group are the pachystermines (*1*), the wild-fire toxin (*2*), and a related anti-metabolite (*3*). The bleomycins and phleomycins (*4*) were also originally thought to possess a β-lactam unit, but this is now known not to be the case (*5*).

In the period up to 1970 β-lactam research was mainly concerned with the penicillin and cephalosporin groups of compounds (*6*). Since that time a major extension in the area has taken place with the discovery of several new natural β-lactam structures. In many cases these have features quite different from those present in the penicillins and cephalosporins. This sudden acceleration in discovering new structures has undoubtedly resulted from the design and development of new and sensitive screening procedures. Furthermore new expertise in handling somewhat unstable molecules together with the use of more sophisticated physical, analytical and spectroscopic methods has led to a more rapid determination of structural variations.

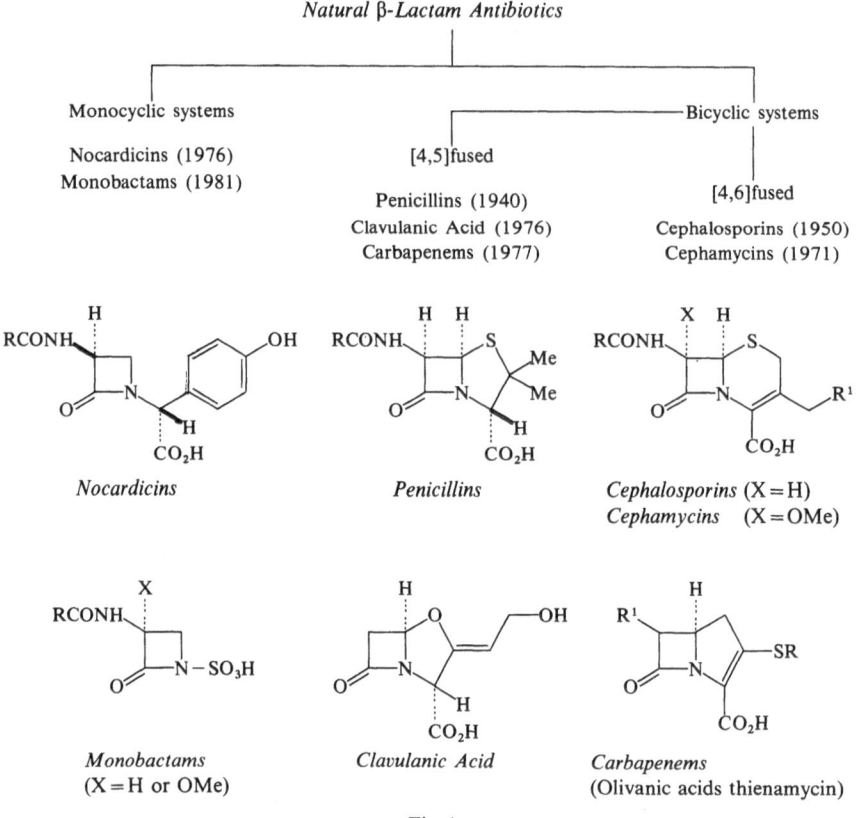

Natural β-Lactam Antibiotics

Monocyclic systems

Nocardicins (1976)
Monobactams (1981)

Bicyclic systems

[4,5]fused

Penicillins (1940)
Clavulanic Acid (1976)
Carbapenems (1977)

[4,6]fused

Cephalosporins (1950)
Cephamycins (1971)

Nocardicins

Penicillins

Cephalosporins (X = H)
Cephamycins   (X = OMe)

Monobactams
(X = H or OMe)

Clavulanic Acid

Carbapenems
(Olivanic acids thienamycin)

Fig. 1

In addition to the discovery of these new natural products, a substantial effort has been devoted to the synthesis of numerous β-lactams of varying structures over the last decade. Many of these have not been found in nature and therefore do not fall within the scope of this review, although a few of the more significant structures of biological interest have been included.

The extent and nature of the various β-lactams classified as antibiotics are shown in Fig. 1. Rather than review these in the more traditional historical way starting with the penicillins, a more systematic approach has been adopted which progresses from monocyclic structures through to bicyclic compounds of varying forms.

II. The Nocardicins

In 1976 the Fujsawa Company reported the isolation and characterisation of the new monocyclic β-lactam nocardicin A (**1**) from the fermentation broth of a strain of *Nocardia uniformis* (*7*). The antibiotic was

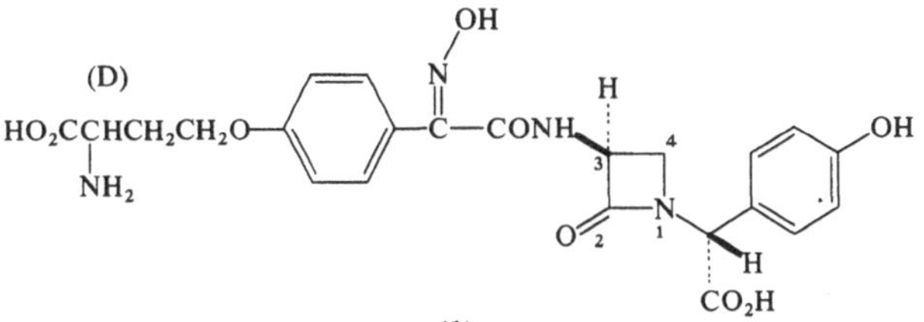

(**1**)

detected by a screening procedure using mutant strains of *Escherichia coli* supersensitive to β-lactam antibiotics. Nocardicin A shows a modest level of antibacterial activity *in vitro* against Gram negative bacteria. It was also revealed that several related compounds known as nocardicins B to G had been isolated (*8, 9*) and shown to have the structures depicted in Table 1. The structural determination of the nocardicins was based on a number of chemical interconversions together with spectroscopic evidence. The final confirmation of the full structure of nocardicin A was obtained from acid degradation experiments which gave (**2**), (**3**) and (**4**). The latter was further degraded to the L-diaminopropionic acid (**5**) and D-*p*-hydroxyphenylglycine (**6**).

Table 1. *Structures of Naturally Occurring Nocardicins (8, 9)*

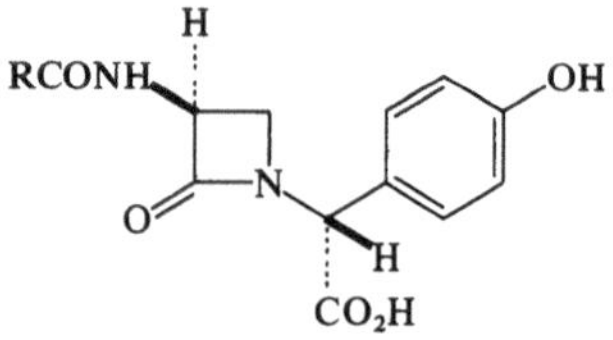

Nocardicin	R
A	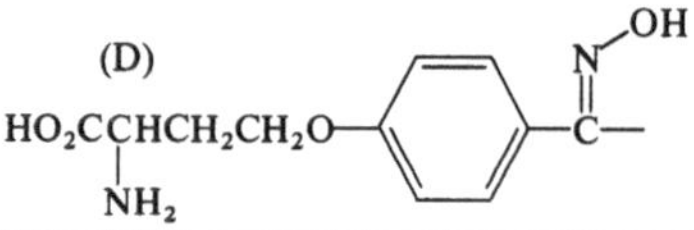 (D) $HO_2CCHCH_2CH_2O-$ (with NH_2) — phenyl ring — C with $=N-OH$
B	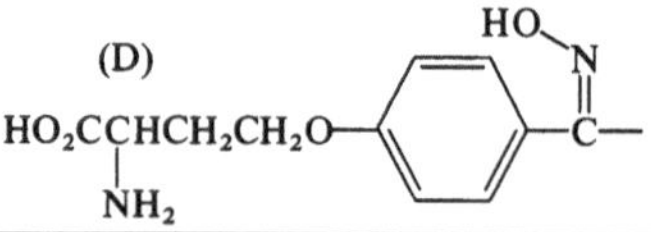 (D) $HO_2CCHCH_2CH_2O-$ (with NH_2) — phenyl ring — C with $=N$ (HO)
C	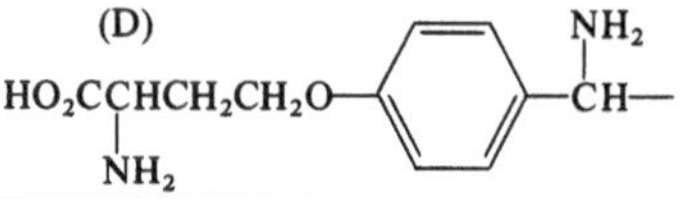 (D) $HO_2CCHCH_2CH_2O-$ (with NH_2) — phenyl ring — $CH-$ (with NH_2)
D	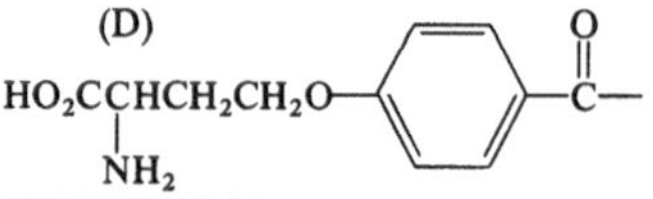 (D) $HO_2CCHCH_2CH_2O-$ (with NH_2) — phenyl ring — C with $=O$
E	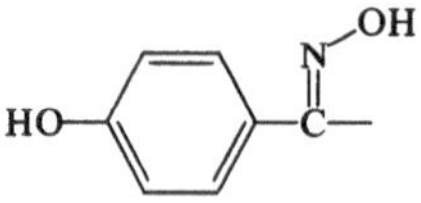 $HO-$ phenyl ring — C with $=N-OH$
F	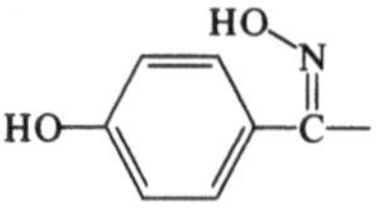 $HO-$ phenyl ring — C with $=N$ (HO)
G	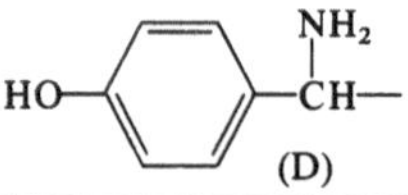 $HO-$ phenyl ring — $CH-$ (with NH_2) (D)

$HO_2CCHCH_2CH_2O-$ (with NH_2) — phenyl ring — $C(=X)-CO_2H$

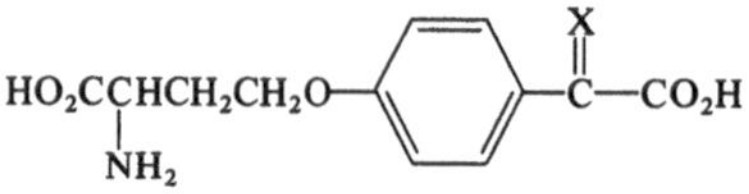

(2) X=NOH
(3) X=O

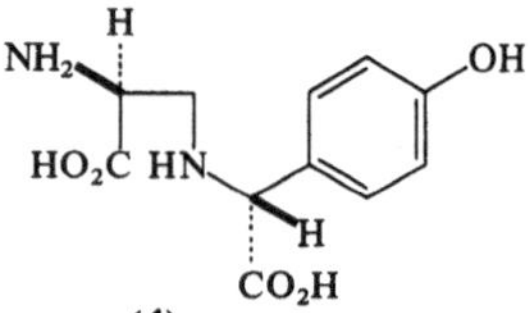

structure **(4)**: NH_2, H, HO_2C HN—, azetidinone ring, phenyl with OH, H, CO_2H

References, pp. 89—106

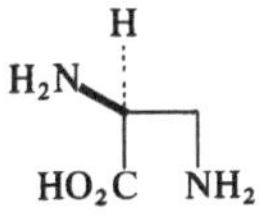

(5)

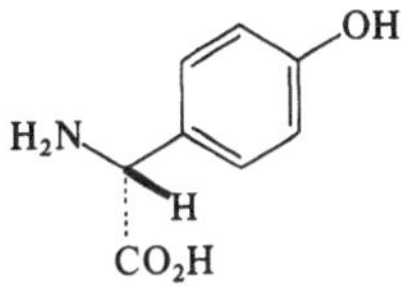

(6)

The nocardicin nucleus 3-aminonocardicinic acid (3-ANA) (7) has not been found in nature, but can be prepared by deacylation of nocardicin C using microbial amidases (10). The first chemical method leading to (7) involved acid treatment of the bisthiourea derivative (8) of nocardicin C (11). A more recent method (12) makes use of the reaction of the oxime grouping of nocardicin A with di-t-butyl dicarbonate. This results in (9) which on treatment with trifluoroacetic acid gives (7). Further confirmation of the structure and stereochemistry of the nocardicins was the identification of the acylamino-derivative (10) derived from 3-ANA, with a compound obtained by partial synthesis from penicillin G (11). Another semi-synthetic approach (13) to the nocardicins from penicillin is by way of the thiazoline (11), the final step being Raney nickel desulphurisation of (12).

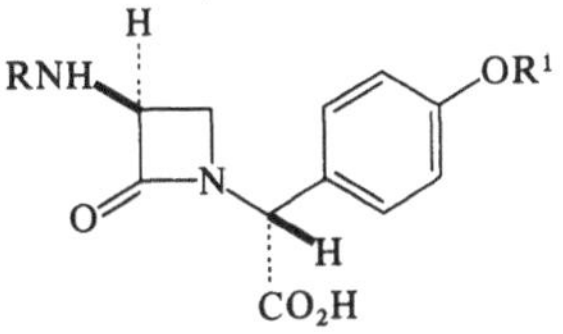

(7) R=R^1=H (3-ANA)
(9) R=R^1=CO$_2$Bu^t

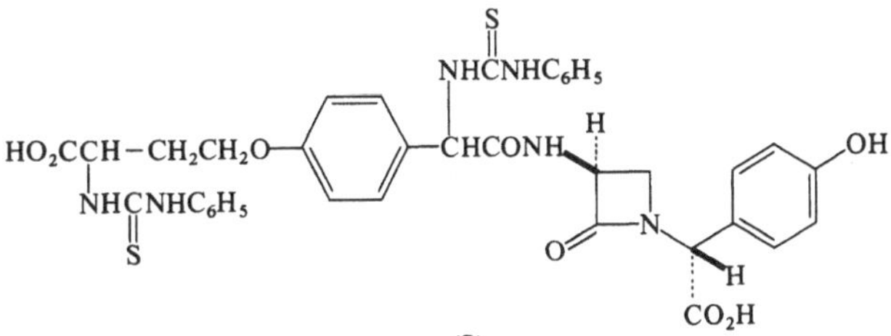

(8)

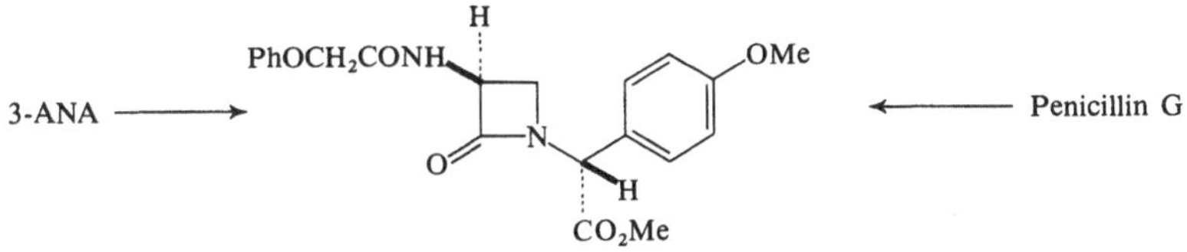

3-ANA $\longrightarrow$

$\longleftarrow$ Penicillin G

(10)

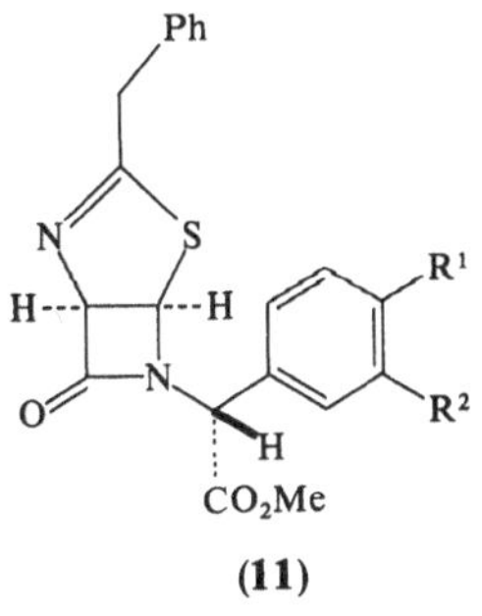

(11)

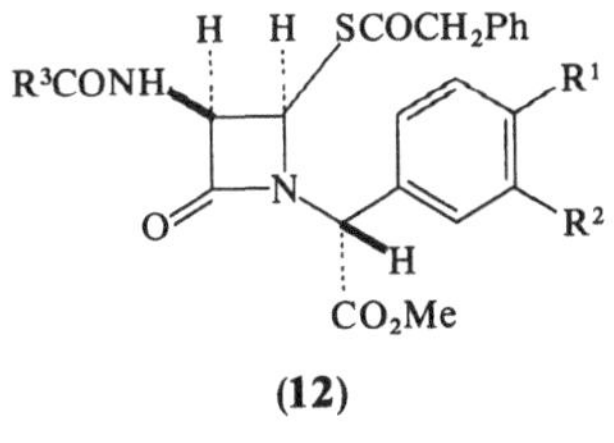

(12)

The total synthesis of nocardicin A and its analogues has received considerable attention. One of the first approaches (*11, 14*) made use of the classical keten-imine reaction for construction of the β-lactam ring. Reaction of phthalimido-acetyl chloride with the thioimidate (**13**) in the presence of triethylamine gave the β-lactam (**14**). Removal of the sulphur grouping and deprotection afforded 3-ANA, which could be acylated with the appropriate side-chain acid to give nocardicins D, E and G. Reaction of nocardicin D with hydroxylamine produced nocardicin A. Alternative routes (*15, 16*) utilise triazines such as (**15**) to provide a source of the imines of type (**16**). This sequence also removes the necessity of a desulphurisation step.

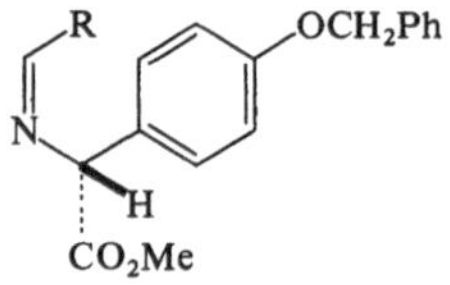

(**13**) R = SMe
(**16**) R = H

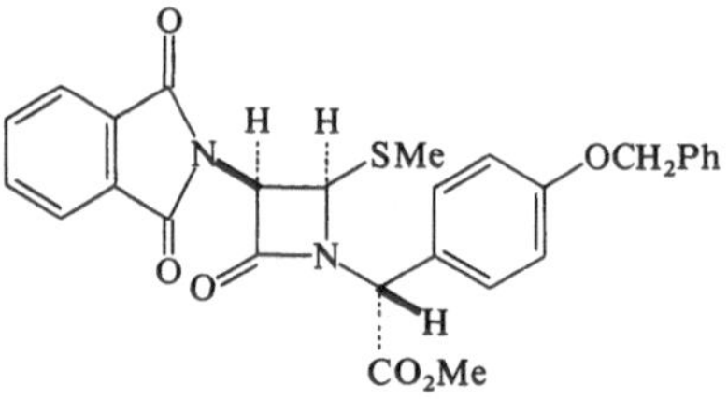

(**14**)

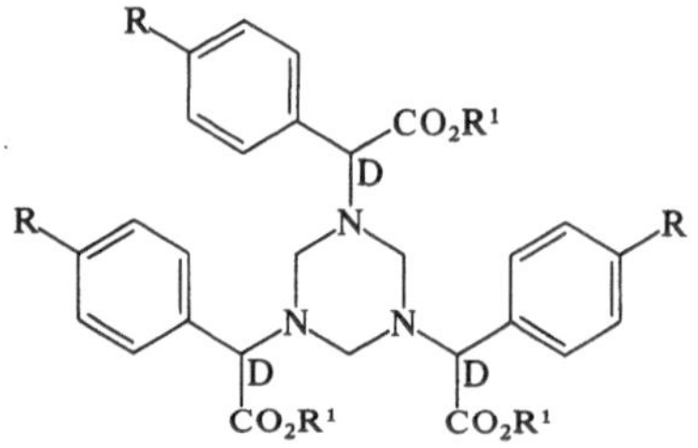

(**15**)

In 1978 the Lilly group reported (*17*) a synthesis of nocardicin A starting from the thiazolidine (**17**) derived from L-cysteine. Formation of the β-lactam (**19**) was by intramolecular cyclisation of the chloride (**18**).

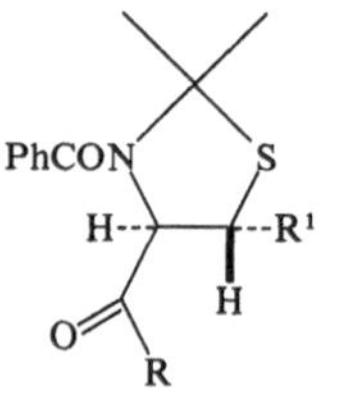

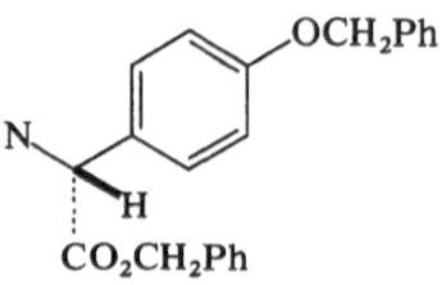

(**17**) R = OH; R¹ = H

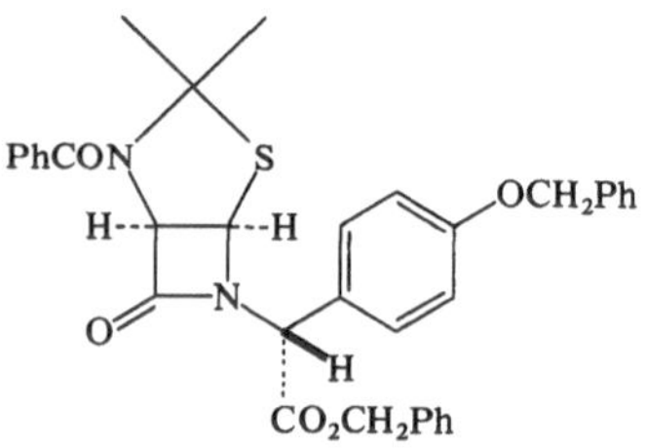

(**18**) R = HN⟨...⟩   ; R¹ = Cl

(**19**)

Elaboration of (19) involved conversion to an oxazoline, ring-opening and chlorination at the 4-position followed by reduction to eventually give the benzyl ester of 3-ANA benzyl ether. Subsequent acylation and deprotection led to nocardicin A.

A more direct synthesis makes use of the L-serine hydroxamate (20) (18). Cyclisation of chloride (21) followed by removal of the N-substituent afforded the β-lactam (22). Progression to (23) necessitated introduction of the protected phenylglycyl residue either by alkylation or diazo-insertion. Deprotection then led to 3-ANA in good overall yield.

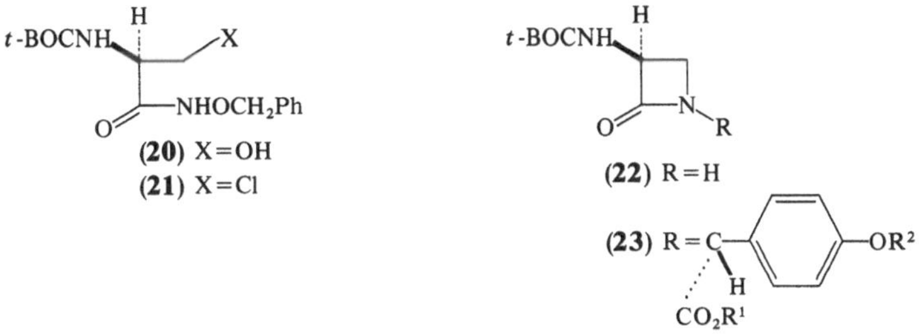

WASSERMANN has reported (19–21) the synthesis of (±)-3-ANA by way of the β-lactam (25). Various methods were used to prepare (25), such as cyclopropane ring expansion of (24), cyclisation of (26) or acylation of the azetidine carboxylate (27). In another approach (22) the α-methylene β-lactam (28) has also been elaborated to 3-ANA via the use of (29).

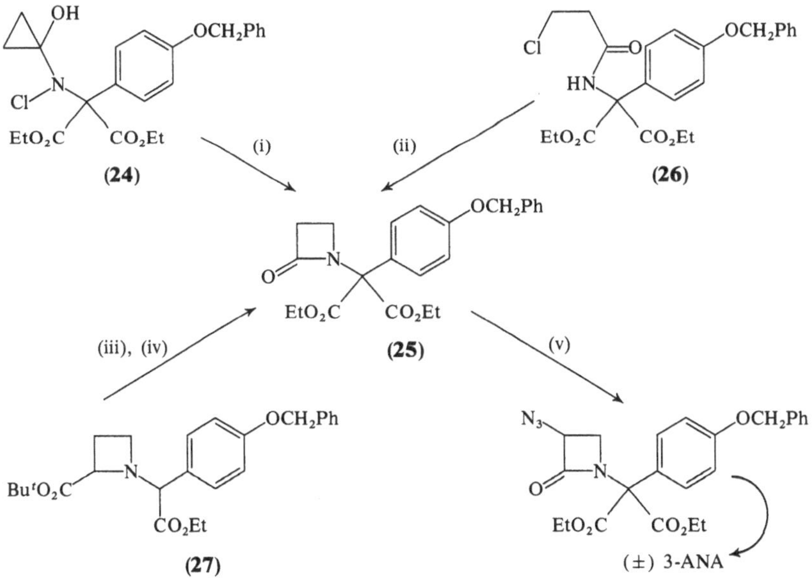

Reagents: (i) AgNO₃; (ii) NaH; (iii) (COCl)₂-HClO₄; (iv) ArCO₃H + Pyridine; (v) Base-p-toluenesulphonyl azide.

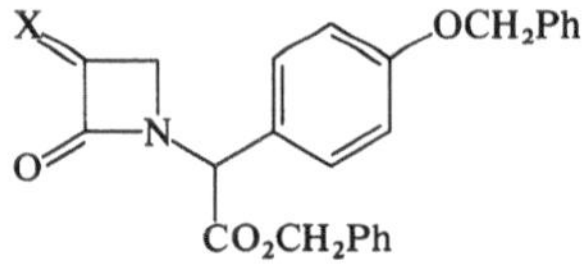

(28) X=CH₂
(29) X=O

III. The Monobactams

In 1981 workers of the Takeda Company described (*23*) the isolation of the monocyclic β-lactams sulfazecin (*24*) and isosulfazecin (*25*) from various species of Pseudomonas. This was the first reported case of β-lactam antibiotics being produced by bacteria. Shortly afterwards chemists at the Squibb Research Institute also reported a number of similar monocyclic structures of bacterial origin (*26*). All these compounds were characterised by the presence of a sulphonic acid grouping in the *N*(1)-position of the β-lactam ring as in (**30**). Generically called the monobactams, the known structural types are shown in Table 2.

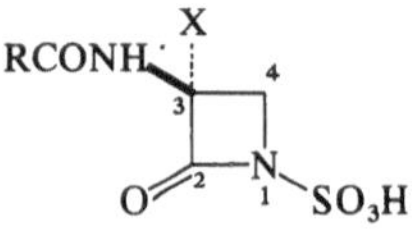

$X = OMe$ or H

(**30**)

Table 2. *Structures of the Monobactams*

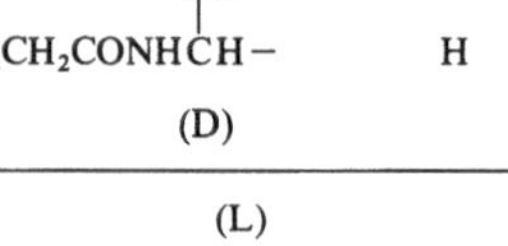

Metabolite	R		M	X	References
Sulfazecin (SQ 26,445, M = Na)	$\underset{(D)}{HO_2C}\overset{NH_2}{\underset{|}{C}}HCH_2CH_2CONH\underset{(D)}{\overset{Me}{\underset{|}{C}}}H-$		H	OMe	(*23, 24, 26, 27*)
Isosulfazecin	$\underset{(D)}{HO_2C}\overset{NH_2}{\underset{|}{C}}HCH_2CH_2CONH-\underset{Me}{\underset{|}{C}}H-$ (L)		H	OMe	(*23, 25*)
SQ 26,180	Me		K	OMe	(*26, 27*)

Table 2 *(continued)*

Metabolite	R	M	X	References
SQ 26,823	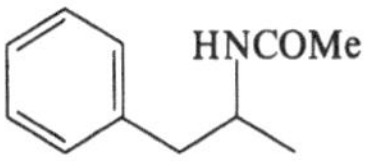 (phenyl)CH₂CH(CH₃)–HNCOMe	Na	OMe	(*26, 28*)
SQ 26,975	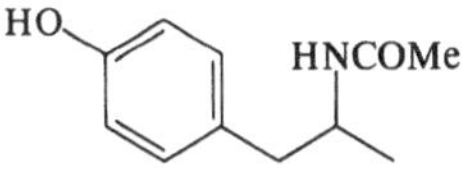 HO–(C₆H₄)–CH₂CH(CH₃)–HNCOMe	K	OMe	(*26, 28*)
SQ 26,700	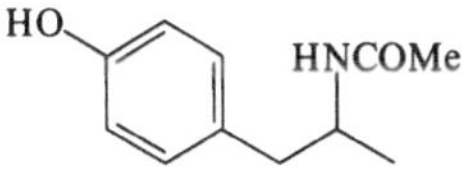 HO–(C₆H₄)–CH₂CH(CH₃)–HNCOMe	K	H	(*26, 28*)
SQ 26,970	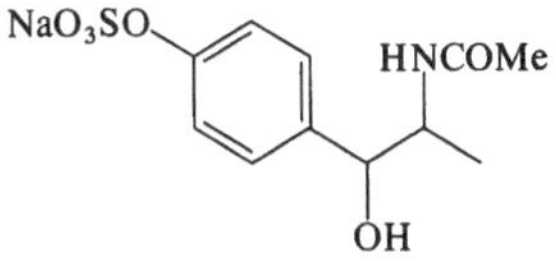 NaO₃SO–(C₆H₄)–CH(OH)CH(CH₃)–HNCOMe	Na	OMe	(*26, 28*)
SQ 26,812	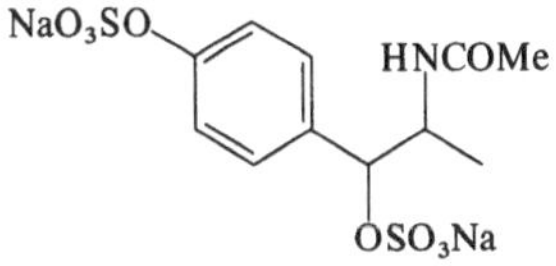 NaO₃SO–(C₆H₄)–CH(OSO₃Na)CH(CH₃)–HNCOMe	Na	OMe	(*26, 28*)
SQ 28,332	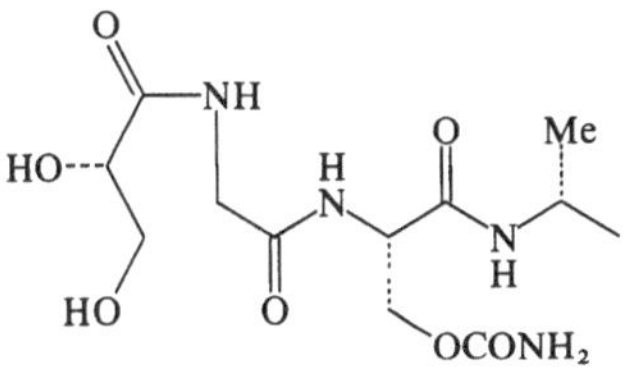 (structure: HO—, HO— substituted chain with —NH, C=O, N–H, Me, OCONH₂)	Na	H	(*283, 284*)

The structure of the simplest member of the series SQ 26,180 was deduced fairly readily from its spectroscopic properties (*27*). Confirmation and assignment of configuration were made by synthesis from the methoxylated cephalosporin (**31**) of known configuration, using the sequence (**31**)→(**32**)→(**33**)→(**34**). In the case of SQ 26,445 (sulfazecin) the side-chain was identified by acid hydrolysis which gave D-glutamic acid and D-alanine; milder hydrolysis gave γ-D-glutamyl-D-alanyl amide and 2-oxo-3-sulphoaminopropionic acid. The structures of other members of the family were derived by a comparison of spectroscopic properties and degradation to determine the side chain components.

Isolation of the monobactams has resulted in the synthesis of a large number of acylamino derivatives of the 3-aminomonobactamic acid (3-AMA) (**35**). Deacylation of the natural products was not satisfactory;

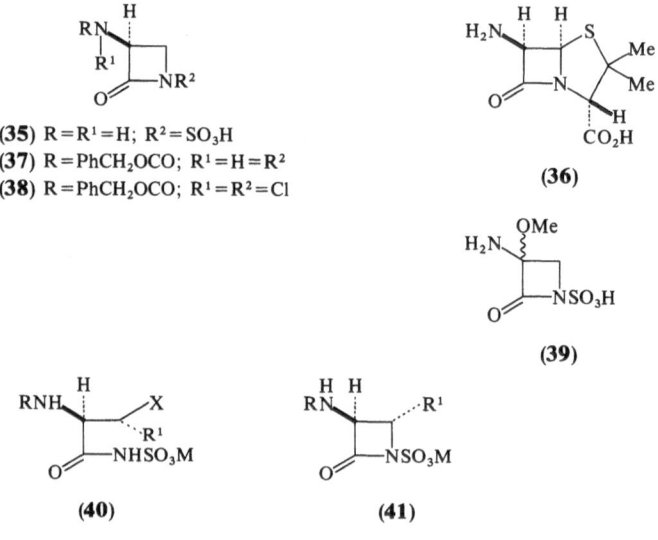

Reagents: (i) Raney Ni;   (ii) Pb(OAc)$_4$-Cu(OAc)$_2$;   (iii) Base hydrolysis;   (iv) DMF-SO$_3$.

therefore (35) was prepared (29) by degradation of 6-aminopenicillanic acid (6-APA) (36) or by total synthesis (30). The penicillin nucleus (36) was readily converted by known methods to (37). Sulphonation and hydrogenation then gave (35). Methoxylation was by way of the dichloride (38) and lithium methoxide, the N(1)-chloro grouping being subsequently removed by reduction. The final product (39) in this case was racemic, although acylation with optically active amino-acids on the C(3)-amino group could lead to diastereoisomer separation.

(35) R=R^1=H; R^2=SO$_3$H
(37) R=PhCH$_2$OCO; R^1=H=R^2
(38) R=PhCH$_2$OCO; R^1=R^2=Cl

(36)

(39)

(40)

(41)

Total synthesis was achieved by base catalysed cyclisation of acyclic acylsulfamates (40) to (41). Thus the serine derived mesylate (40; X = OMes, R^1 = H) gave (41; R^1 = H) and the L-threonine derivative (40; X = OMes, R^1 = Me) the 4-methyl product (41; R^1 = Me). It was from this series of compounds that the highly bioactive synthetic monobactam SQ 26,776 (42) emerged.

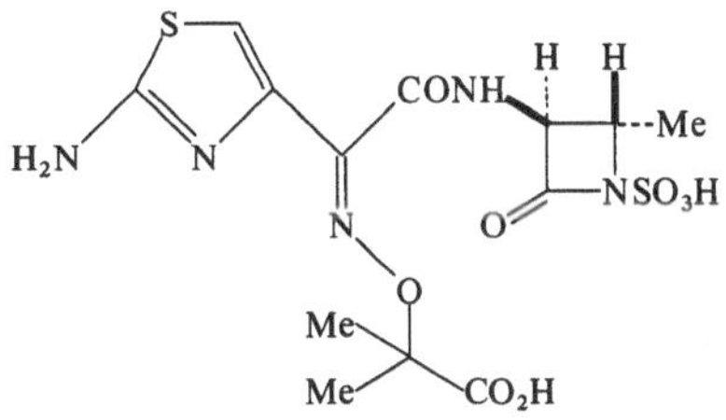

(42) SQ 26,776

IV. The Penicillins

Undoubtedly the best known family of β-lactam antibiotics comprises the penicillins, the discovery, history, chemistry and biology of which have been well reviewed (6, 32, 33, 34). All members of the family possess the fused β-lactam-thiazolidine ring structure (43) and differ only in the nature of the acylamino-substituent (R) present at the C(6)-position of the parent nucleus 6-β-aminopenicillanic acid (44). The naturally occurring penicillins are listed in Table 3. The predominant source of these substances are the true fungi, while the quantity and quality of the various penicillins produced depends on the ingredients of the fermentation medium. Addition of side-chain precursors can lead to a variety of other penicillins. Modern commercial production is almost solely concerned with Penicillin G or V using deep culture methods and high yielding mutants of *Penicillium chrysogenum*.

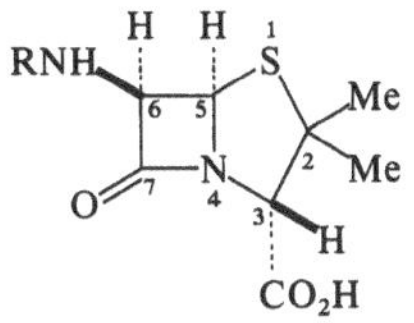

(43) R = R^{1}CO
(44) R = H

Table 3. *Natural Penicillins of Unaided Fermentations (33)*

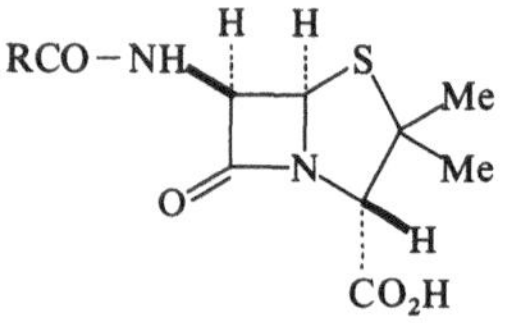

Penicillin	Side-chain (R)
F (pentenylpenicillin)	$CH_3CH_2CH=CHCH_2-$
Dihydro F (pentylpenicillin)	$CH_3(CH_2)_4-$
K (Heptylpenicillin)	$CH_3(CH_2)_6-$
G (Benzylpenicillin)	$PhCH_2$
X (p-Hydroxybenzylpenicillin)	$p\text{-HO}-C_6H_4CH_2-$
N (D-4-Amino-4-carboxybutylpenicillin)	$HO_2C-CH{\overset{NH_2}{\underset{(D)}{\diagdown}}}(CH_2)_3-$
Iso-N (L-4-Amino-4-carboxybutylpenicillin)	$HO_2C-CH{\overset{NH_2}{\underset{(L)}{\diagdown}}}(CH_2)_3-$

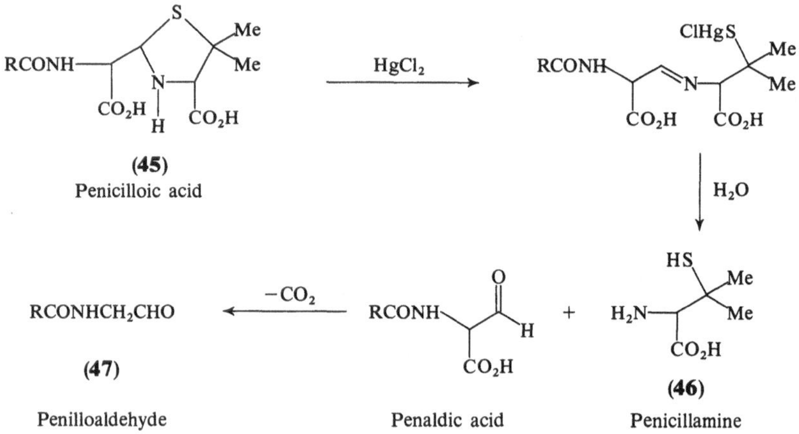

(45)
Penicilloic acid

(47)
Penilloaldehyde

(46)
Penicillamine

Penaldic acid

Elucidation of the structure of the penicillin nucleus involved extensive degradation studies and X-ray analysis. Reaction of a penicillin under hydrolytic conditions followed by further degradation with mercuric chloride was shown to give D-penicillamine (46) and the penilloaldehyde (47). From this evidence it was concluded that the primary hydrolysis product was the penicilloic acid (**45**), the penicillin being a dehydration product of this acid. The choice of structure therefore lay between the oxazolone (**48**) and the β-lactam structure (**49**). Evidence for the latter came with the isolation of the desthiobenzylpenicillin (**50**) as one of the products from the Raney nickel desulphurisation of benzylpenicillin (*35*). Final confirmation of the structure and stereochemistry of the penicillin molecule was provided by X-ray crystallographic studies on various penicillin salts carried out at Oxford (*36*).

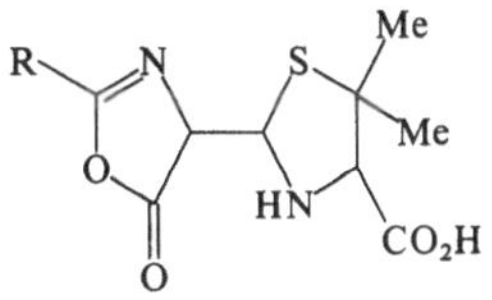

(**48**) 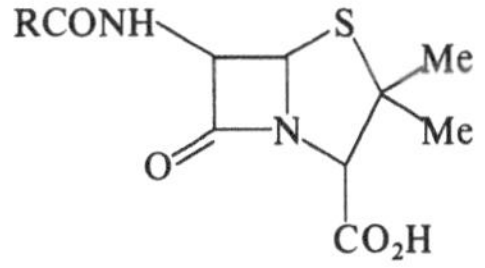   (**49**)

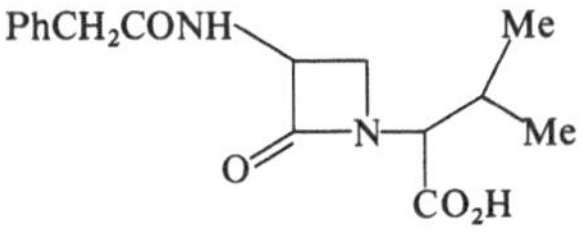

(**50**)

The preparation of penicillins by fermentation is limited to relatively few compounds with side chains derived from mono-substituted acetic acids. The isolation of 6-aminopenicillanic acid (**44**) either by direct fermentation (*37*) or enzymic deacylation of penicillin G (*38*) radically altered this situation. Reacylation of the nucleus with a variety of side chain acids has lead to the preparation of several thousands of new "semi-synthetic" penicillins, many with improved chemotherapeutic properties. In recent years some attention has been directed towards chemical removal of the penicillin side chain, the first successful method being that described by WEISSENBURGER (*39*) in 1970. A solution of the silyl ester of benzyl-penicillin was converted to the imidate (**51**); aqueous work-up then provided 6-APA (**44**).

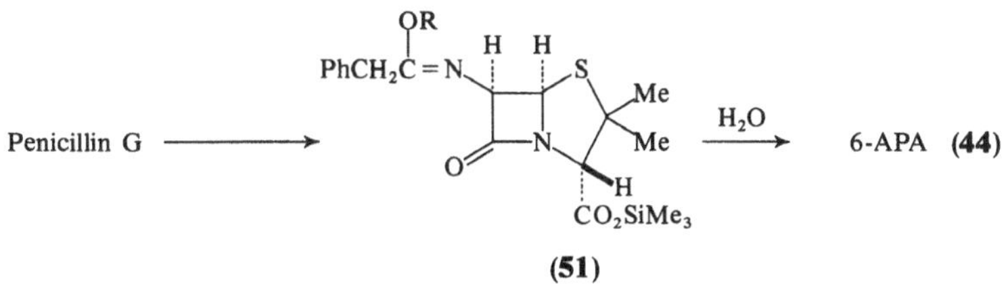

(**51**)

Prior to the complete structural elucidation of penicillin some effort was directed towards the preparation of the erroneous oxazolone-thiazolidine structure (48) using (52) and penicillamine (53). Surprisingly trace amounts of bioactive penicillins corresponding to the fused β-lactam structure (49) were obtained (40). The first cyclisation of a penicilloate derivative was described by Sheehan (41) using the acid chloride of the phthalimido derivative (54) to give (55).

Unfortunately such cyclisation methods were not appropriate to acylamino derivatives since formation of the five-membered oxazolone ring was preferred. Further work by Sheehan using the newly introduced N,N¹-dicyclohexylcarbodiimide (DCC) as cyclisation reagent for the formation of the β-lactam ring resulted in the first rational synthesis of a natural penicillin (potassium penicillin V) in 1957 (42). This was achieved by the sequence (56)→(57)→(58)→(59). Subsequent reports (43) outlined the application of such methods to a general synthesis of penicillins.

Later synthetic approaches by Bose (44) utilised a ketene-imine cycloaddition reaction for construction of the β-lactam ring. Correction of the stereochemistry is required, however, since the thermodynamically more favoured trans-configuration of β-lactam protons is obtained. A method of epimerisation was developed by the Merck group (45) and is illustrated in their total synthesis of sodium penicillin G (64). Using the optically active thiazoline (60), azidoacetyl chloride added in the presence of triethylamine gave the 6α-epimer (61) in 98% yield. A deprotonation-protonation sequence using (62) provided a 2:1 ratio of cis to trans products, leading to the 6β-amino-derivative (63a) after regeneration of the free base and separation by crystallisation. Subsequent acylation and deprotection afforded the penicillin (64).

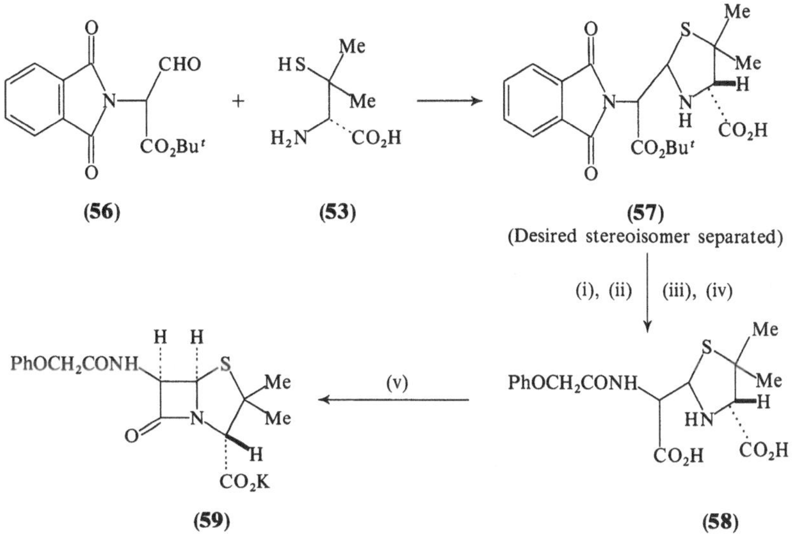

(56) + **(53)** → **(57)**
(Desired stereoisomer separated)

(i), (ii) | (iii), (iv)

(59) ← (v) ← **(58)**

Reagents: (i) N₂H₄-HCl; (ii) PhOCH₂COCl-Et₃N; (iii) HCl, pyridine; (iv) KOH (1 equivalent); (v) DCC-aqueous dioxan.

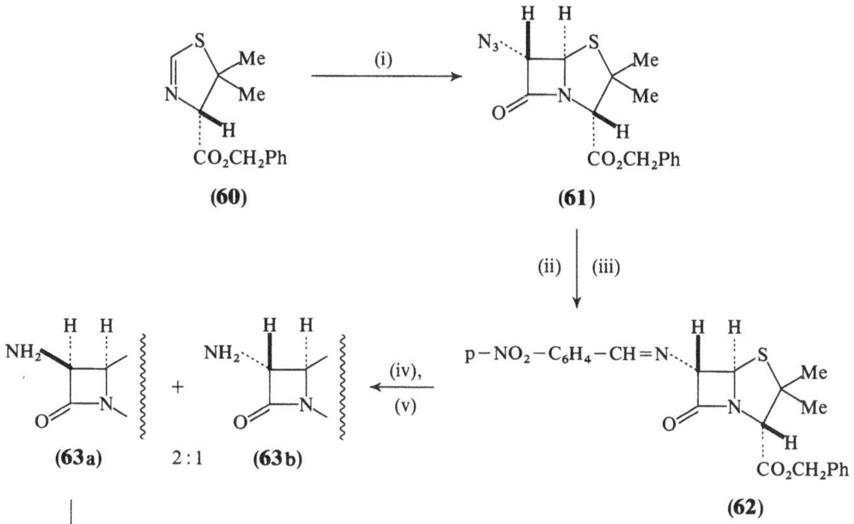

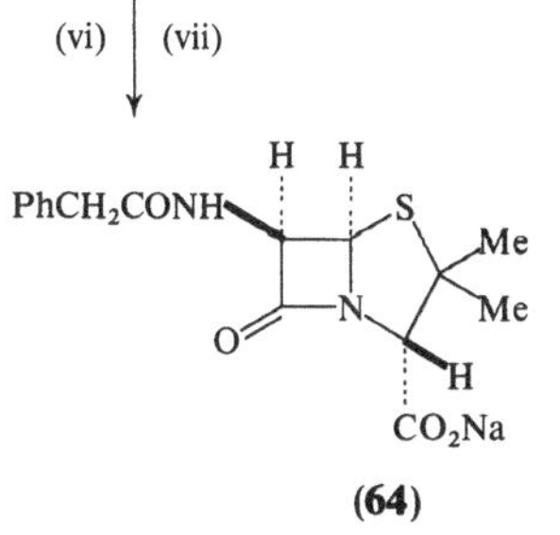

(60) → (i) → **(61)**

(ii) | (iii)

(63a) 2:1 **(63b)** ← (iv), (v) ← **(62)**

(vi) | (vii)

(64)

Reagents: (i) $N_3CH_2COCl\text{-}Et_3N$;
(ii) H_2/Pt_2O;
(iii) $p\text{-}NO_2C_6H_4CHO$;
(iv) PhLi, DMF: AcOH, H_2O;
(v) DNPH, TsOH;
(vi) $PhCH_2COCl$-pyridine;
(vii) $H_2\text{-}10\%$ Pd/C; $NaHCO_3$

Although other approaches to modified penicillin structures, together with many other aspects of the chemistry of the natural substrate, have been described (46), only one stereospecific synthesis of the penicillin molecule has so far emerged (47). Condensation of the D-isodehydrovaline methyl ester (65) with the cysteine-derived thiazolidine acid (66) to give (67) was followed by functionalisation α to the sulphur atom to give the chloride (68). Smooth closure to the β-lactam (69) was achieved using sodium hydride. The thiazolidine β-lactam (69) was converted by a sequence of five steps to a diastereoisomeric mixture of the sulphoxides (70), from which a single stereoisomeric sulphenic acid (71) was generated on thermolysis. Cyclisation then occurred to give the penicillin sulphoxide (72), which on deoxygenation provided the penicillin methylester (73). Modification (48) to this sequence later led to a shorter stereocontrolled entry into the penam ring system, but only to compounds having a 2β-chloromethyl or 2β-bromomethyl substituent (74).

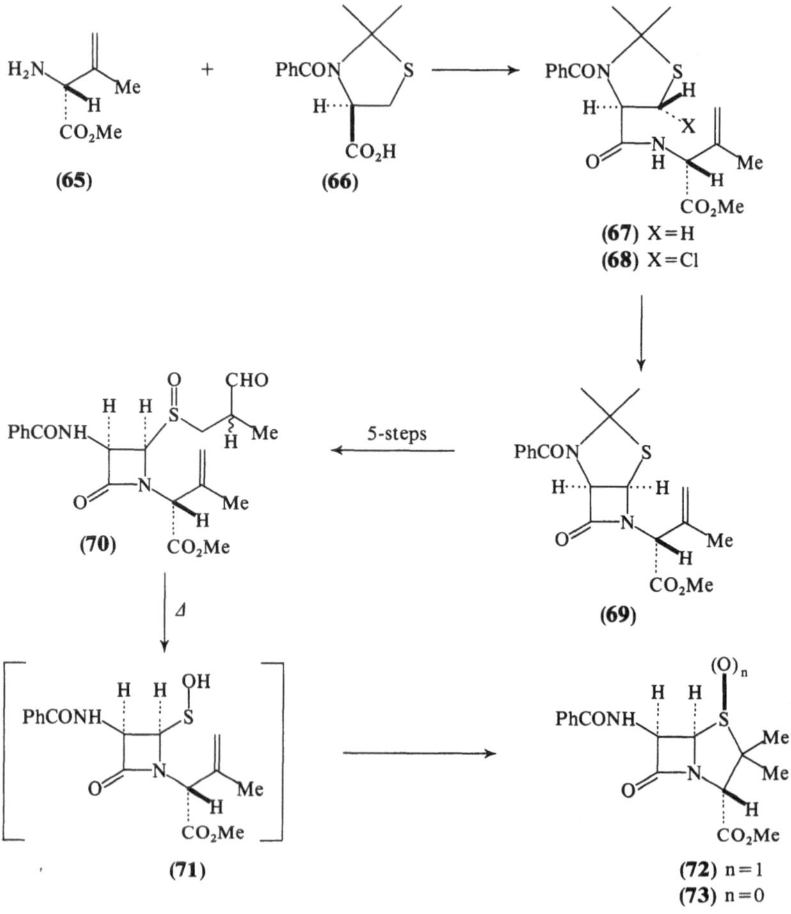

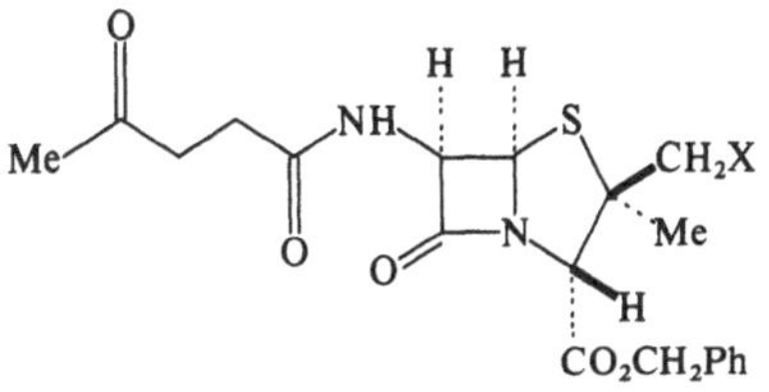

(74) X=Cl or Br

V. Clavulanic Acid

The antibacterial activity associated with β-lactam antibiotics is dependent on the presence of the intact β-lactam ring. Many bacteria possess the ability to produce β-lactamase enzymes capable of hydrolysing the β-lactam leading to inactivation of and bacterial resistance to the antibiotic. Co-administration of a β-lactamase inhibitor with antibiotics sensitive to such enzymes affords one method of overcoming such resistance. During the course of screening (49) for natural inhibitors of β-lactamases the novel bicyclic β-lactam clavulanic acid (75) was isolated from *Streptomyces clavuligerus*. The structure determination (50, 51) of this metabolite, based on spectroscopic and X-ray analysis, revealed a quite distinct difference from other known fused β-lactam structures. Unlike the penicillins, clavulanic acid possesses no acylamino side-chain while its β-lactam is fused to an oxazolidine ring. This 7-oxo-4-oxa-1-azabicyclo[3.2.0]heptane nucleus is more often referred to as the 1-oxadethiapenam or clavam ring system using the numbering historically associated with the penicillins [as in structure (75)]. A β-hydroxyethylidene grouping is present at C(2) of the nucleus, while the carboxy function at C(3) and the hydrogen at C(5) possess the same configuration as found in the penicillins.

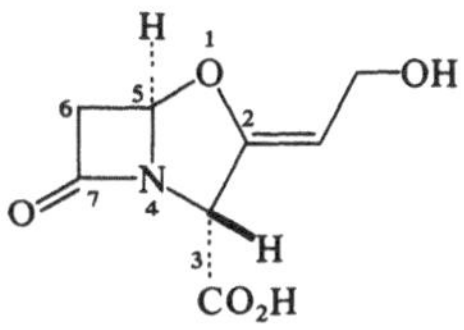

(75)

Methanolysis of sodium or lithium clavulanate provides a high yield of the pyrrole (79) (52, 53, 54). Since the enamine (78) has also been isolated a fairly well defined degradative pathway for the formation of (79) can be postulated further verifying the fused β-lactam-oxazolidine structure of clavulanic acid. Rupture of the β-lactam followed by fragmentation of the oxazolidine (76) would lead to the keto acid (77); decarboxylation provides

the enamine (78) which on intramolecular cyclisation and dehydration leads
to the pyrrole (79).

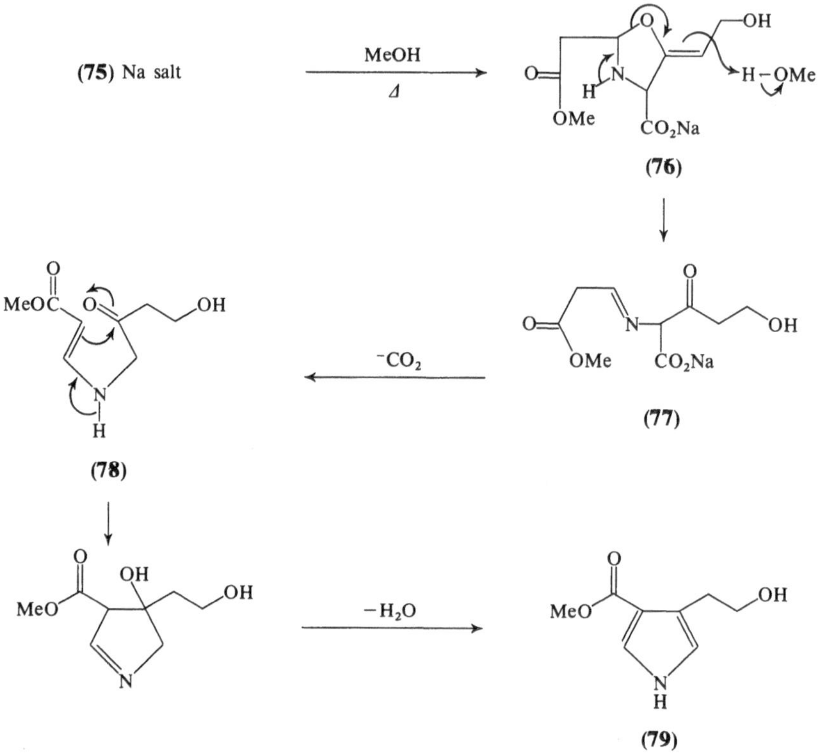

(75) Na salt $\xrightarrow[\Delta]{\text{MeOH}}$

(76)

(77)

(78) $\xleftarrow{-CO_2}$

(79) $\xleftarrow{-H_2O}$

Clavulanic acid is a potent inhibitor of a number of β-lactamases and is
capable of protecting those β-lactam antibiotics normally sensitive to these
enzymes. The biological properties of clavulanic acid alone, and in
combination with various penicillins, have been well reviewed (55). An
extensive investigation of the chemistry of clavulanic acid has been
undertaken since its discovery. Normally the benzyl or p-nitrobenzyl ester
has been used since the free acid can be regenerate readily by hydrogenolysis. Reactions of the double bond include isomerisation to the
(E)-isomer (56), oxidation to the epoxide (53) and ozonolysis (57) to give the
lactone (80). Various reduction products of the double bond have also been
obtained (54).

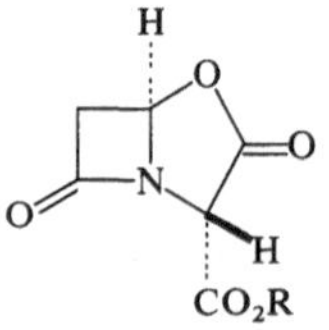

(80) 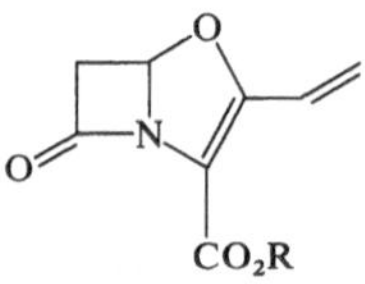   (81) 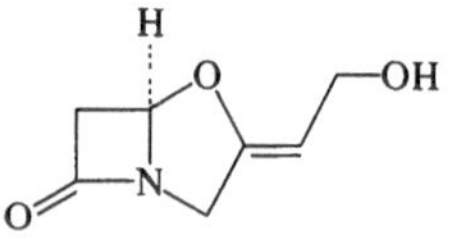   (82)

Oxidation (58) of the allylic hydroxyl group of benzyl clavulanate gave the corresponding aldehyde when using pyridinium chlorochromate as oxidant, but under the Pfitzner-Moffat conditions elimination took place resulting in the formation of the diene (81). The latter can also be obtained from the allylic chloride (59). Other aspects of the chemistry of the hydroxyl group include the preparation of esters and ethers and replacement reactions using sulphur, carbon and nitrogen nucleophiles (54). Clavulanic acid can also be decarboxylated (60) to (82). Other interesting aspects of the chemistry of clavulanic acid include its conversion into the penem ring system (83) (61), while reaction of the β-lactam carbonyl with methoxycarbonylmethylenetriphenylphosphorane gives rise to the olefin (84) (62).

(83)   (84)

Following the isolation of clavulanic acid a number of other structurally related bicyclic β-lactams have been isolated from various *Streptomyces* species. These are shown in Table 4. Interestingly in some cases (65, 66) the stereochemistry at C(5) is assigned the opposite configuration to that found in clavulanic acid itself. The compounds lacking the C(3)-carboxyl group are mainly active as antifungal agents.

Interest in the β-lactamase inhibitory activity of clavulanic acid has led to the synthesis of a number of analogues (53, 54). A formal total synthesis of racemic clavulanic acid itself was reported in 1977 (67). Alkylation of the synthetic azetidinone (84a) using the 2-bromo derivative of dimethyl 3-oxoglutarate provided (85) which exists largely in the enolised form. Replacement of the thiomethyl substituent with chlorine followed by potassium carbonate catalysed cyclisation gave (86) as a single double bond isomer. Irradiation resulted in a 3:2 mixture of (86) and (87), which was reduced using di-isobutylaluminium hydride to give a low yield of (88) and (89). These were separated to provide (±) methyl clavulanate (88) and the iso-derivative (89). As the methyl ester of natural clavulanic acid can be hydrolysed to the parent compound the preparation of (88) constitutes a formal total synthesis of racemic (75).

An alternative synthesis has also been described by the same group (68). In this case acylation of the ester enolate of (90) with vinylacetyl chloride gave (91). Chlorinolysis and cyclisation as before produced the diene (92). Careful ozonolysis of (92) followed by hydrogenolysis of the ozonide formed the racemic ester (88) together with the aldehyde (93).

Table 4. *Natural Products Based on the 1-Oxadethiapenam (Clavam)* Ring System*

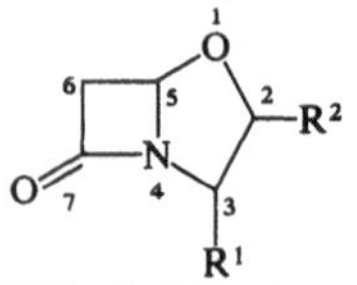

R¹	R²	Stereochemistry	References
CO_2H	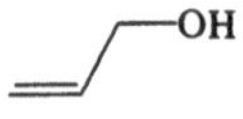 —OH	3R, 5R, Z double bond	(50, 51)
CO_2H	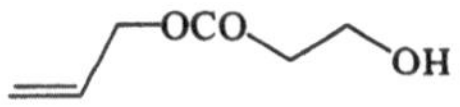 —OCO—OH	3R, 5R, Z double bond	(63)
H	—CO_2H	Not determined	(64)
H	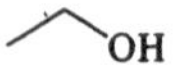 OH	Not determined	(64)
H	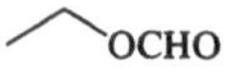 OCHO	Not determined	(64)
H	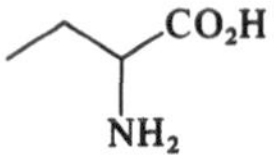 CO_2H NH_2	2S, 5S (L-alanyl)	(65)
H	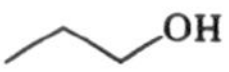 OH	2S, 5S	(66)

(* Numbering based on the systematic 7-oxo-4-oxa-1-azabicyclo [3.2.0] heptane ring system is rarely used in the literature)

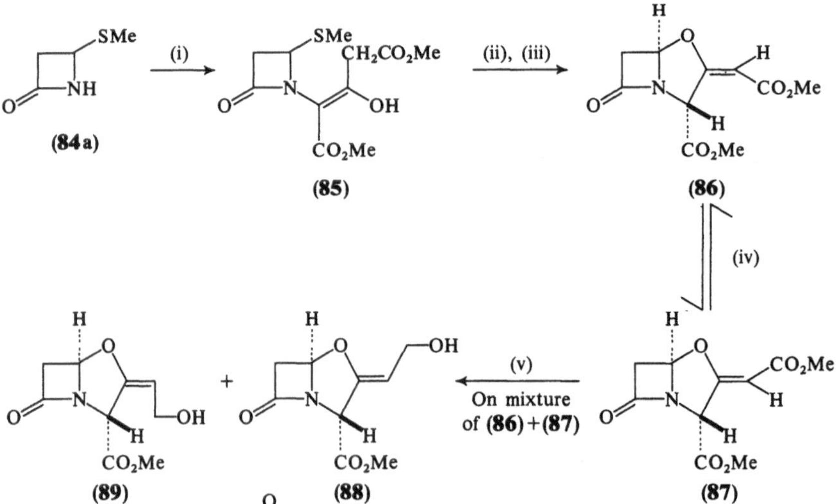

Reagents: (i) NaH-Br CO$_2$Me-DMF; (ii) Cl$_2$-CCl$_4$;

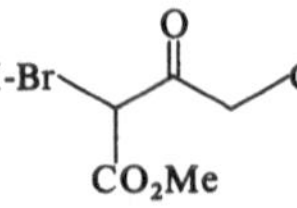

(iii) K$_2$CO$_3$-DMF; (iv) hν; (v) Diisobutylaluminium hydride-toluene ($-70°$)

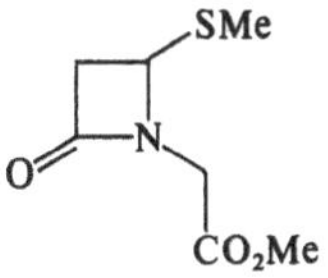

(90) 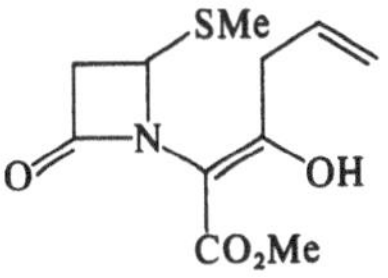   **(91)** 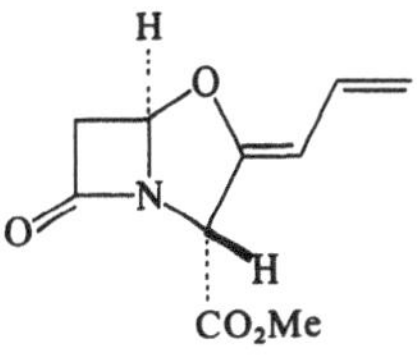   **(92)** 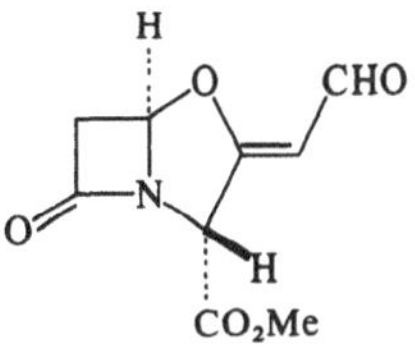   **(93)**

VI. Olivanic Acids, Thienamycin and Other 1-Carbadethiapen-2-ems

The microorganism *Streptomyces olivaceus* produces a number of β-lactam antibiotics and β-lactamase inhibitors which have collectively become known as the olivanic acids (*49, 69–75*). Together with other streptomycete metabolites such as thienamycin (*80, 81*), PS-5 (*85, 86*), the carpetimycins (*90, 91*) and the asparenomycins (*92*), they form a family of natural products characterised by the presence of the 1-carbadethiapen-2-em (7-oxo-1-azabicyclo[3.2.0]hept-2-ene-2-carboxylic acid) ring system (**94**). The simplest member of the family is the completely unsubstituted nucleus (**94**), recently isolated from various species of *Serratia* and *Erwinia* (*98*). All the other natural products possess substituents at C(2) and C(6) of this nucleus, the former being attached by way of a sulphur linkage. Differences in the nature and stereochemistry of these substituents have provided a wide variety of structures. Almost forty natural variations of this ring system are now known and these are listed in Table 5.

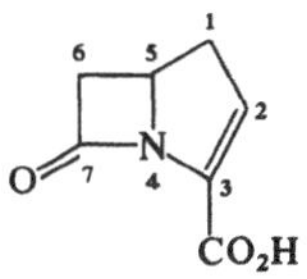

(94) 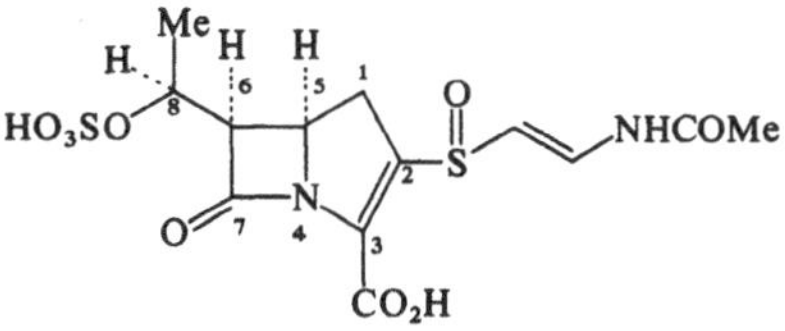   **(95)** MM 4550

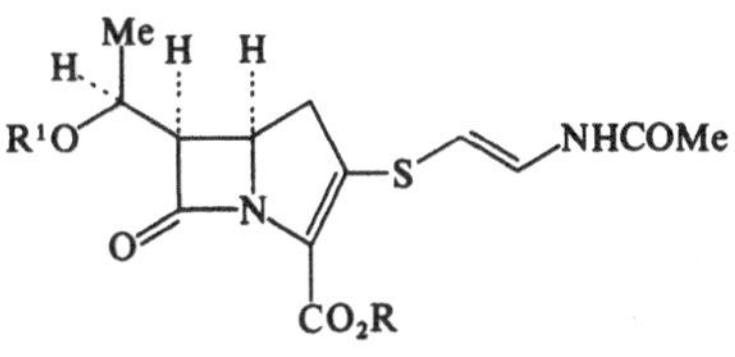

(96) R=H; R^1=SO_3H; MM 13902
(98) R=Me; R^1=SO_3H
(103) R=H; R^1=H; MM 22382

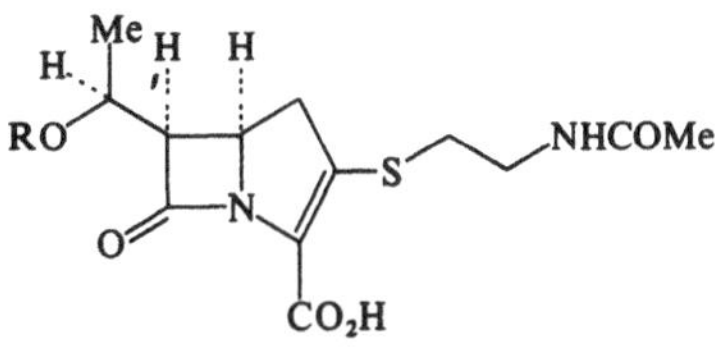

(97) R=SO_3H; MM 17880
(104) R=H; MM 22380

Table 5. *Structural Variations of Known Naturally Occurring 1-Carbadethiapen-2-ems*

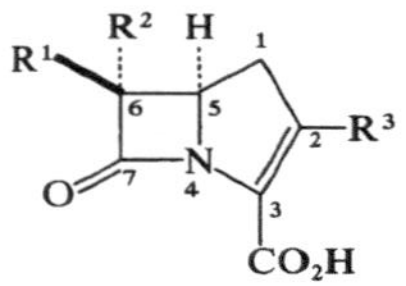

Metabolite	R^1	R^2	R^3	References
MM 4550 (MC696-SY2-A)	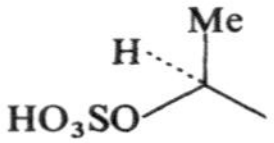 H—⟨Me⟩ HO₃SO	H	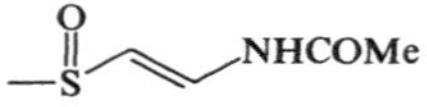 S(O)—CH=CH—CH₂—NHCOMe	(49, 69, 71, 72, 75, 76)
MM 13902 (Epithienamycin E)	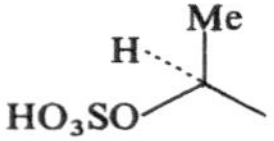 H—⟨Me⟩ HO₃SO	H	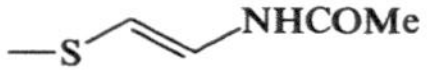 S—CH=CH—CH₂—NHCOMe	(49, 69, 71, 72, 77)
MM 17880 (Epithienamycin F)	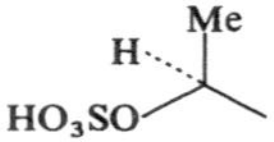 H—⟨Me⟩ HO₃SO	H	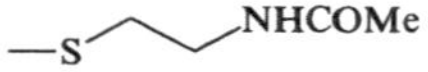 S—CH₂—CH₂—NHCOMe	(70, 71, 72, 75, 77)
MM 22380 (Epithienamycin A)	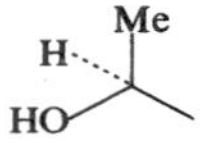 H—⟨Me⟩ HO	H	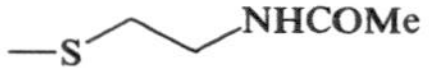 S—CH₂—CH₂—NHCOMe	(73, 74, 75, 77)
MM 22381 (Epithienamycin C)	H	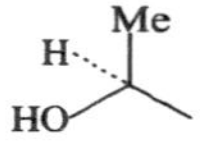 H—⟨Me⟩ HO	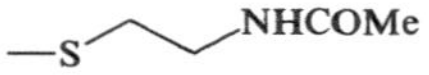 S—CH₂—CH₂—NHCOMe	(73, 74, 75, 77)
MM 22382 (Epithienamycin B)	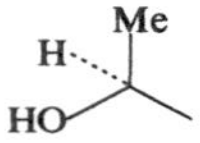 H—⟨Me⟩ HO	H	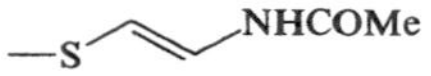 S—CH=CH—CH₂—NHCOMe	(73, 74, 75, 77)
MM 22383 (Epithienamycin D)	H	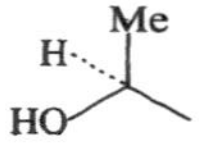 H—⟨Me⟩ HO	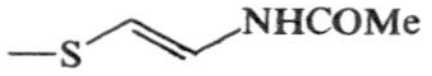 S—CH=CH—CH₂—NHCOMe	(73, 74, 75, 77)

Compound			Ref.
MM 27696	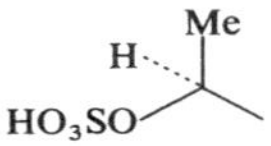 (HO$_3$SO, Me, H)	H	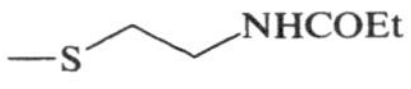 —S$\sim\sim$NHCOEt	(78)
8U-207	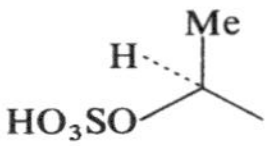 (HO$_3$SO, Me, H)	H	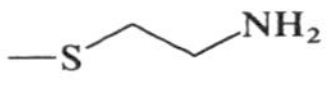 —S$\sim\sim$NH$_2$	(79)
Thienamycin	H	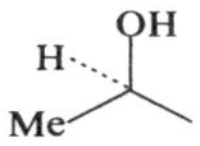 (H, OH, Me)	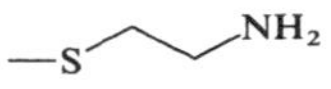 —S$\sim\sim$NH$_2$	(80, 81)
N-Acetylthienamycin	H	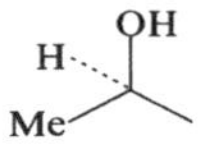 (H, OH, Me)	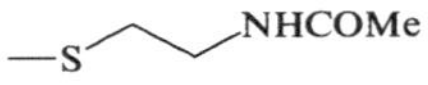 —S$\sim\sim$NHCOMe	(82)
N-Acetyldehydrothienamycin	H	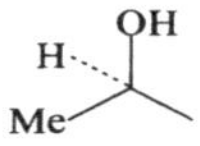 (H, OH, Me)	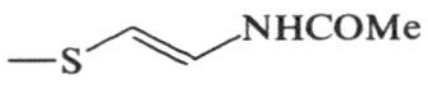 —S$\sim$=$\sim$NHCOMe	(83)
Nor-thienamycin	H	HOCH$_2$—	—S$\sim\sim$NH$_2$	(84)
PS-5 (MM 22744)	H	CH$_3$CH$_2$—	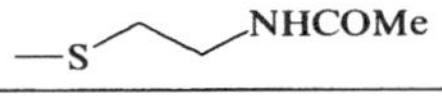 —S$\sim\sim$NHCOMe	(85, 86)
PS-6	H	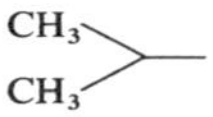 (CH$_3$)$_2$CH—	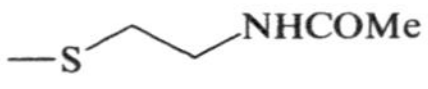 —S$\sim\sim$NHCOMe	(87)
PS-7	H	CH$_3$CH$_2$—	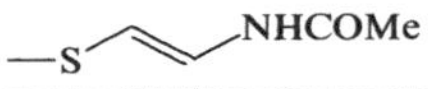 —S$\sim$=$\sim$NHCOMe	(87)
PS-8	H	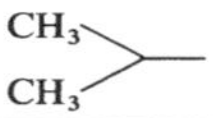 (CH$_3$)$_2$CH—	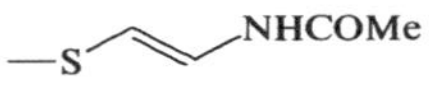 —S$\sim$=$\sim$NHCOMe	(88)
NS-5	H	CH$_3$CH$_2$—	—S$\sim\sim$NH$_2$	(89)

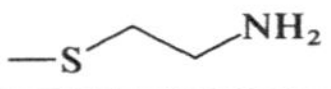

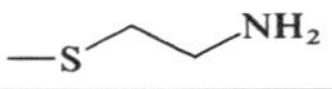

Table 5 (continued)

Metabolite	R^1	R^2	R^3	References
Carpetimycin A (C-19393H₂; KA-6643A)	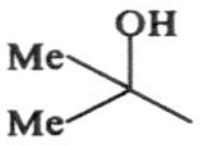 Me–C(OH)–Me	H	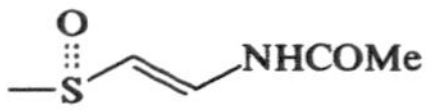 –S(=O)–CH=CH–CH₂–NHCOMe	(90, 91)
Carpetimycin B (C-19393S₂; KA-6643B)	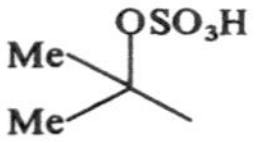 Me–C(OSO₃H)–Me	H	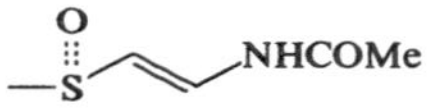 –S(=O)–CH=CH–CH₂–NHCOMe	(90, 91)
Asparenomycin A		$R^1, R^2 =$ 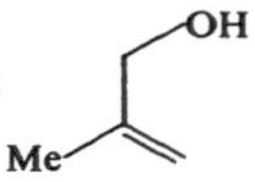 (CH₂=C(Me)–CH₂–CH(OH)–)	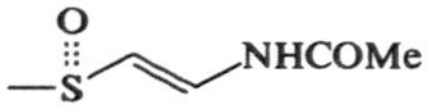 –S(=O)–CH=CH–CH₂–NHCOMe	(92)
Asparenomycin B		$R^1, R^2 =$ 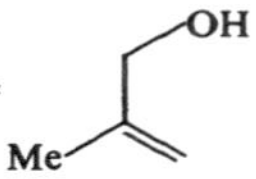 (CH₂=C(Me)–CH₂–CH(OH)–)	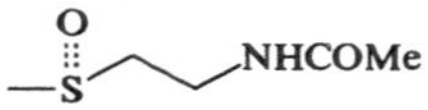 –S(=O)–CH₂CH₂–NHCOMe	(92)
Asparenomycin C		$R^1, R^2 =$ 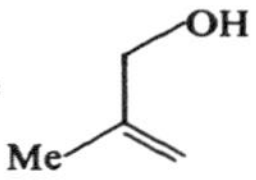 (CH₂=C(Me)–CH₂–CH(OH)–)	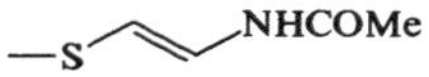 –S–CH=CH–CH₂–NHCOMe	(92)
SF 2103A (Pluracidomycin A)	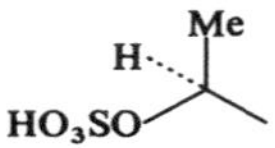 HO₃SO–CH(Me)–H	H	–SO₃H	(93, 94)
Pluracidomycin B	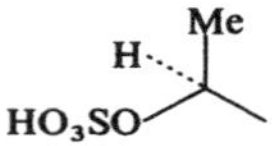 HO₃SO–CH(Me)–H	H	 –S(=O)–CH₂CO₂H	(94)
Pluracidomycin C	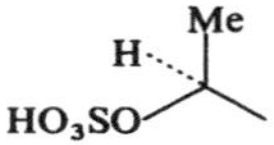 HO₃SO–CH(Me)–H	H	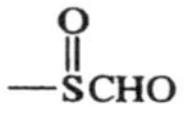 –S(=O)–CHO or –S(=O)–CH(OH)OH 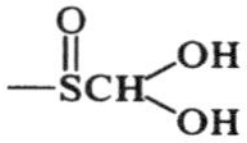	(94)

Compound	Substituent (left)	Substituent (middle)	S-side chain	Ref.
MM 22383 (Epithienamycin D) S-oxide	H	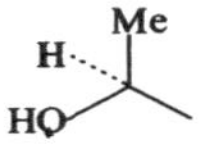 Me, H, HO (CH(OH)CH₃)	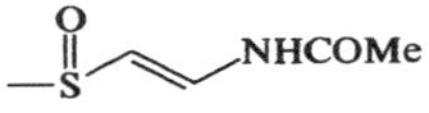 —S(=O)—CH=CH—NHCOMe	(94)
OA-6129A	H	CH₃CH₂—	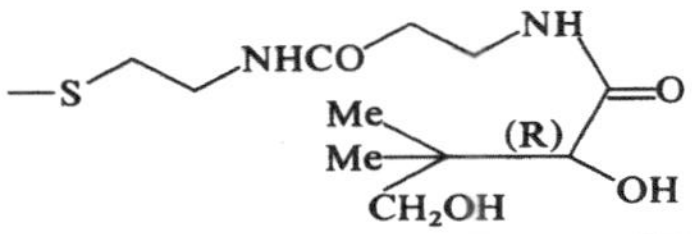 —S—CH₂CH₂—NHCO—...—NH—...—(R), OH, Me Me, CH₂OH	(95, 96)
OA-6129B₁	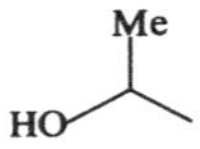 HO, Me	H	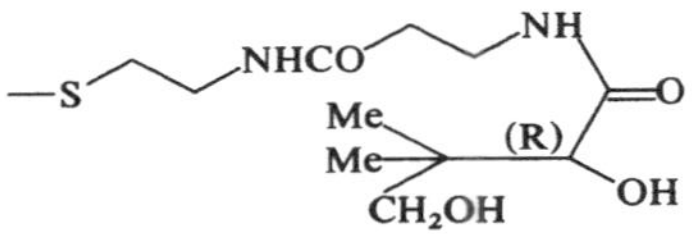 —S—NHCO—...—NH—(R), OH, Me Me, CH₂OH	(95, 96)
OA-6129B₂	H	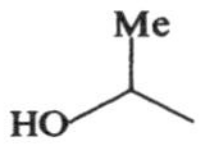 Me, HO	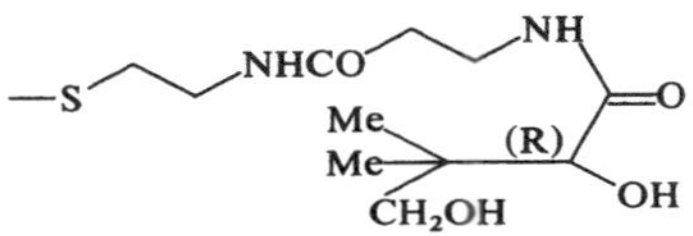 —S—NHCO—...—NH—(R), OH, Me Me, CH₂OH	(95, 96)
OA-6129C	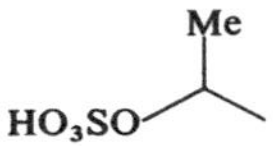 HO₃SO, Me	H	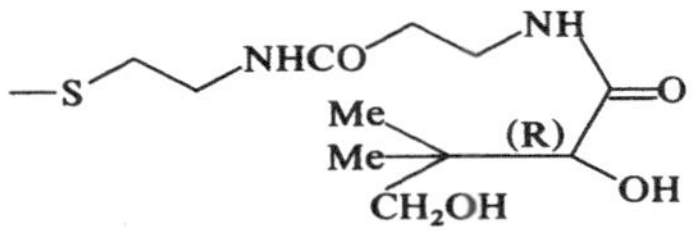 —S—NHCO—...—NH—(R), OH, Me Me, CH₂OH	(95, 96)
Epithienamycin G	H	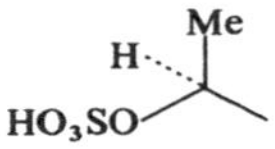 Me, H, HO₃SO	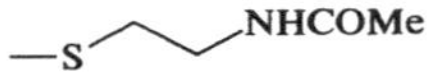 —S—NHCOMe	(77)
SQ 27,860	H	H	H	(77)
8-Epithienamycin	H	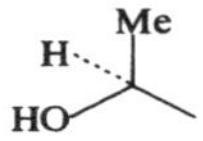 Me, H, HO (CH(OH)CH₃)	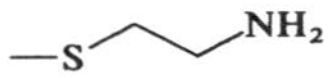 —S—NH₂	(98)

Table 5 (continued)

Metabolite	R¹	R²	R³	References
MM 22382 S-Oxide (C-19393 E₅; PA-3108811; Epithienamycin B-S-oxide)	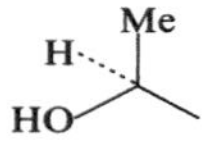 Me–CH(H)(OH)	H	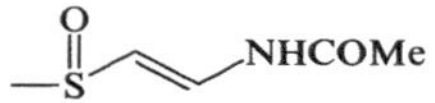 O=S–CH=CH–CH₂–NHCOMe	(94, 99, 100)
KA-6643D	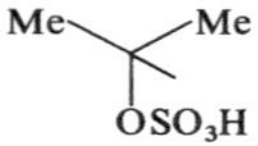 Me₂C(OSO₃H)	H	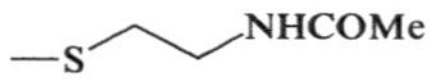 S–CH₂CH₂–NHCOMe	(101)
KA-6643F	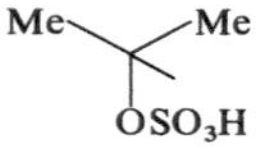 Me₂C(OSO₃H)	H	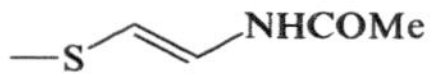 S–CH=CH–NHCOMe	(101)
KA-6643G	H	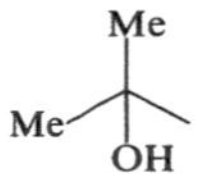 Me₂C(OH)	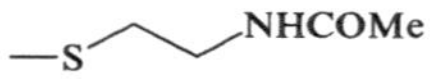 S–CH₂CH₂–NHCOMe	(101)
KA-6643X	R¹ R²= C(=CH₂)(CH₂OH)(Me)	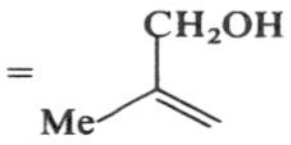	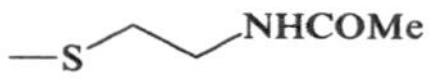 S–CH₂CH₂–NHCOMe	(101, 102)
KA-6643I Carpetimycin C	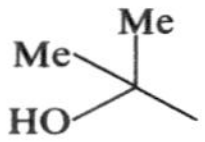 Me₂C(OH)	H	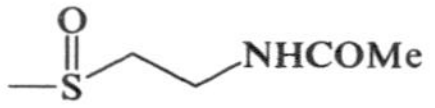 O=S–CH₂CH₂–NHCOMe	(205)
KA-6643J Carpetimycin D	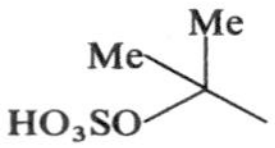 Me₂C(OSO₃H)	H	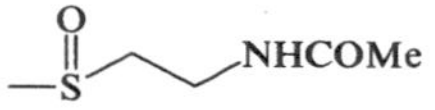 O=S–CH₂CH₂–NHCOMe	(205)

The first three metabolites isolated from *Streptomyces olivaceus* were the sulphated derivatives designated MM 4550 (**95**), MM 13902 (**96**) and MM 17880 (**97**). All have the *cis*-arrangement of protons about the β-lactam ring with the configuration at C(8) being (*S*) (*69, 70, 75*). A key reaction in confirming these structures was the rearrangement of the monomethyl ester (**98**) of MM 13902 to the pyrrole (**99**) on heating in dimethyl sulphoxide. Subsequent esterification followed by elimination of the sulphate residue gave a mixture of isomers of the olefin (**100**) (*70, 75*). This rearrangement also occurred with MM 17880 monoester, but not with MM 4550. Elimination of the sulphate moiety of the di-ester (**101**) of MM 13902 was stereospecific, giving solely the (*E*)-isomer of the ethylidene derivative (**102**), suggesting that the configuration in the side-chain was (*S*). This was subsequently confirmed by X-ray analysis (*75*).

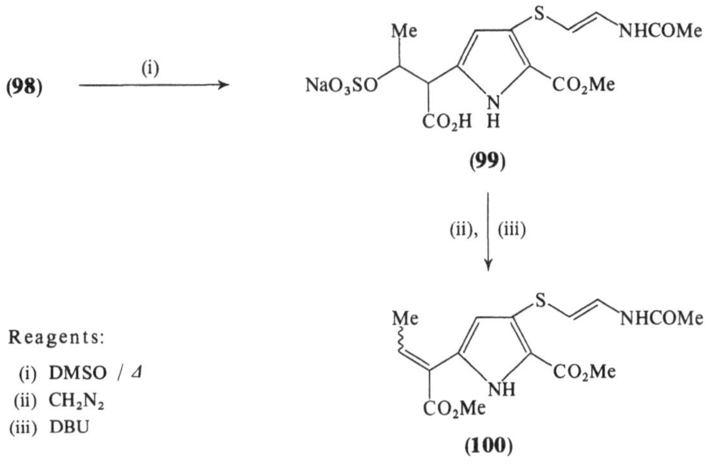

Reagents:

(i) DMSO / Δ
(ii) CH₂N₂
(iii) DBU

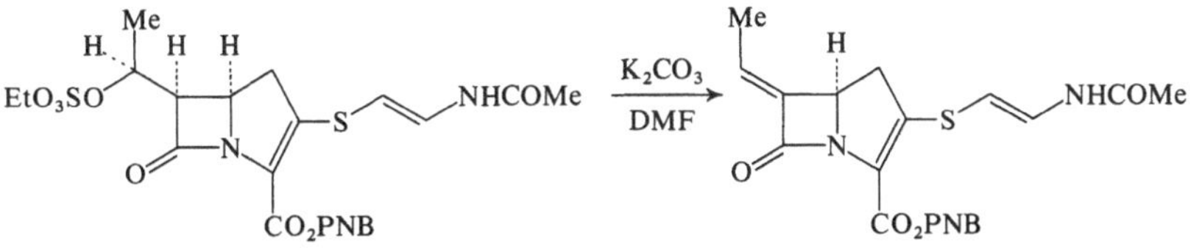

Later the non-sulphated metabolites MM 22380 (**104**) and MM 22382 (**103**) were obtained from *S. olivaceus*, together with the corresponding isomers MM 22381 (**105**) and MM 22383 (**106**) having the *trans*-arrangement of protons about the β-lactam ring.

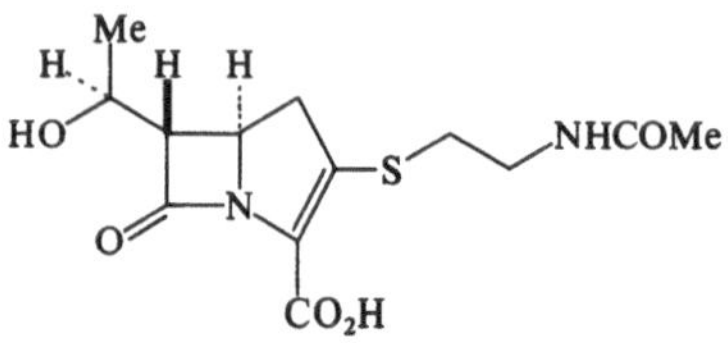

(105) MM 22381 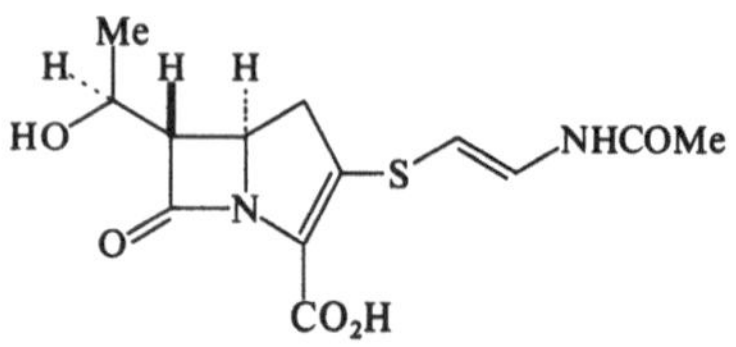   **(106)** MM 22383

Independently of the work on the olivanic acids, researchers at the Merck Company had isolated, and determined the structure of, the related compound thienamycin **(107)** produced by *Streptomyces cattleya* (*80, 81*). Here a free amino group is present, while the configuration at C(8) is (*R*). An extensive review of all aspects of the chemistry of thienamycin has recently been published (*103*). Two derivatives of thienamycin found in the same fermentation broth were *N*-acetyl thienamycin **(108)** (*81, 82*) and *N*-acetyldehydro-thienamycin **(109)** (*81, 83*). Workers at Merck have also isolated from *Streptomyces flavogriseus* various epithienamycins (A to F), which correspond to the olivanic acids (*77*). Subsequently discovered by numerous Japanese workers were PS-5 **(110)** (*85, 86*), carpetimycin A **(111)** (*90, 91*), asparenomycin A **(112)** (*92*) and the many other compounds listed in Table 5.

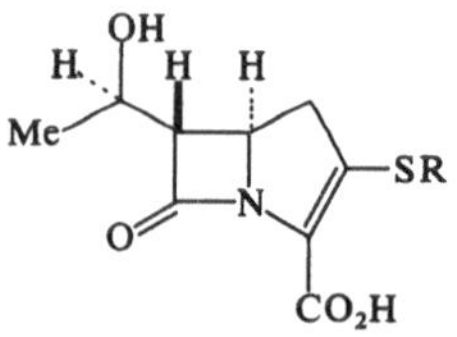

(107) R=CH₂CH₂NH₂
(108) R=CH₂CH₂NHCOMe
(109) R=CH=CH−NHCOMe

$$(107)\ R=CH_2CH_2NH_2$$
$$(108)\ R=CH_2CH_2NHCOMe$$
$$(109)\ R=CH{=}CH-NHCOMe$$

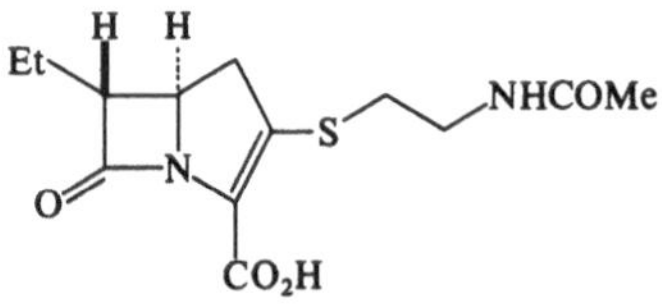

(110) 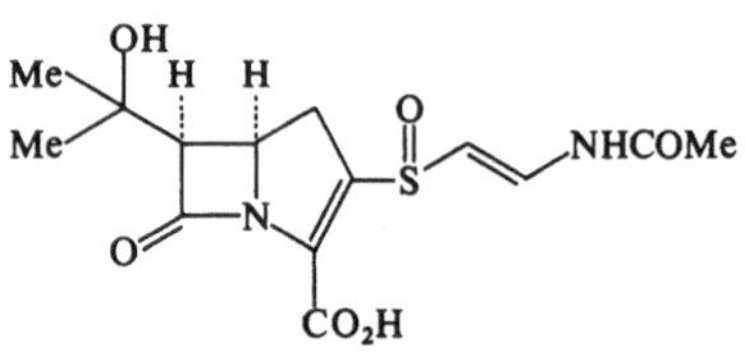   **(111)**

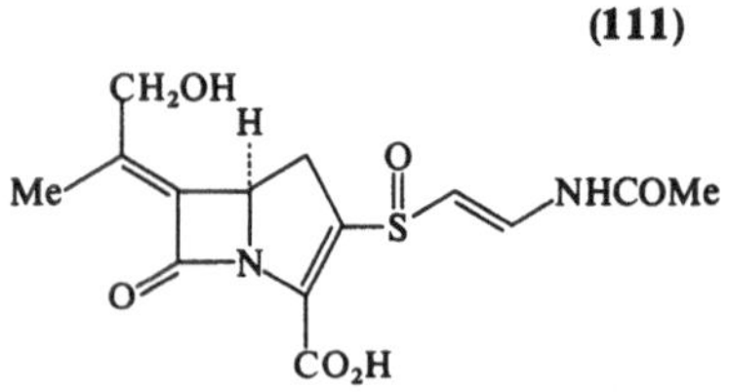

(112)

The novelty and potent biological properties of the carbapenems has resulted in an extensive synthetic effort, both to produce analogues as well as the natural products themselves. Impetus for this effort is the generally low amount of material available by fermentation processes. For the purpose of this review only syntheses of the natural products will be described. In many cases these methods have been readily adapted for the preparation of purely synthetic analogues.

The first synthesis (*104, 105*) of racemic thienamycin by the Merck group made use of the azetidinone (**113**) derived from acetoxybutadiene and chlorosulphonyl isocyanate (CSI). Reduction, hydrolysis and condensation with acetone gave the 1,3-tetrahydrooxazine (**114**). Introduction of the hydroxyethyl side chain by way of an aldol condensation produced predominantly the thermodynamically more stable *trans*-β-lactam (**115**) as a mixture of (*R*)- and (*S*)-isomers in the side chain. On removal of the acetone residue a proportion of the unwanted (*S*)-isomer crystallised from the mixture. The synthesis was continued with the mixture by way of the aldehyde (**116**) and the thio-acetal (**117**). Bromination, elimination and introduction of the *N*(1) malonate residue gave (**119**) ready for an intramolecular alkylation reaction.

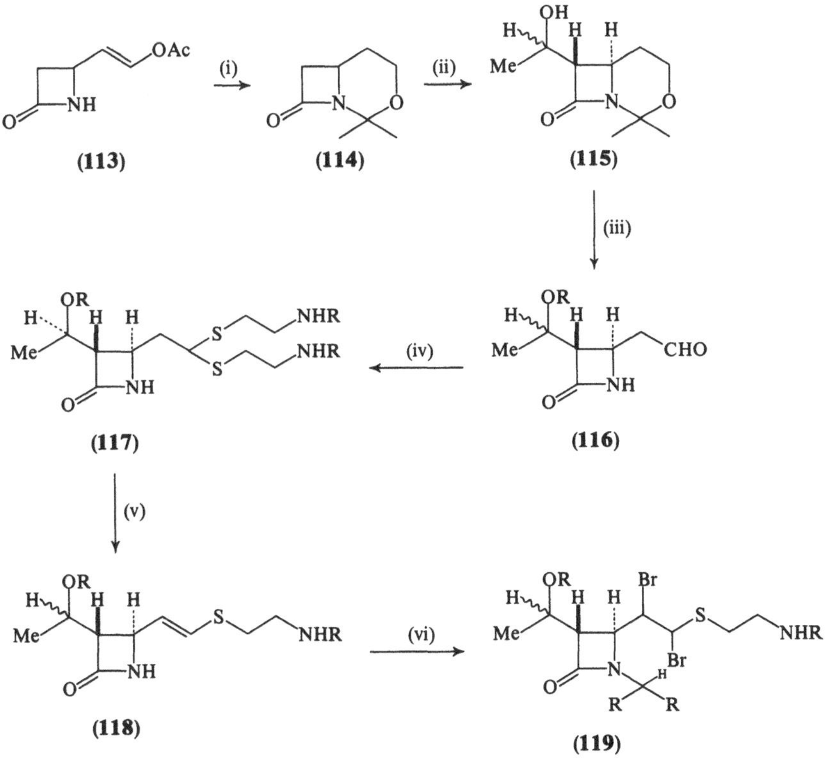

(113) (114) (115)

(117) (116)

(118) (119)

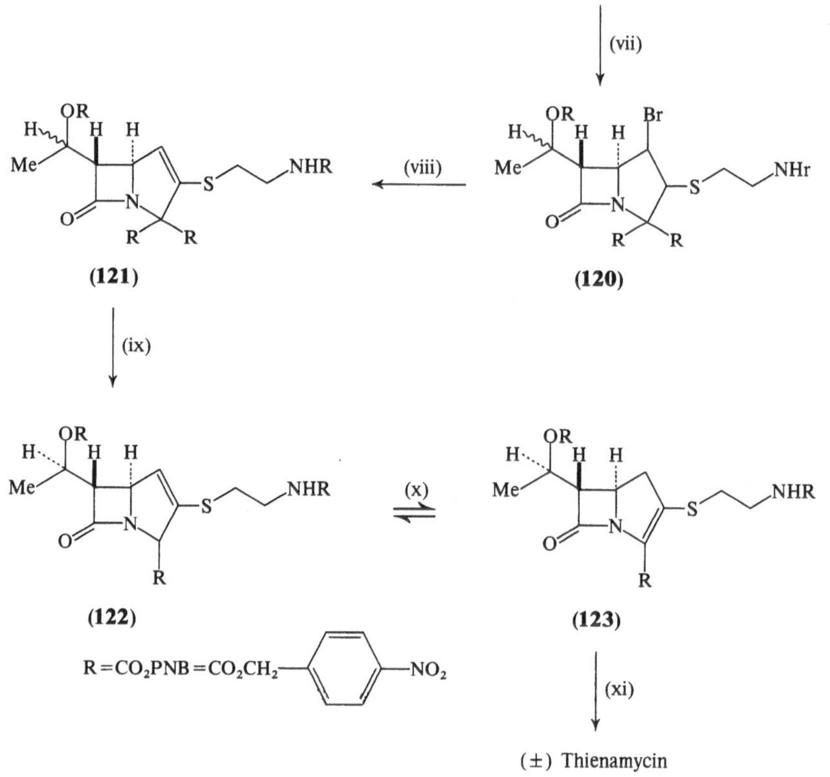

$R = CO_2PNB = CO_2CH_2$——⟨ ⟩——NO_2

(±) Thienamycin

Reagents: (i) H_2, Pd/C, NaOMe, $Me_2C(OMe)_2$; (ii) LDA, MeCHO; (iii) Base, $ClCO_2PNB$, H^+, CrO_3, pyridine; (iv) HS⌒⌒NHR, BF_3-Et_2O; (v) Br_2, Et_3N;

(vi) $O=C(CO_2PNB)_2$, $SOCl_2$, P(n-Bu)$_3$, Br_2; (vii) Et_3N;

(viii) AgF, pyridine; (ix) Separation of isomers, Collidine, LiI;

(x) Diisopropylamine, DMSO; (xi) H_2, Pd/C.

After formation of the fused 4 − 5 ring system (120) another elimination step led to the Δ-1 carbapenem (121). At this stage the two diastereoisomers could be separated, decarboxylated and the requisite 8(R)-isomer iso-merised to (123). Interestingly the equilibrium lies largely on the side of the unconjugated ester (122). Removal of the p-nitrobenzyl protecting groups by hydrogenolysis led to the isolation of racemic thienamycin.

Prior to the discovery of the naturally occurring 1-carbapenem nucleus (94), a synthesis of the racemic material had been described (106). The 4-acetoxyethyl substituted β-lactam (124) derived from (113) was converted by standard procedures to the phosphorane (125). Hydrolysis to the alcohol (126) followed by oxidation resulted in cyclisation to the bicyclic ester (127).

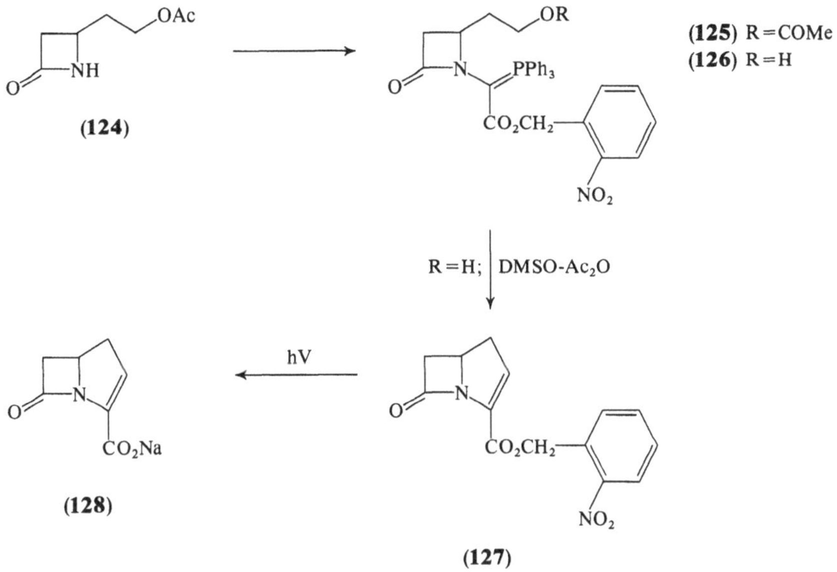
OAc
NH
(124)
OR
PPh3
CO2CH2
NO2
(125) R=COMe
(126) R=H
R=H; DMSO-Ac2O
hV
CO2Na
(128)
(127)

Removal of the photolabile acid protecting group afforded the unstable sodium salt (**128**). Various esters (**131**) of (**128**) have been prepared (*107*) using the phosphorane (**130**) derived from the 4-allylazetidinone (**129**); in this case oxidation of the terminal methylene grouping by ozonolysis provided the aldehyde for the intramolecular Wittig reaction. Both the *p*-nitrobenzyl ester (*108*) and acetonyl ester (*109, 110*) of (**131**) have been used in this way to prepare (**128**).

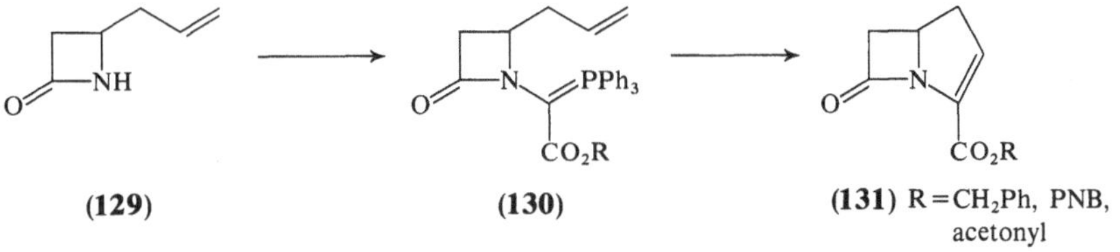
NH
PPh3
CO2R
(129)
(130)
(131) R=CH2Ph, PNB, acetonyl

The intramolecular Wittig cyclisation procedure has been used extensively to prepare derivatives related to the olivanic acids, particularly since it was discovered that thiol esters would participate in the ring closure reaction (*108*). This methodology provided the basis for the total synthesis of (±)-MM 22383 and (±)-*N*-acetyldehydrothienamycin (*108, 111*).

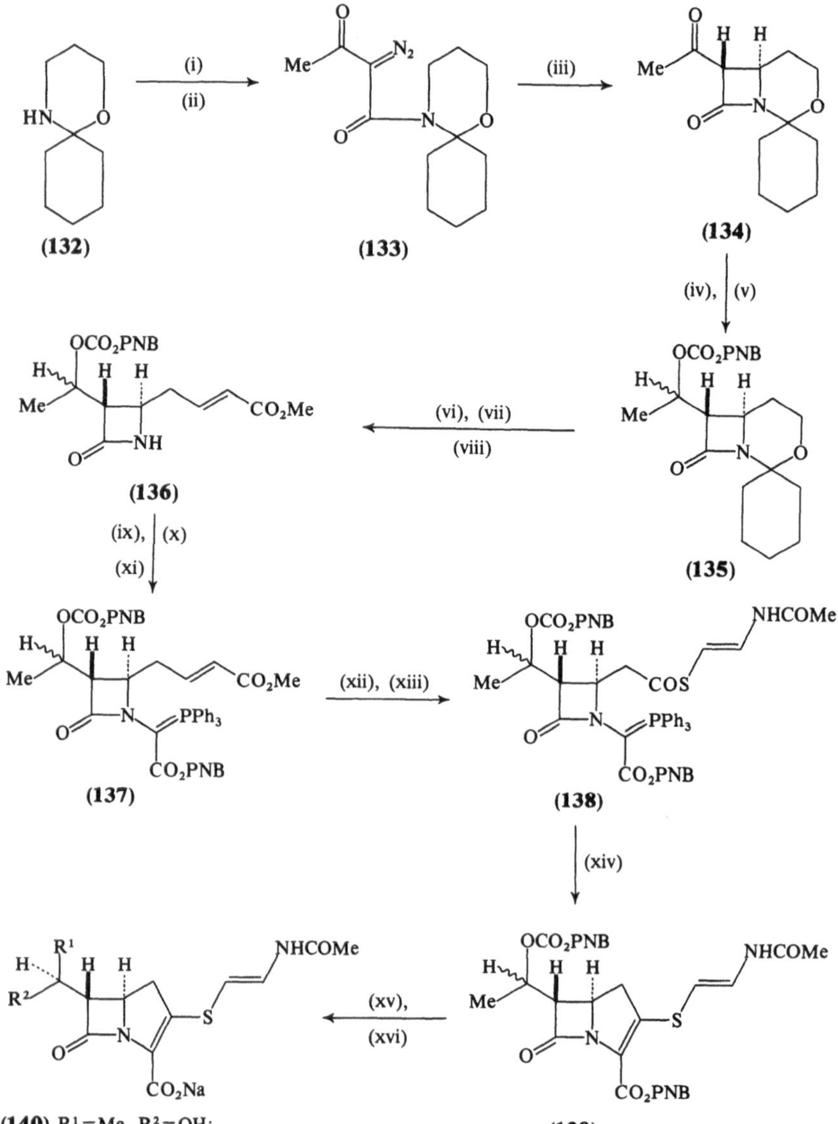

(132)   **(133)**   **(134)**

(136)   **(135)**

(137)   **(138)**

(139)

(140) R¹=Me, R²=OH;
(±)-MM 22383 (5*RS*, 6*SR*, 8*SR*)

(141) R¹=OH, R²=Me;
(±)-N-Acetyldehydrothienamycin (5*RS*, 6*SR*, 8*RS*)

Reagents: (i) Diketen; (ii) p-Me-C$_6$H$_4$SO$_2$N$_3$, Et$_3$N; (iii) Rh$_2$(OAc)$_4$; (iv) NaBH$_4$; (v) n-BuLi, ClCO$_2$PNB; (vi) H$^+$; (vii) Pyridinium chlorochromate; (viii) Ph$_3$P=CHCO$_2$Me; (ix) CHO–CO$_2$PNB; (x) SOCl$_2$,2,6-lutidine; (xi) Ph$_3$P,2,6-lutidine; (xii) O$_3$, MCPA; (xiii) SOCl$_2$, pyridine, AgS⁀⁀NHCOMe ; (xiv) Δ, toluene, 48 h; (xv) Separation of isomers; (xvi) H$_2$, Pd/C, NaHCO$_3$.

The tetrahydro-1,3-oxazine (132) from cyclohexanone and aminopropanol was converted to the diazo-intermediate (133) and cyclised using rhodium acetate to the *trans*-β-lactam (134). Non-selective reduction of the ketone gave both hydroxy epimers. Progression through the sequence as outlined provided the thiol-ester phosphorane (138) possessing the required (*E*)-acetamidoethenyl substituent. Cyclisation in boiling toluene gave the two epimers of (139) which were separated, and deprotected to afford ($\pm$)-MM 22383 (140) and ($\pm$)-*N*-acetyldehydrothienamycin (141).

An alternative approach to the synthesis of thienamycin and analogues also makes use of a rhodium acetate catalysed carbene insertion reaction. In this case formation of the C(3)-*N*(4) bond leads to the key ketone intermediate (149). Using this procedure and starting from L-aspartic acid a stereocontrolled synthesis leading to (+)-thienamycin has been achieved (*112*). The β-lactam precursor was formed from the protected amino-acid (142) and elaborated to the *N*-silylated azetidinone (145). Introduction of the hydroxyethyl substituent to form (146) was by way of a stereocontrolled potassium selectride reduction of the corresponding acetyl compound. Oxidation to the acid (147) was followed by conversion to the diazo-intermediate (148), and a high yielding cyclisation step to the ketone (149). Activation as the enol phosphate and reaction with protected cysteamine gave the protected material (150), which on hydrogenolysis provided (+)-thienamycin.

Another approach described by the Merck group (*113*) makes use of the lactone (151) derived from acetone dicarboxylic acid. Conversion to the amino-acid followed by cyclisation to the β-lactam resulted in (152) having the (*S*)-stereochemistry in the side chain. This was corrected to the (*R*)-configuration by an inversion step after conversion to the β-keto-ester. In the olivanic acid series inversion at C(8) has also been achieved to give members of the thienamycin series (*114*).

A semi-synthetic approach by the Merck group (*115*) makes use of the 4-chloro-azetidinone (153) which can be obtained from 6-aminopenicillanic acid. Reaction with the silylated diazo-intermediate (154) in the presence of silver tetrafluoroborate followed by desilylation provides a one-step process to (148) from (153). A similar displacement with a 4-acetoxy substituted azetidinone derived from aspartic acid provides another route (*116*) to (148). The amidine derivative (155) of thienamycin is chemically more stable than the amine and a direct introduction of this side-chain by way of the keto-ester has also been achieved (*117*). It is this amidine (MK 0787) that is undergoing clinical investigation rather than thienamycin itself.

The 1,3-dipolar cycloaddition reaction of nitrones and crotonate esters gives isoxazolines, which are readily converted to β-amino-acids and then β-lactams (*118*). KAMETANI (*119, 120*) has made extensive use of these

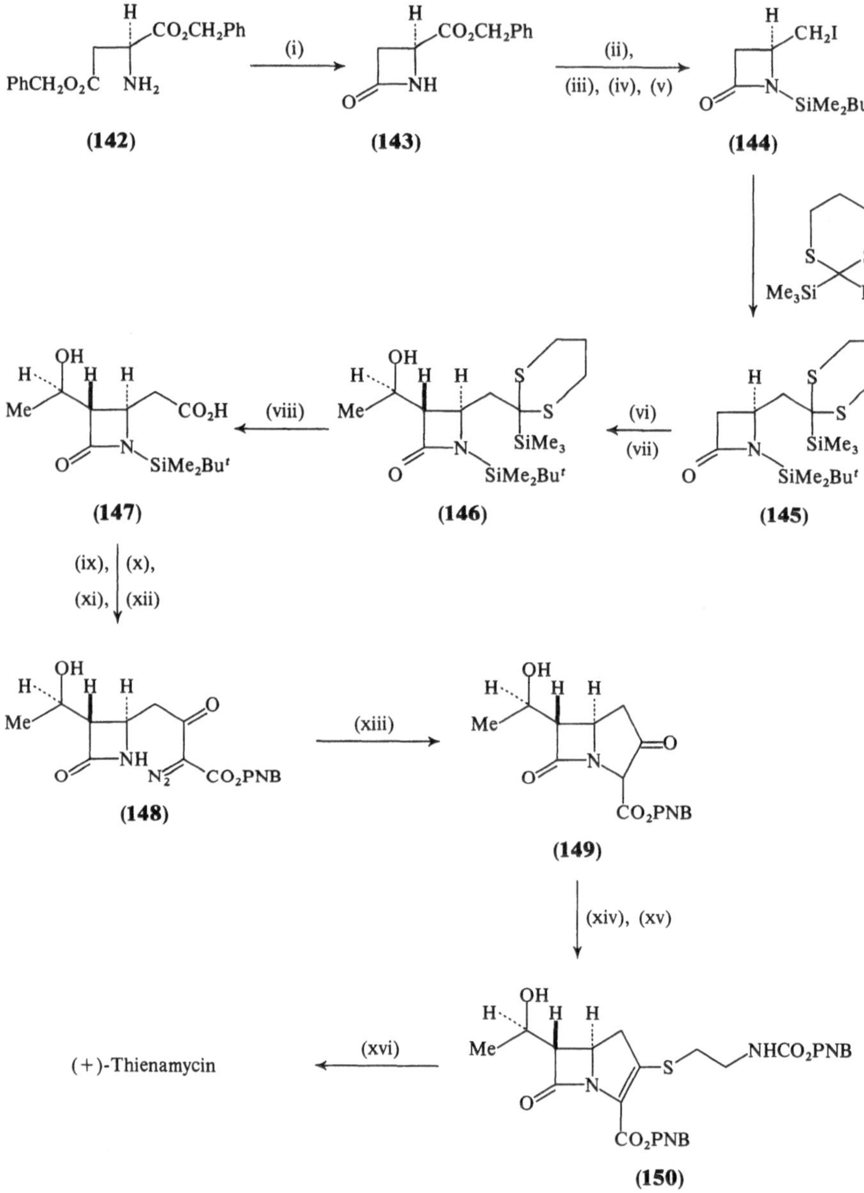

Reagents: (i) TMSCl, Bu^{t}MgCl, H$^+$; (ii) NaBH$_4$; (iii) MeSO$_2$Cl; (iv) NaI; (v) Bu^{t}Me$_2$SiCl; (vi) LDA-Acetylimidazole; (vii) K-Selectride; (viii) HgCl$_2$, H$_2$O$_2$; (ix) Carbonyl diimidazole; (x) Mg(O$_2$CCH$_2$CO$_2$PNB)$_2$; (xi) H$^+$; (xii) ArSO$_2$N$_3$-Et$_3$N; (xiii) Rh$_2$(OAc)$_4$; (xiv) ClPO(OPh)$_2$-iPr$_2$NEt; (xv) HS⁀NHCO$_2$PNB, i-Pr$_2$NEt; (xvi) H$_2$-Pd/C.

References, pp. 89—106

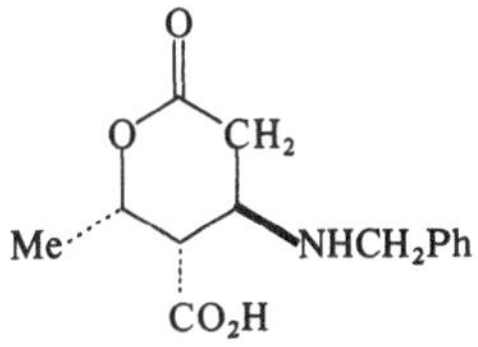

(151)

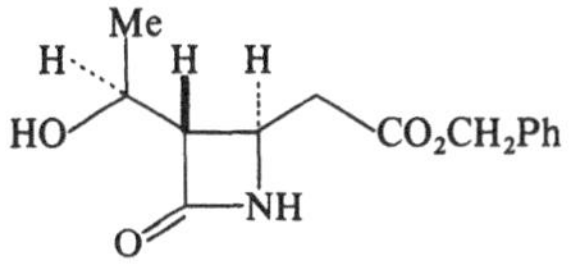

(152)

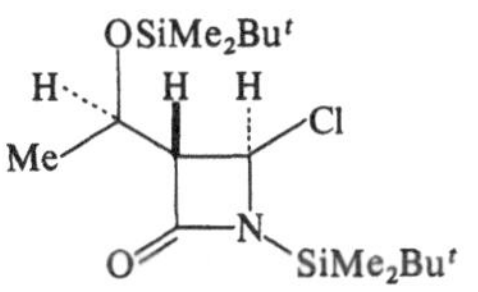

(153)

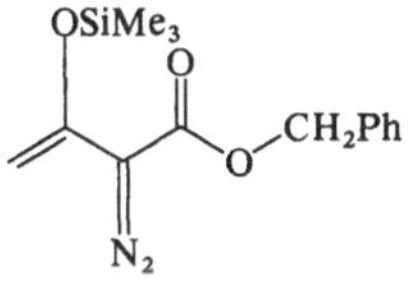

(154)

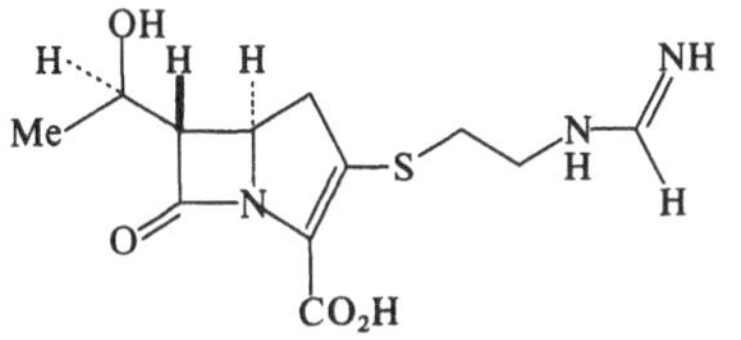

(155)

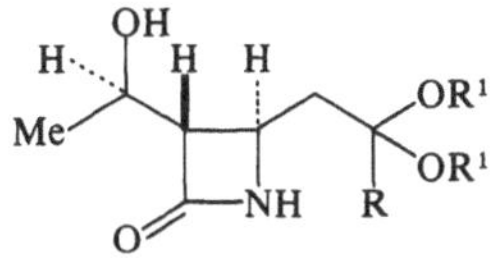

(156) R=H or CO₂Et

reactions to prepare intermediates for thienamycin synthesis. Thus he has obtained ketals of type (156) which are precursors for compounds such as (117) and (148). KAMETANI has also reviewed (*121*) the various approaches to the synthesis of carbapenem antibiotics. Several other formal synthesis of thienamycin have been reported (*122, 129*), including asymmetric approaches from carbohydrate precursors (*123, 125, 126*).

The Beecham group found that thiols add readily to the double bond of C(2)-unsubstituted 1-carbapenems, and this approach has been used to synthesise racemic PS-5 (*130*). The N-silylated 4-allylazetidinone (157) was alkylated with ethyl iodide and the product (158) transformed to the phosphorane (159). Cyclisation to (160) was followed by reaction with acetamidoethane thiol to form three isomers of the addition product (161). These could be converted to the carbapenem (162) on reaction with iodobenzene dichloride in the presence of pyridine. Isomerization to (163) and deprotection afforded the racemic natural product. The ester (163) has also been prepared *via* the diazo-intermediate (164) derived from the 4-acetoxy azetidinone (165) (*131*). A total synthesis of chiral PS-5 has been achieved using the resolved acid (166) (*132*). This was converted to (164) and then to optically pure PS-5. It has also been possible to synthesise PS-5 and PS-7 from the olivanic acid derivatives MM 17880 and MM 13902 (*133*). The benzyl ester of (±)-MM 22381 was obtained from the azabicycloheptane (167) derived from the addition of acetamidoethane thiol to the appropriate C(2)-unsubstituted nucleus (*108*).

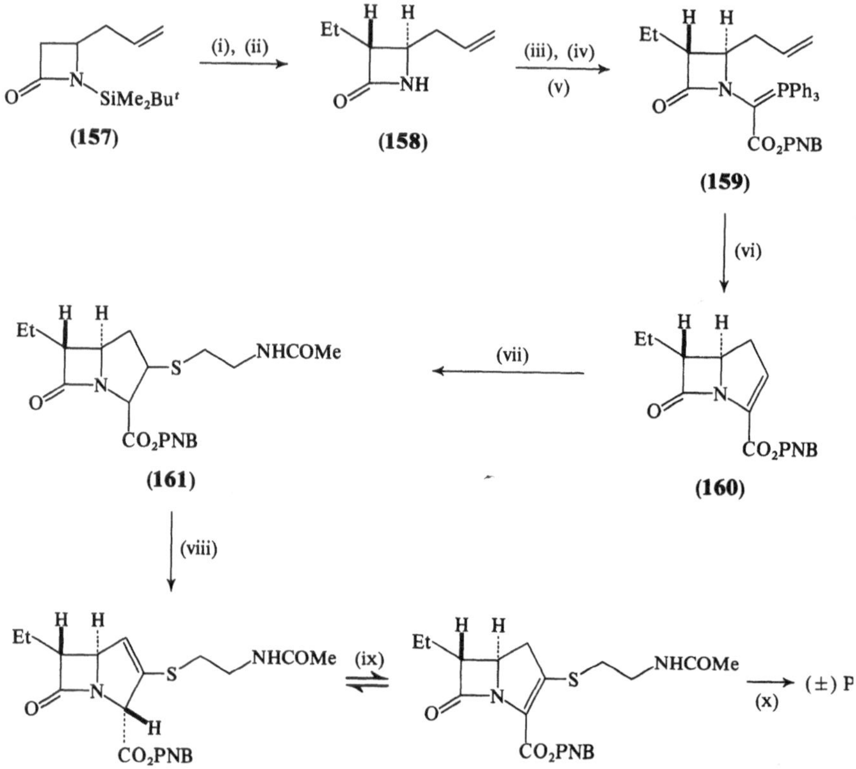

Reagents: (i) (Cyclo-C$_6$H$_{11}$)Pr^{i}NLi, EtI; (ii) KF, MeOH; (iii) CHOCO$_2$PNB; (iv) SOCl$_2$, 2,6-lutidine; (v) Ph$_3$P, 2,6-lutidine; (vi) TFA, O$_3$,Ph$_3$P,NaHCO$_3$; (vii) HS$\diagdown$NHCOMe, K$_2$CO$_3$; (viii) PhICl$_2$, pyridine; (ix) DBU, (x) H$_2$-Pd/C.

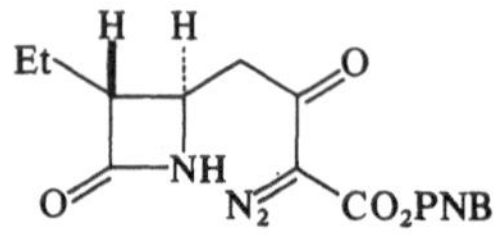

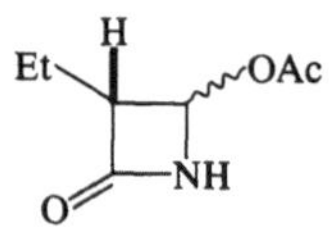

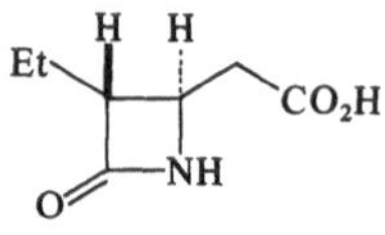

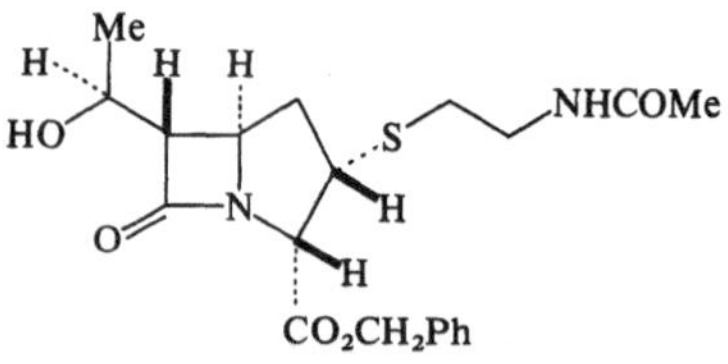

Two independent syntheses of the naturally occurring sulphoxide carpetimycin A (C-19393 H_2) have been published (*134, 135*). A major problem in this series is to obtain the *cis*-stereochemistry of β-lactam protons, while introducing the hydroxyisopropyl substituent at C(6). Most methods of introducing such side-chains lead to the more thermodynamically favoured *trans*-compounds. Sulphenylation of the oxazine (**114**) to (**168**) was followed by the normal aldol reaction with acetone to give (**169**).

(**171**) R=CH₂O(CH₂)₂OMe
(**174**) R=TMS

(**172**)

(**173**)

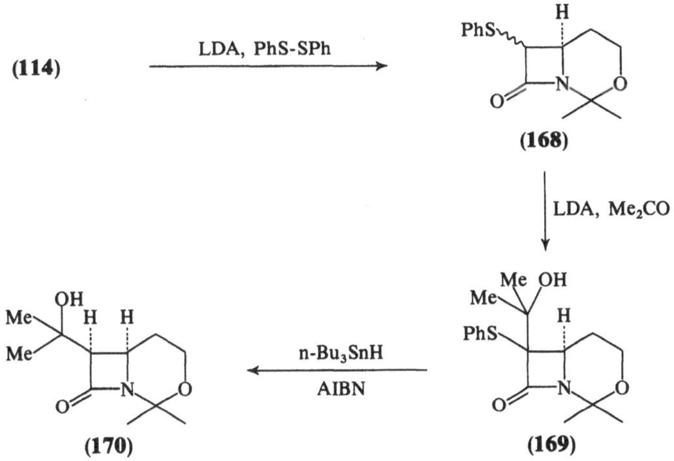

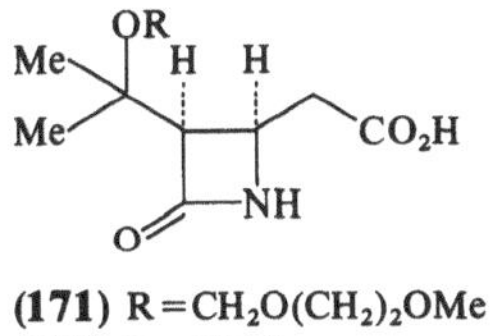

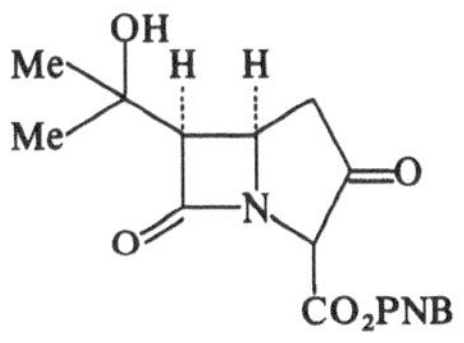

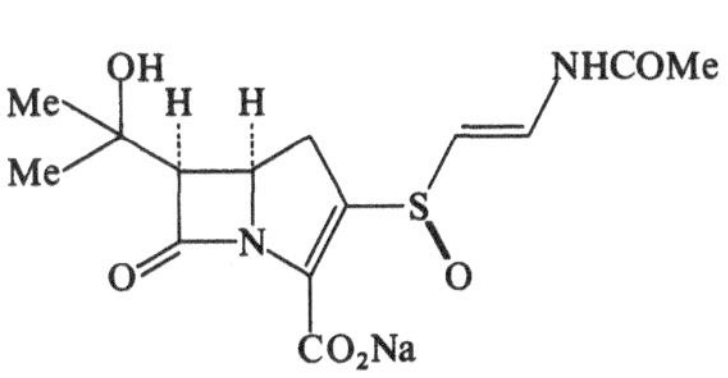

Radical desulphurisation was found to give the *cis*-product (**170**) irrespective of the stereochemistry of (**169**). Progression to the acid (**171**) was fairly straightforward, although protection of the alcohol was required to prevent lactone formation. Subsequent conversion to (**172**) was followed by introduction of the (E)-acetamidoethenylthio-substituent, oxidation and deprotection to form ($\pm$)-carpetimycin A (173) (134). The second approach (135) was also by way of (172), which in this case was derived from the chiral acid (174). Initially chirality was introduced into the synthetic sequence by enzymic hydrolysis of the prochiral ester (175) giving (176). Borohydride reduction gave the lactone (177) which could be converted to the substituted amino acid (178). Cyclisation to the β-lactam gave predominantly the *cis*-substituted product which was converted to (174) and (172). Conversion to optically active carpetimycin A (173) then followed the procedure of the previous scheme.

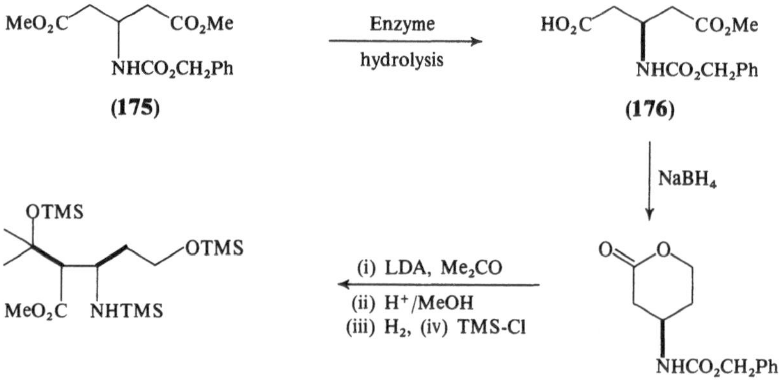

(175)   Enzyme hydrolysis   (176)

NaBH$_4$

(178)   (i) LDA, Me$_2$CO (ii) H$^+$/MeOH (iii) H$_2$, (iv) TMS-Cl   (177)

VII. Cephalosporins and Cephamycins

The cephalosporins form a large group of antibiotics in which the β-lactam is fused to a six membered dihydrothiazine ring. The first member of the group to be identified was cephalosporin C (179) isolated from a species of *Cephalosporium* (136, 137). Chemical studies established the presence of the α-aminoadipic acid side chain, but unlike the penicillins penicillamine could not be obtained on hydrolytic degradation. An important advance was made towards structural determination by use of nuclear magnetic resonance spectroscopy, which revealed the absence of the geminal dimethyl group characteristic of the thiazolidine ring of penicillins. Further the presence of an O-acetyl grouping was indicated. At this juncture

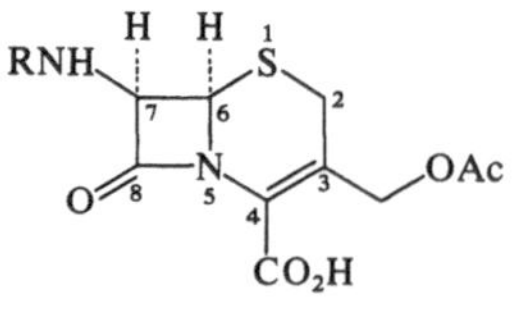

(179) R =

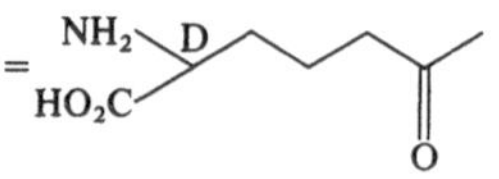

(180) R = H

structure (179) was proposed (137), and this was shortly afterwards confirmed by completion of an X-ray crystallographic study (138) on the sodium salt of cephalosporin C.

Early work on the reactivity of the acetoxy grouping in (179) showed that displacement reactions occurred fairly readily with certain heterocyclic tertiary bases (139), while hydrolysis to deacetylcephalosporin C was readily effected using an acetyl esterase (140). The structure elucidation of this first cephalosporin derivative in 1961 has been followed by the discovery of a number of other natural metabolites produced by fungi and various actinomycete species. The various structures are listed in Table 6. Aspects of the history, chemistry and biology of the group are covered in the extensive review edited by FLYNN (6).

Cleavage of the amide bond of the aminoadipyl side chain of cephalosporin C affords the 7-aminocephalosporanic acid (7-ACA) nucleus (180). This was first achieved in small yield by direct acid hydrolysis (151). The inability to cleave the side chain enzymatically as in the penicillins (6) resulted in much effort to find an efficient chemical method to provide (180), since this is the source of many clinically important semi-synthetic cephalosporins. Many of these compounds have advantages over the penicillins in terms of acid stability and resistance to β-lactamases.

The first effective method of side chain cleavage made use of the reaction of (179) with nitrosyl chloride (152). Loss of nitrogen from the intermediate diazo compound (181) gives imine (182) which on hydrolysis provides 7-ACA (180) and the lactone (183). This provided a route to substantial quantities of (180) thus allowing the preparation of many side chain derivatives. Later a more efficient procedure made use of phosphorus pentachloride followed by cleavage of an imino ether (153). The utility of other phosphorus halogen compounds for cleavage of the amide bond has also been described (154).

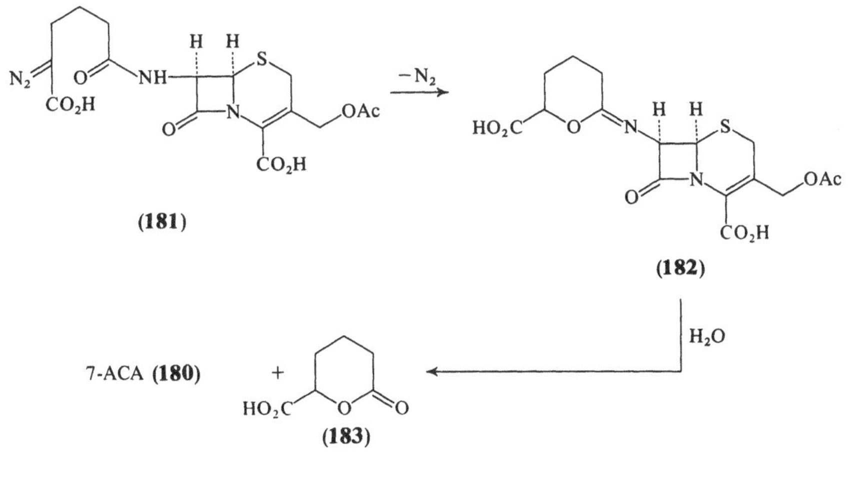

Table 6. *Naturally Occurring Cephalosporins*

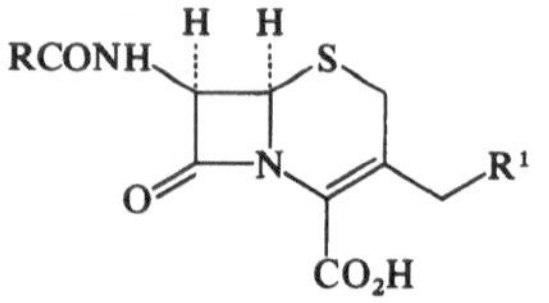

Cephalosporin	R	R^1	References
Cephalosporin C	$HO_2C-CH(CH_2)_3-$ $\quad\quad\quad NH_2$	$-OCOMe$	(136, 137)
Deacetylcephalosporin C	$HO_2C-CH(CH_2)_3-$ $\quad\quad\quad NH_2$	$-OH$	(141, 142)
Deacetoxycephalosporin C	$HO_2C-CH(CH_2)_3-$ $\quad\quad\quad NH_2$	$-H$	(143, 144)
F-1	$HO_2C-CH(CH_2)_3-$ $\quad\quad\quad NH_2$	$-SMe$	(145)
C-43-219	$HO_2C-CH(CH_2)_3-$ $\quad\quad\quad NH_2$	$-SC\begin{smallmatrix}Me\\Me\end{smallmatrix}-CH\begin{smallmatrix}CO_2H\\NH_2\end{smallmatrix}$	(146)
C-1778a	$HO_2C-(CH_2)_3-$	$-H$	(147)
C-1778b	$HO_2C-(CH_2)_3-$	$-OH$	(147)
C-1778c	$HO_2C-(CH_2)_3-$	$-OCOMe$	(147)
N-Acetyl-deacetoxy- cephalosporin C	$HO_2C-CH(CH_2)_3-$ $\quad\quad\quad NHCOMe$	$-H$	(148)
–	$HO_2C-CH(CH_2)_3-$ $\quad\quad\quad NH_2$	$-OCONH_2$	(149)
–	$HO_2C-CH(CH_2)_3-$ $\quad\quad\quad OH$	$-OH$	(150)

A large number of nuclear analogues of the cephalosporin ring system have been synthesised and comprehensively reviewed (46, 155, 156). Approaches to the natural products themselves have been limited. The Roussel (157) and Squibb (158) groups have utilised suitably protected intermediates of type (184) to make the amino acid (185). Cyclisation as in SHEEHAN's penicillin synthesis followed by deprotection afforded the amino lactone (186).

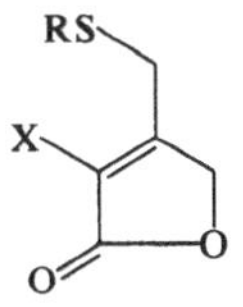

(184) X=OH or NH₂   (185) 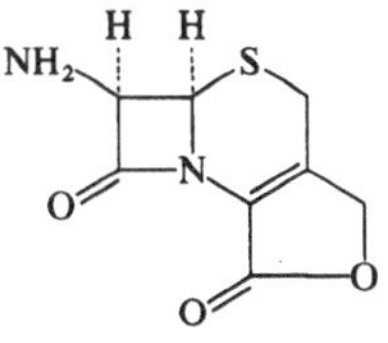   (186)

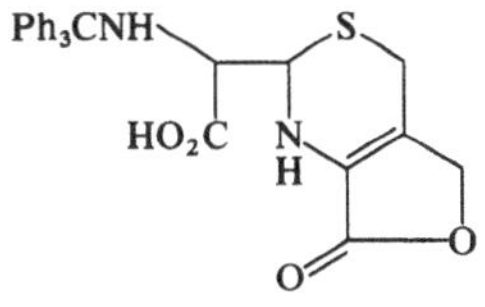

Acylation to the cephalothin analogue (187) is fairly straightforward. Ring opening of the lactone ring of (187) to provide deacetylcephalothin (188) is possible, but the yield is poor (159). Another synthesis of (±)-deacetylcephalothin lactone (187) makes use of the cycloaddition reaction of azidoacetyl chloride with 4H-furo[3,4-d]-1,3-thiazine (160). One other important process for obtaining cephalosporins is to start from a penicillin sulphoxide. Thermal rearrangement of the latter in the presence of acid gives the cephalosporin ring system by way of a sulphenic acid intermediate (161 – 163).

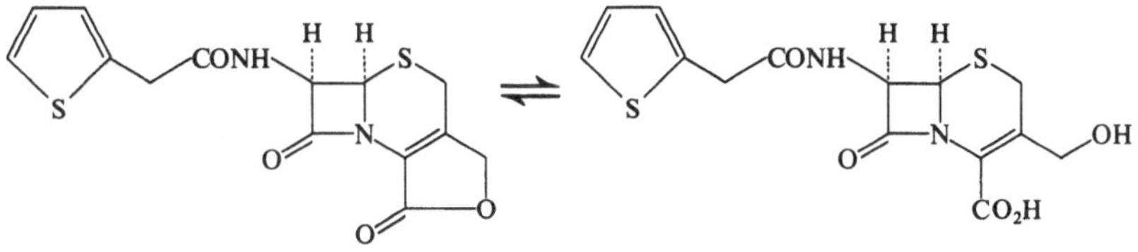

(187)   (188)

The only complete synthesis of cephalosporin C provides a classic example of the synthetic art of the late R. B. WOODWARD, and formed the subject of his Nobel Prize winning lecture of 1965. The synthesis, published in 1966, starts from L(+)-cysteine (189) (164, 165). Protection of the nitrogen, sulphur and acid functional groups provided the cyclic intermediate (191) ideally suited for the introduction of the amino function which was to become the nitrogen atom of the crucial β-lactam intermediate (195). This was achieved in a completely stereocontrolled manner by a novel substitution method. Thermal introduction of the hydrazo-substituent to form (192) was followed by oxidation and conversion to the trans-hydroxy ester (193). Inversion by displacement of the mesylate with azide and subsequent reduction gave the cis-amino ester (194) which afforded the β-lactam (195) on cyclisation.

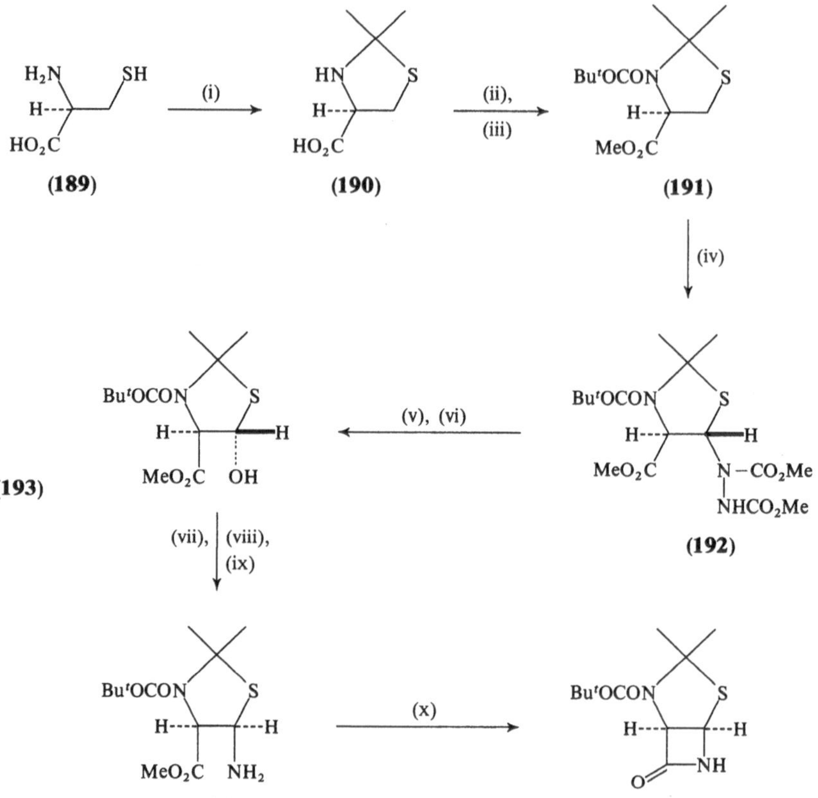

Reagents: (i) Me$_2$CO, Δ; (ii) Bu^{t}OCOCl, pyridine; (iii) CH$_2$N$_2$;
(iv) MeO$_2$C$-$N$=$N$-$CO$_2$Me, Δ; (v) Pb(OAc)$_4$; (vi) NaOAc, MeOH;
(vii) MeSO$_2$Cl, diisopropylamine; (viii) NaN$_3$; (ix) Al/Hg; (x) triisobutylaluminium.

The β-lactam (195) readily reacted with the dialdehyde (196), derived from the aldol product of trichloroethyl glyoxylate and malondialdehyde. The N- and S-protecting groups could now be removed from (197) with trifluoroacetic acid, a process which also effected ring closure to the amino aldehyde (198). This was acylated to give the D-α-aminoadipic acid derivative (199) protected on both the nitrogen and carboxylic acid groupings. Reduction, acetylation, and equilibration provided the cephalosporin C ester (201). Removal of the protecting groups using zinc and acetic acid gave cephalosporin C (179) identical with authentic material. The use of the trichloroethyl residue as a protecting group was another innovation introduced by WOODWARD and was subsequently employed in many other areas of β-lactam chemistry.

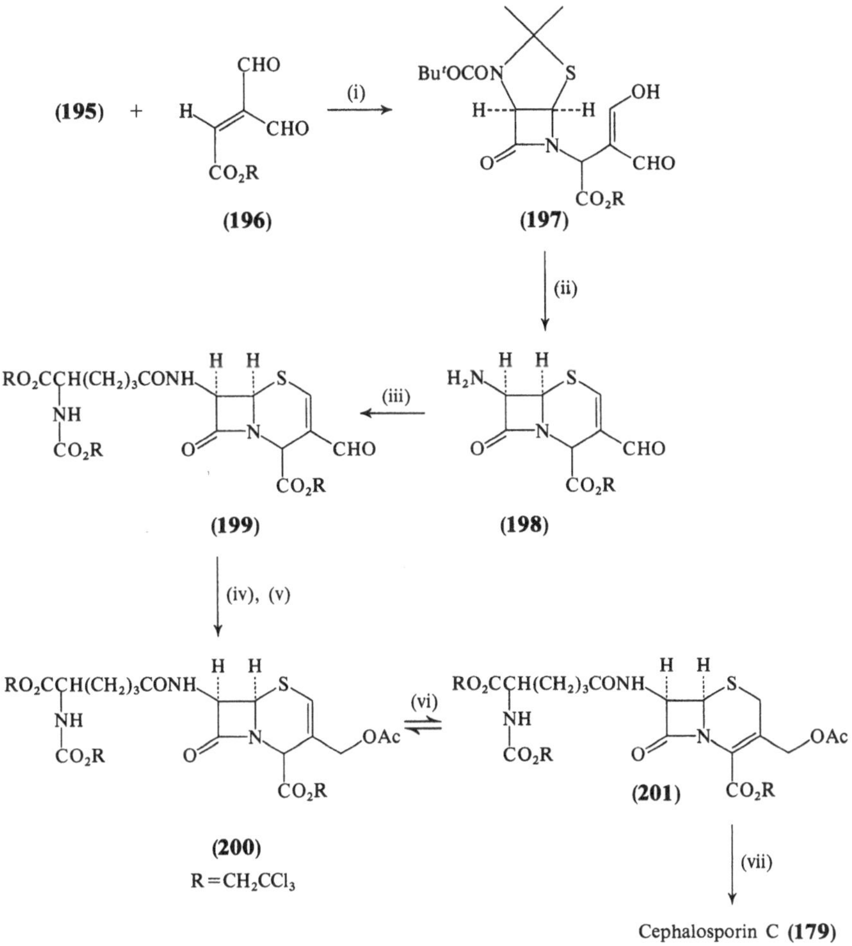

Reagents: (i) Δ, 80°; (ii) CF₃CO₂H; (iii) Acylation; (iv) Diborane; (v) Ac₂O, pyridine; (vi) Pyridine; (vii) Zn − acetic acid.

Some ten years after the structure of cephalosporin C had been established, the isolation of two naturally occurring cephalosporins possessing a 7(α)-methoxy group was reported (*149*). These were identified as 7-methoxycephalosporin C (**202**) and the C(3)-carbamate (**203**). Further examples of this type of natural product have been reported and are listed in Table 7. All possess the α-aminoadipic acid side chain, but vary in the substitution pattern at C(3). Collectively they are known as the cepha-mycins (*174*). A major point of interest with the cephamycin type of antibiotic is their intrinsically higher resistance to hydrolysis by β-

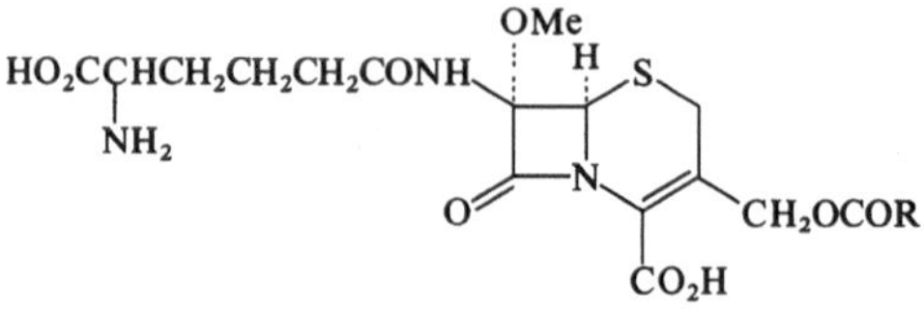

$$HO_2CCHCH_2CH_2CH_2CONH \quad \overset{OMe}{\underset{}{\overset{H}{\underset{NH_2}{|}}}} S$$

(structure with O, N, CH_2OCOR, CO_2H)

(202) R = Me
(203) R = NH$_2$

Table 7. *Naturally Occurring Cephamycins*
(7-Methoxycephalosporins)

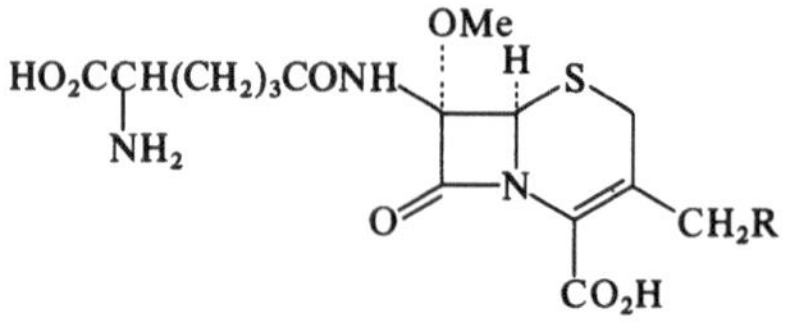

$$HO_2CCH(CH_2)_3CONH \quad \overset{OMe}{\overset{H}{}} S$$

(structure with NH_2, O, N, CH_2R, CO_2H)

Compound	R	References
Cephamycin C	$-OCONH_2$	(*149*)
7-Methoxycephalosporin C	$-OCOMe$	(*149*)
Cephamycin A	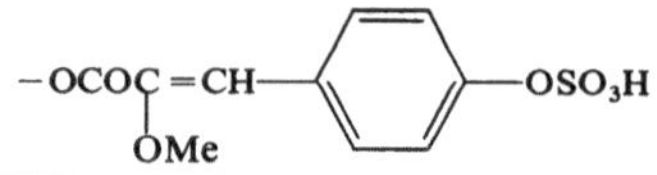 $-OCOC=CH-\!\!\!\bigcirc\!\!\!-OSO_3H$ with OMe	(*166, 167, 168*)
Cephamycin B	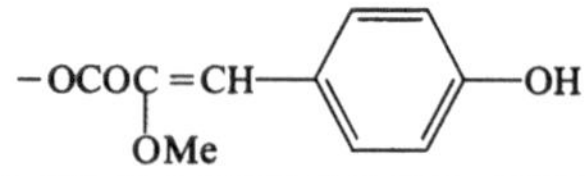 $-OCOC=CH-\!\!\!\bigcirc\!\!\!-OH$ with OMe	(*166, 167, 168*)
C-2801X	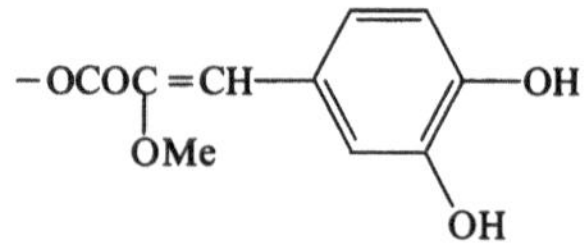 $-OCOC=CH-\!\!\!\bigcirc\!\!\!-OH$ with OMe and OH	(*169*)
7-Methoxydeacetoxy-cephalosporin C (WS 3442D)	H	(*170*)
7-Methoxydeacetyl-cephalosporin C (Y-G19ZD3)	OH	(*171, 173*)
Oganomycin A	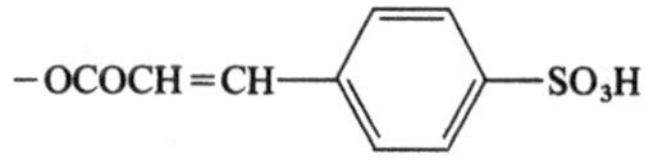 $-OCOCH=CH-\!\!\!\bigcirc\!\!\!-SO_3H$	(*172*)
SF-1623	$-S-SO_3H$	(*173*)

lactamases compared to the unsubstituted compounds (*165*). This also applies to certain 6-methoxy substituted penicillins, although in this case no naturally occurring metabolites have yet been discovered (*175*).

As for the cephalosporins, one of the first areas of cephamycin chemistry to be investigated concerned the replacement of the amidoadipic side chain with new acylamino substituents. This was first achieved by a novel acyl exchange reaction (*176, 177*). Under appropriate conditions the cephamycin C ester (**204**) can be converted to diacyl derivatives such as (**205**). Removal of the trichloroethyl protecting groups leads to formation of the piperidone (**206**) and the new cephamycin ester (**207**). Deprotection of the latter affords the important semi-synthetic derivative cefoxitin. Other routes making use of the chemistry of various derived imines and imidoyl chlorides have been described (*178 – 181*). Another approach makes use of the reaction of (**204**) with oxalyl chloride to give the oxamic acid (**208**); with diphenylcarbodiimide (**208**) produced the amino ester (**209**) and the trione (**210**) (*182*).

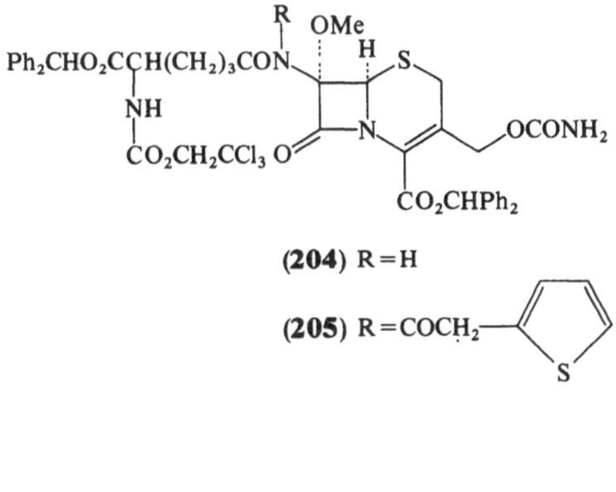

(**204**) R = H

(**205**) R = COCH$_2$—

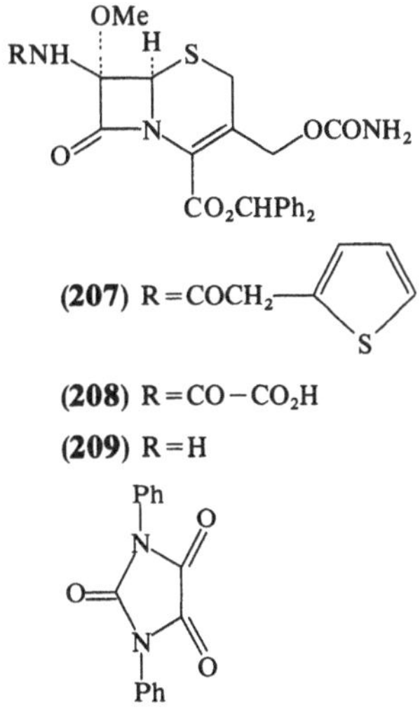

(**206**)

(**207**) R = COCH$_2$—

(**208**) R = CO – CO$_2$H

(**209**) R = H

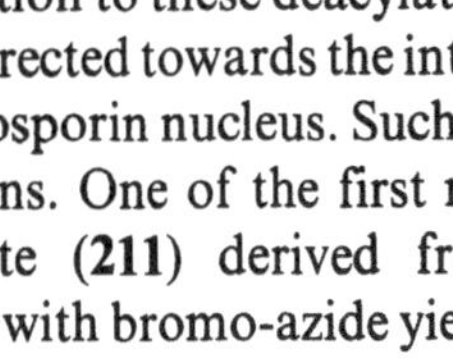

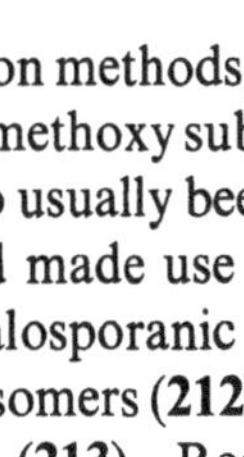

(**210**)

In addition to these deacylation or transacylation methods, much effort has been directed towards the introduction of a 7α-methoxy substituent into the cephalosporin nucleus. Such methods have also usually been applicable to penicillins. One of the first methods developed made use of the diazo intermediate (**211**) derived from 7-aminocephalosporanic acid (*183*). Treatment with bromo-azide yielded a mixture of isomers (**212**), which with silver tetrafluoroborate and methanol formed (**213**). Reduction and acylation readily followed. The stereoselective addition of methoxide to acylimine intermediates such as (214) is a method much used (*184, 185*), while addition to sulfenimines (215) is also possible (*186—188*).

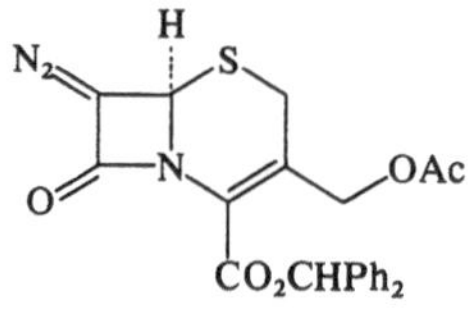

(211)

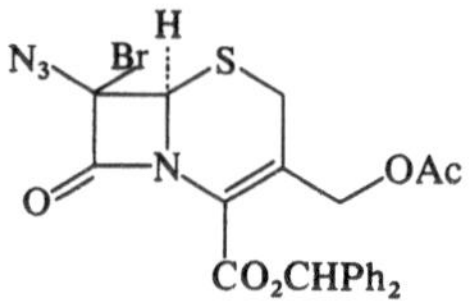

(212)

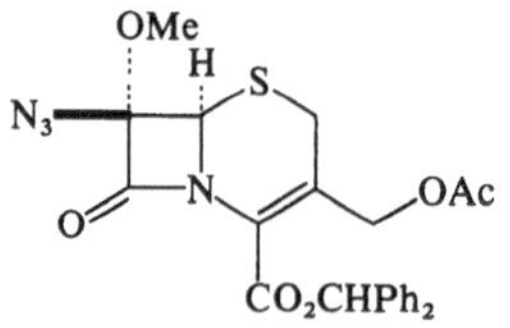

(213)

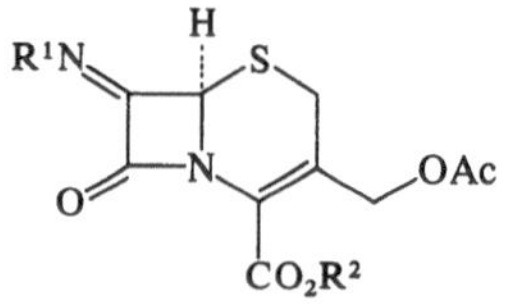

(214) $R^1 = RCO$
(215) $R^1 = RS$

Another approach uses imino derivatives of type (216), which form methoxy ketenimines (218) by way of addition of methoxide to (217) (*189*). Conversion to the acylamino side chain is then possible. Other variations using ketenimines have been reported (*190, 191*). One other method using imines involves addition to the quinonoid intermediate (219), derived by oxidation of the corresponding phenolic Schiff's base (*192*).

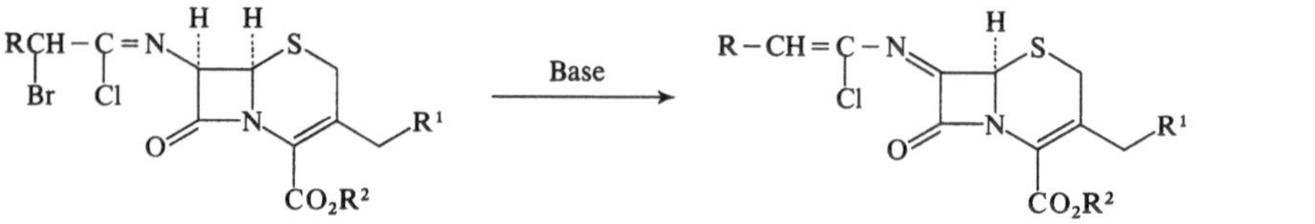

(216)

(217)

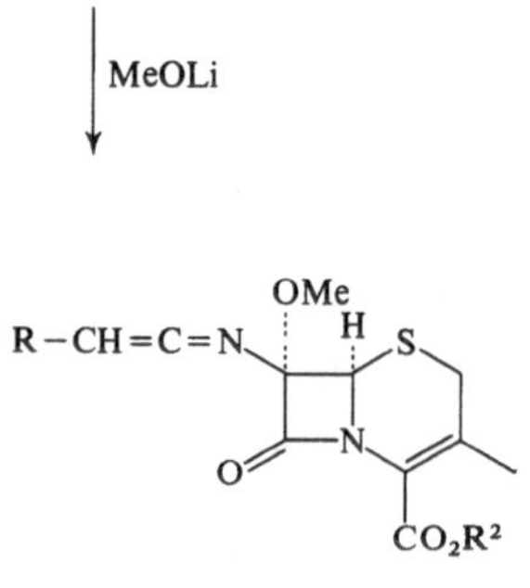

| MeOLi

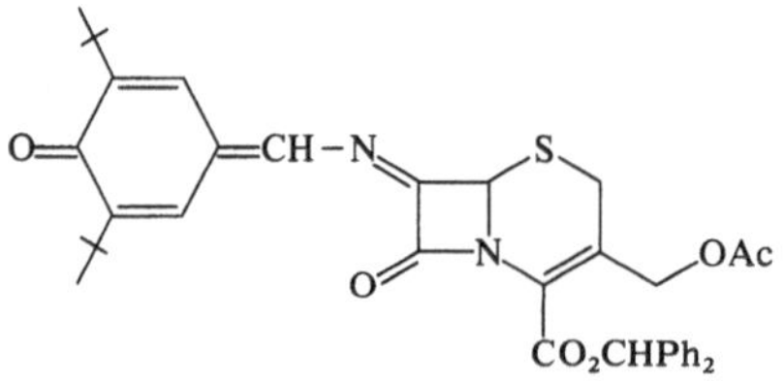

(219)

(218)

A second major approach to semi-synthetic cephamycin synthesis utilises carbanion formation α to the β-lactam carbonyl. Reaction with electrophiles then provides substrates suitable for the introduction of the methoxy group. Several papers have described how the anion generated from the Schiff's base (**220**; $R^1 = Bu^t$) reacts readily with methyl methan-ethiosulfonate to form (**221**; $R^1 = Bu^t$). Solvolysis in methanol in the presence of mercury salts leads to (**222**; $R^1 = Bu^t$) (*193 – 196*). A similar method uses the *p*-nitrobenzylidene Schiff's base and methoxycarbonyl-methyl disulphide as the electrophile (*197*). Bromination using the anion from (**220**; $R^1 = CHPh_2$) affords (**223**; $R^1 = CHPh_2$) in which the bromine can be replaced by methoxy to form (**222**; $R^1 = CHPh_2$) (*198*).

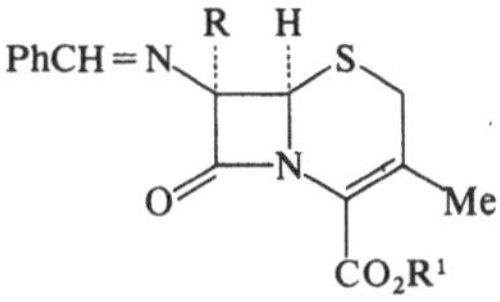

(**220**) R=H
(**221**) R=SMe
(**222**) R=OMe
(**223**) R=Br

The azido intermediate (**224**) can be prepared by a cycloaddition reaction using the thiazine (**225**) and azidoacetyl chloride in the presence of triethylamine (*199*). Reduction and Schiff's base formation gives (**226**) which can be converted to (±)-cephalothin. On the other hand introduction of the thiomethyl group followed by acylation gives (**227**), ideally suited for conversion to (**228**) and ultimately (±)-cefoxitin (*200*).

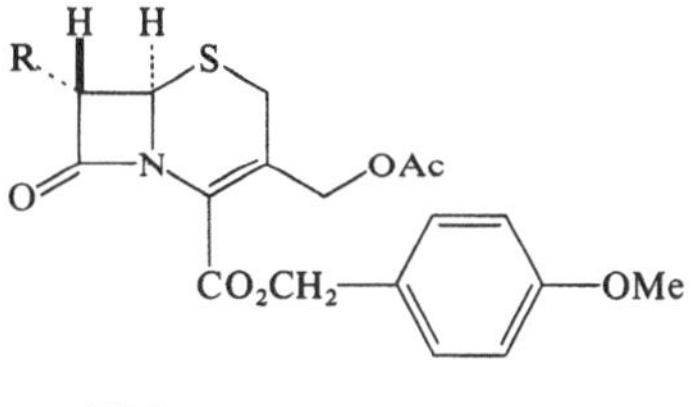

(**224**) R=N₃
(**226**) R=*p*-NO₂−C₆H₄CH=N

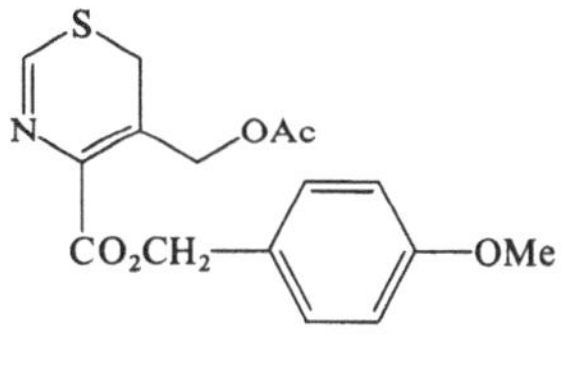

(**225**)

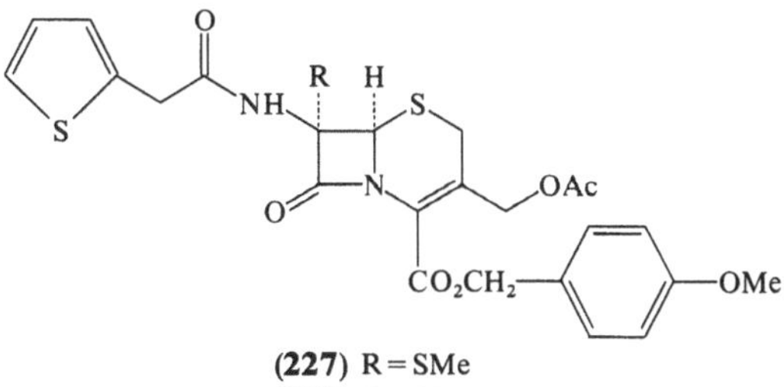

(**227**) R=SMe
(**228**) R=OMe

One other approach to the cephamycin ring system is that reported by
Kishi (*201, 202*) using a synthetic scheme based mainly on biosynthetic
pathways suggested by Cooper (*203*). *N*-Acetylbromodehydroalanine *t*-
butyl ester (**229**) was converted to the methoxydibromo acid (**230**), and then
to the dehydrovaline derivative (**231**). The amide (**231**) was converted to the
thioamide (**232**), which without purification was successfully used in a
double cyclisation reaction to give the β-lactam thiazoline (**233**).

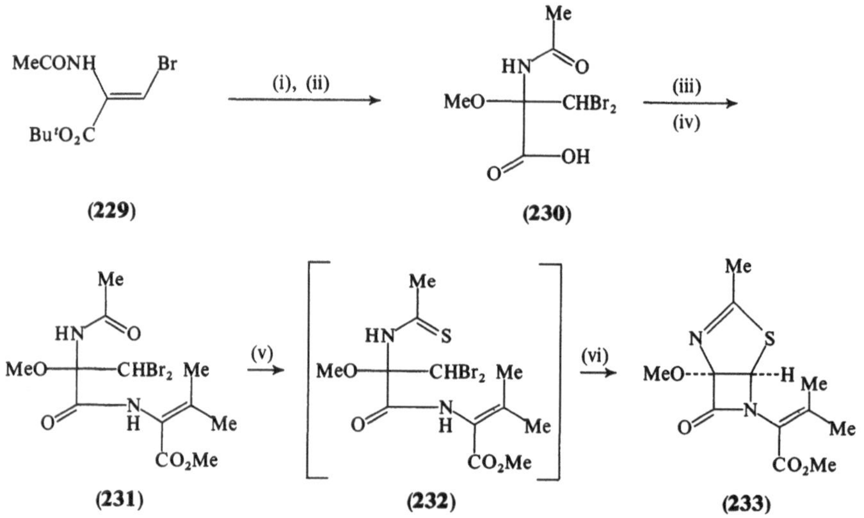

(**229**)   (**230**)

(**231**)   (**232**)   (**233**)

Reagents: (i) Br$_2$, MeOH−CH$_2$Cl$_2$; (ii) H$^+$; (iii) DCC, NH$_2$⟨...⟩; 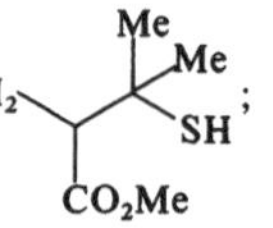
(iv) Tosyl chloride-pyridine, Et$_3$N; (v) P$_2$S$_5$; (vi) NaH.

Further elaboration of (**233**) involving allylic bromination and reduction gave the deconjugated ester (**234**) with the "natural" penicillin
configuration together with the "unnatural" isomer. Oxidation of (**234**) in
the presence of trifluoroacetic acid followed by reduction gave a complex
mixture from which could be isolated the methoxy cephalosporin ester (**235**)
and the penicillin (**236**). Alternatively the bromide (**237**) can be induced to
undergo cyclisation to (**235**) by merely allowing a methylene chloride
solution of (**237**) to evaporate to dryness at room temperature over three
days.

Although naturally occurring cephalosporins in which the sulphur atom
is replaced by oxygen are not known, mention should be made of the semi-
synthetic 7-methoxy-1-oxacephalosporin derivative moxalactam (6059-S)

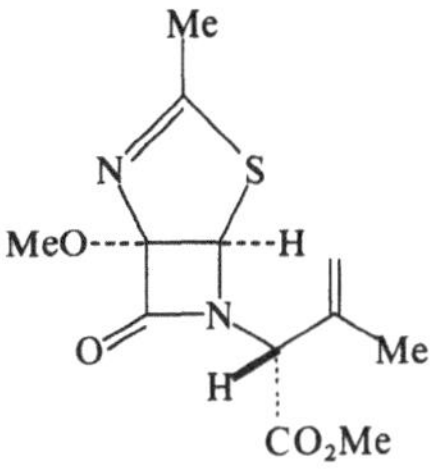

(234) 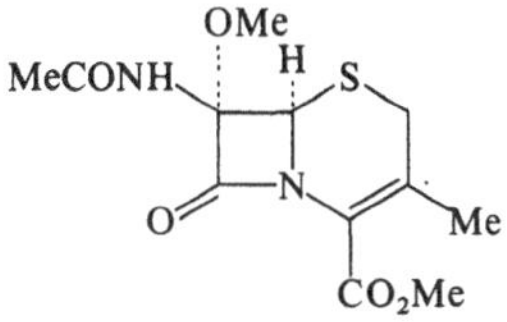   (235)

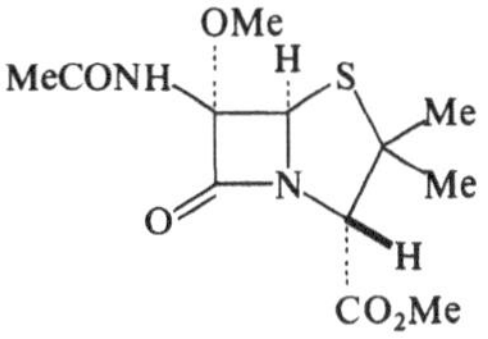

(236) 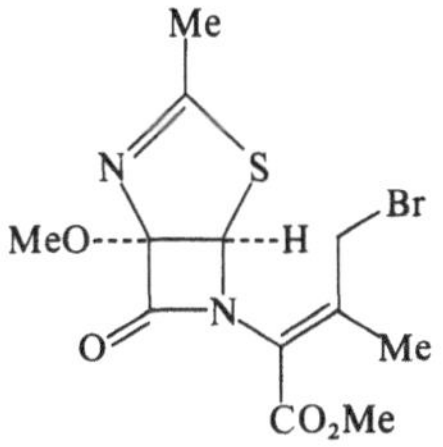   (237)

(238). Developed by the Shionogi group (204), this highly potent antibiotic is prepared by a multistage synthetic sequence starting from the penicillin nucleus. The successful large scale use of such a route to obtain this clinically important substance provides a good illustration of the great advances made in the synthetic manipulation of β-lactams since the first discovery of the penicillins and cephalosporins.

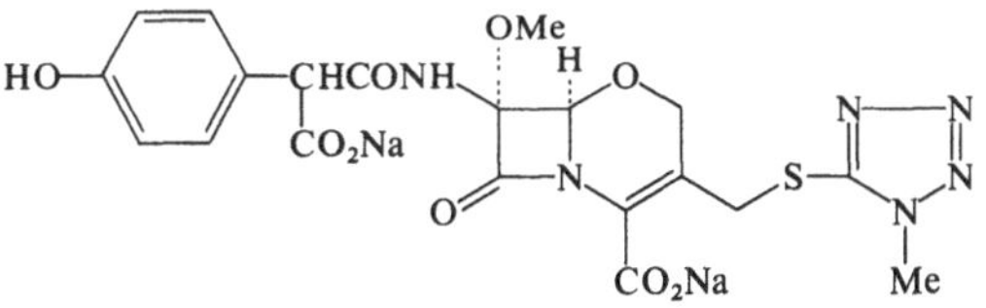

(238) Moxalactam (6059-S)

VIII. Pachystermines, Wild-Fire Toxin and Antimetabolites

The occurrence of the β-lactam ring amongst natural products other than the β-lactam antibiotics is small. In the alkaloid field two examples are known. These are the two steroidal alkaloids pachystermines A (239) and B (240), the structures of which were determined by extensive degradation and spectroscopic studies (1).

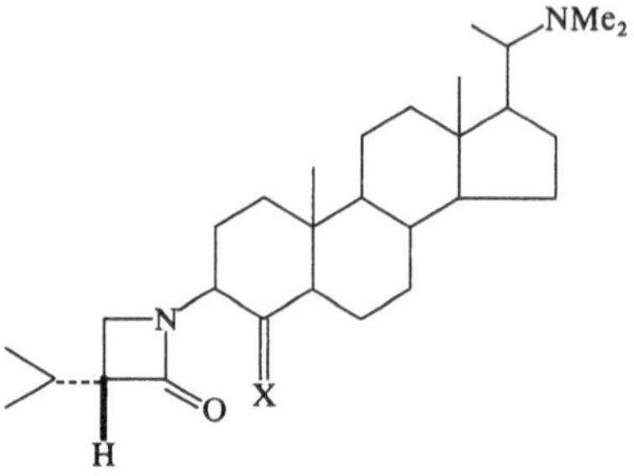

(239) X=O
(240) X=H, OH

Wild-fire toxin is produced by *Pseudomonas tabaci,* an organism which causes a leafspot disease in tobacco plants. The correct structure was determined by Stewart (*2*), and shown to be **(241)**, although no assignment of stereochemistry was made. Wild-fire toxin is an unstable substance rapidly converted at neutral pH to isotabtoxin **(242)** by way of intramolecular attack of the amino-group on the β-lactam. Hydrolysis of wild-fire toxin gives threonine and tabtoxinine **(243)**.

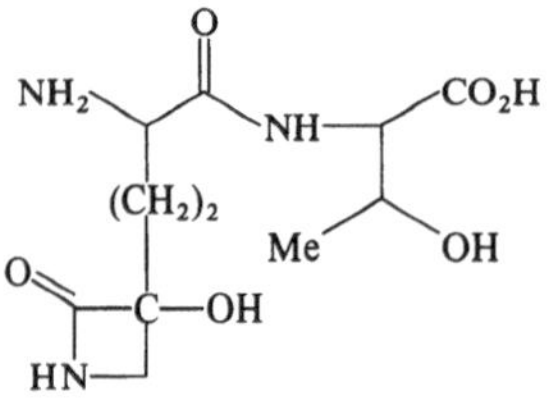

(241) 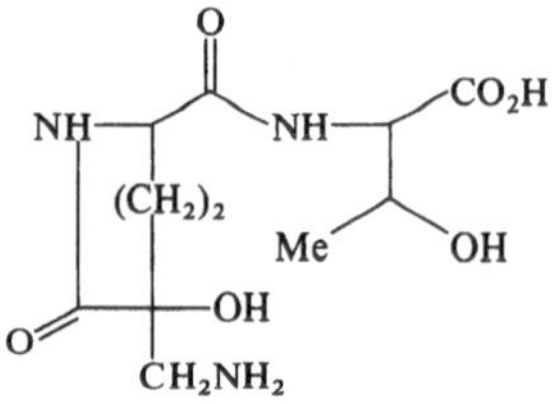   **(242)** 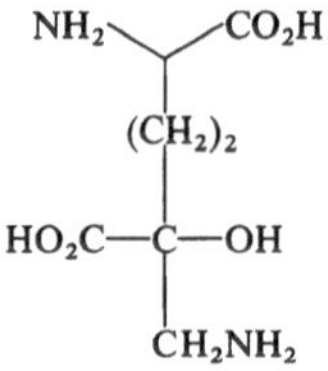   **(243)**

In the course of a screening programme which produced a number of antimetabolites from various microorganisms, Scannell has described the isolation of a monocyclic β-lactam from an unidentified *Streptomyces* species (*3*). The compound was identified as **(244)**, and is somewhat similar to the wild-fire toxin.

A stereospecific synthesis of the wild-fire toxin (tabtoxin) has recently been described (*219*).

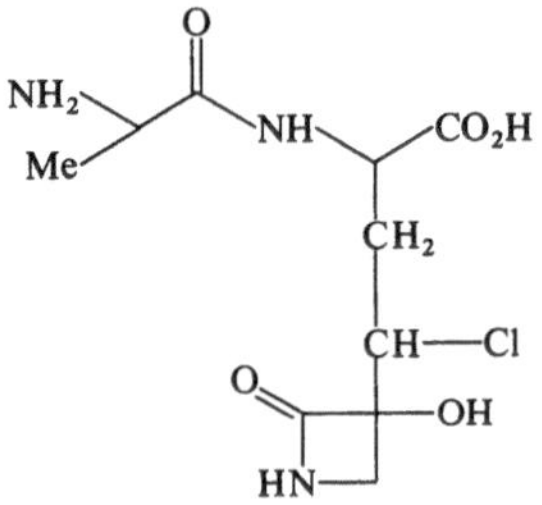

(244)

IX. Biosynthesis

1. Biosynthesis of Nocardicin A

Examination of the structure of nocardicin A (**1**) reveals that the compound contains three α-amino acid units and one α-oximino acid, residue and suggests that the compound is probably derived from a peptide.

The first reported study on the biosynthesis of nocardicin A was by HOSODA *et al.*, who administered a variety of ^{14}C-labelled compounds to shaken fermentations of *Nocardia uniformis* subsp. *tsuyamonensis* (206). Examination of the resulting nocardicin A showed that L-[U-^{14}C]-tyrosine, [G-^{14}C]-shikimic acid, and L-[U-^{14}C]-serine were all incorporated into the antibiotic. [U-^{14}C]-Glycine and L-[U-^{14}C]-homoserine were also incorporated but to a lesser degree, whereas the labels from L-[1-^{14}C]-tyrosine, L-[U-^{14}C]-phenylalanine, L-[U-^{14}C]-alanine and DL-[U-^{14}C]-α-amino-butyric acid were not incorporated to any significant extent.

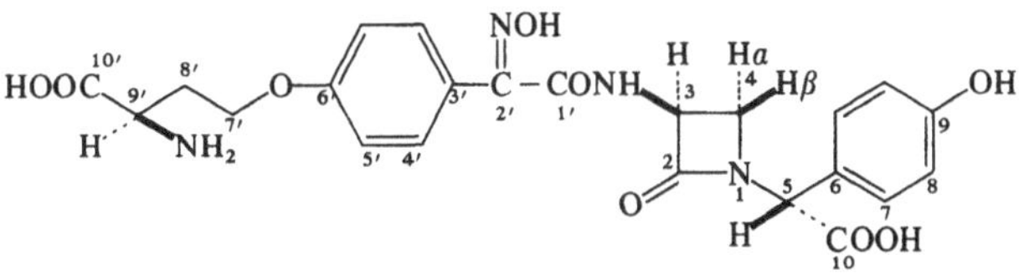

(**1**)

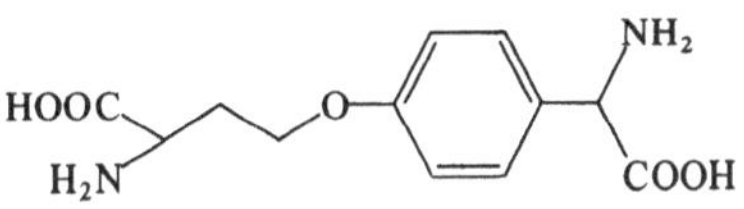

(**245**) 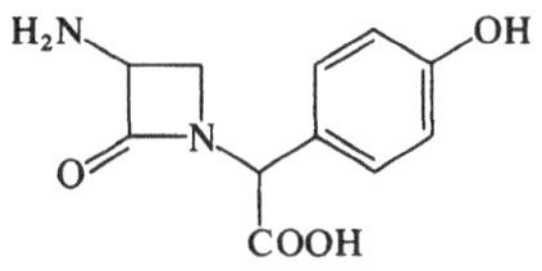   (**7**)

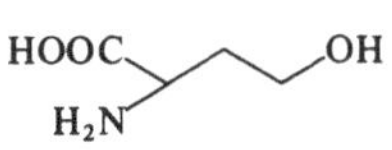

(**246**)

In order to locate the position of label the authors used three degradation techniques. The first involved reduction of the oximino to an amino group by hydrogenation over palladium on charcoal. The resulting nocardicin C was then hydrolysed with an acylase from a *Pseudomonas* sp to give 4-(3-amino-3-carboxypropoxy)-phenylglycine (**245**) and 3-aminono-cardicinic acid (**7**). In the second degradation the *p*-hydroxyphenylglycine (**6**) residue was isolated by hydrolysing nocardicin A with aqueous potassium hydroxide. The third degradation involved hydrogenation of nocardicin A over platinum oxide to release the terminal D-homoserine residue (246).

By use of the above degradations the authors showed that L-[U-^{14}C]-tyrosine was incorporated specifically into the two aromatic residues, L-[U-^{14}C]-homoserine was incorporated specifically into the D-homoserinyl residue and L-[U-^{14}C]-serine was incorporated specifically into the β-lactam ring of nocardicin A.

Townsend and Brown (207) extended these studies by feeding both radiolabelled and stable isotope labelled compounds to cultures of *N. uniformis* subsp. *tsuyamonensis*. They found that the carbon label of L-[2-^{3}H, 1-^{14}C]-*p*-hydroxyphenylglycine was incorporated much more efficiently than that of L-[U-^{14}C]-tyrosine but that there was substantial loss of the tritium label. D-[2-^{3}H, 1-^{14}C]-*p*-hydroxyphenylglycine was relatively poorly incorporated, though the authors did not present evidence that the compound was taken up by the cells. The position of incorporation was determined by feeding DL-[1-^{13}C]-*p*-hydroxyphenylglycine and examining the resulting labelled nocardicin A by ^{13}C-NMR spectrometry, when ^{13}C-enrichment was observed at C-10 and C-1'. Townsend and Brown confirmed the observation of Hosoda *et al.* that L-[U-^{14}C]-homoserine was incorporated but also showed that the incorporation of L-[1-^{14}C]-methionine was substantially better. They also showed that [1-^{14}C]-glycine was incorporated to a lesser extent than [2-^{14}C]-glycine which in turn was less well incorporated than L-[U-^{14}C]-serine. ^{13}C-NMR studies showed that [2-^{13}C]-glycine enriched C-3 and C-4 of nocardicin A (C-3 to a greater extent than C-4) whereas DL-[3-^{13}C]-serine enriched only C-4. The authors suggested that glycine was converted to serine in *N. uniformis* by serine hydroxymethyltransferase. In a double labelling experiment Townsend and Brown showed that both protons on C-3 of L-serine were retained during incorporation into nocardicin A.

Townsend *et al.* extended their studies on serine incorporation by showing that L-[2-^{3}H, 1-^{14}C]-serine was incorporated into nocardicin A with a 19% retention of tritium label which they concluded was adequate to rule out the possibility of a dehydroalanyl intermediate (208). Neither the carbon label nor the tritium label of D-[2-^{3}H, 1-^{14}C]-serine was incorporated. The authors also synthesised serines labelled with a deuterium atom on C-3 in the form of the enantiomeric pairs (247) and (248), and (249) and (250). These pairs were not resolved as only the L-isomer would be

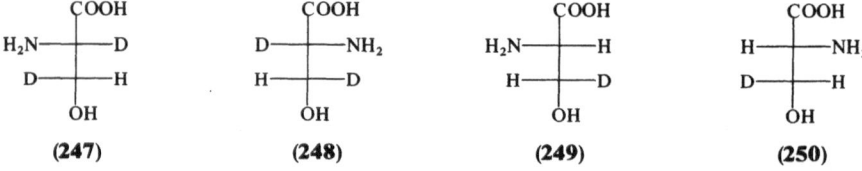

(247)   (248)   (249)   (250)

incorporated in each case. Feeding the mixtures to *N. uniformis* was therefore equivalent to feeding the (2*S*, 3*R*) isomer (**247**) or the (2*S*, 3*S*) isomer (**249**). When the resulting samples of nocardicin A from these experiments were examined by ^{2}H-NMR spectrometry it was found that inversion of stereochemistry had taken place, the (3*R*) deuterium label appearing at the 4α position of nocardicin A and the (3*S*) deuterium label appearing at the 4β position, i.e. (**251**)→(**1**). The authors concluded that

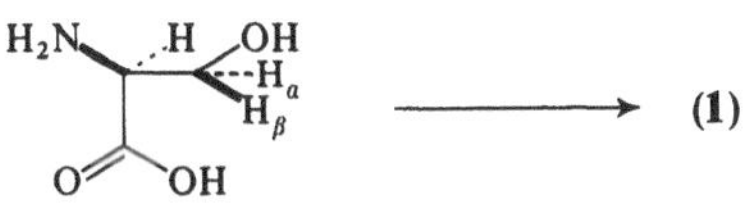

(**251**)

this inversion of stereochemistry probably occurred during the β-lactam ring closure as the result of nucleophilic attack by the amide nitrogen on the β-carbon, which resulted in displacement of serine hydroxyl in a classic S_N2 reaction.

The biosynthesis of nocardicin A is summarised schematically in Fig. 2.

2. Biosynthesis of the Monobactams

As is the case with the nocardicins, the β-lactam ring of the monobactams is derived from the amino acid L-serine. Working with the simplest monobactam (**34**) produced by *Chromobacterium violaceum*, O'SULLIVAN *et al.* (*209*) showed that L-[U-^{14}C]-serine was better incorporated than L-[U-^{14}C]-cysteine, [2-^{14}C]-pyruvate, L-[U-^{14}C]-lactate, L-[U-^{14}C]-alanine or L-[U-^{14}C]-aspartate. The addition of unlabelled L-serine to *C. violaceum* fermentations depressed the incorporation of L-[U-^{14}C]-cysteine, [U-^{14}C]-glycine and [2-^{14}C]-pyruvate, whereas the addition of unlabelled glycine or pyruvate had little effect on the incorporation of L-[U-^{14}C]-serine. Allylglycine, L-cysteine and D-cysteine all stimulated the incorporation of L-[U-^{14}C]-serine. [2-^{14}C]-Glycine was better incorporated than [1-^{14}C]-glycine, and the authors concluded that this probably indicated that glycine is the precursor of serine in *C. violaceum* (*cf.* nocardicin biosynthesis).

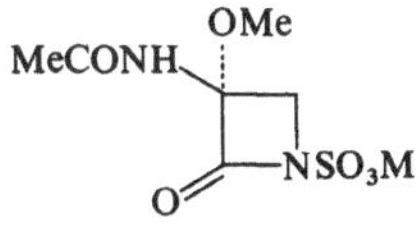

(**34**)

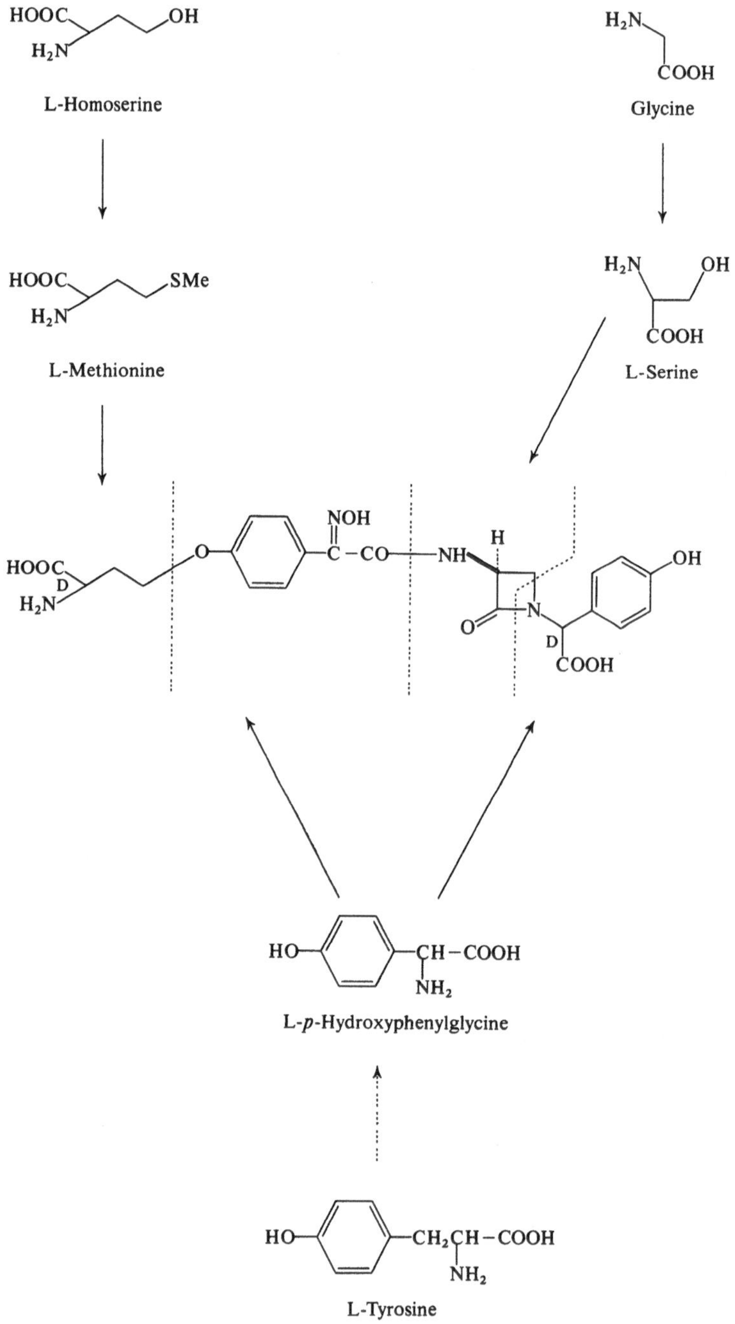

Fig. 2. Biosynthesis of nocardicin A

In order to confirm the role of serine the same workers administered a mixture of L-[U-^{14}C]-serine and L-[3-^{3}H]-serine to a *C. violaceum* fermentation. The resulting ^{14}C-labelled (**34**) showed 101% retention of the tritium label. Similarly, feeding a mixture of L-[3-^{14}C]-serine and L-[3-^{3}H]-serine resulted in 83% tritium retention in the product. However in two experiments where mixtures of L-[U-^{14}C]-cystine and L-[3,3′-^{3}H]-cysteine were fed only 14% and 20% of the tritium label was retained in the products. These experiments clearly indicated that serine is a closer precursor of (**34**) than cysteine. Again, the retention of tritium label from L-[3-^{3}H]-serine is reminiscent of nocardicin biosynthesis and suggests that the β-lactam ring is closed by an S_N2 displacement of the serine hydroxyl.

The investigation of the role of serine in the biosynthesis of the monobactam (**252**) was more complex. This amino acid could possibly be incorporated into the D-alanyl residue of the side chain as well as into the β-lactam ring. A mixture of ^{14}C- and ^{3}H-labelled serine was converted to alanine but with loss of 90% of the tritium label. However, when this mixture was fed to growing cells of *Acetobacter* sp., the tritium label was efficiently retained indicating that the labelled serine was being incorporated into the β-lactam ring rather than into the side chain alanyl residue.

In a further double-isotope labelling experiment O'SULLIVAN *et al.* showed that L-serine was incorporated into (**253**) produced by *Agrobacterium radiobacter*, again without loss of tritium label.

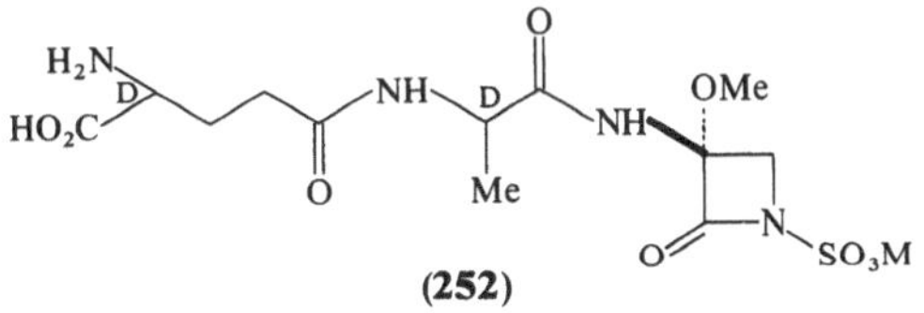

(**252**)

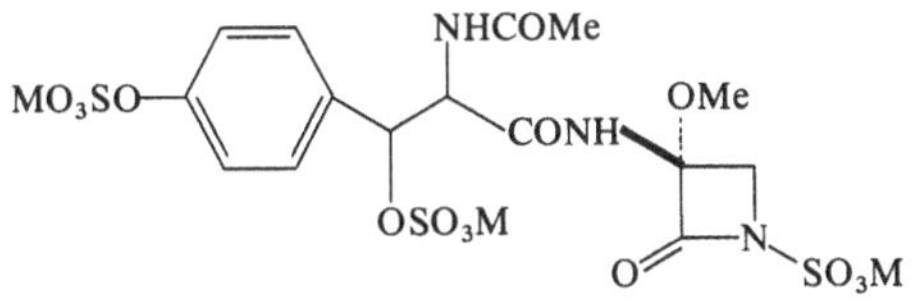

(**253**)

In order to determine the origin of the β-lactam 3-methoxyl group the authors fed L-[methyl-^{14}C]-methionine. The resulting ^{14}C-labelled (**252**) was purified by paper electrophoresis then heated with hydroiodic acid to liberate the methoxyl methyl group as methyl iodide which was trapped and counted. It was found that 87% of the ^{14}C-label of (**252**) was recovered in the methyl iodide showing that the methoxyl methyl group was specifically derived from methionine as is the case in cephamycin biosynthesis.

3. Biosynthesis of the Penicillins

The bicyclic nucleus of the penicillins (43) is derived from the amino acids cysteine (254) and valine (255) as shown.

(254)	(43)	(255)
Cysteine	Penicillin	Valine

The role of cysteine in penicillin biosynthesis was first demonstrated by Arnstein and Grant (210, 211) who synthesised L- and D-cystine labelled in the β-position with ^{14}C or with ^{35}S or ^{15}N. When these compounds were fed to cultures of Penicillium chrysogenum the L-isomer was found to be better incorporated into the resulting benzylpenicillin. Using a series of chemical degradations (Fig. 3), they were able to demonstrate the specific incorporation of cystine into benzylpenicillin. It was concluded that cystine is a direct precursor of penicillin, probably after reduction to cysteine.

In a double-labelled experiment it was shown that when L-[3-^{14}C, ^{35}S] cystine was fed to P. chrysogenum the isotope ratio of the starting material was retained unchanged in the isolated benzylpenicillin. This demonstrated that cystine (rather than penicillamine) was the source of the thiazolidine sulphur atom. Further evidence on the origin of the penicillin sulphur atom was provided by Stevens et al. (212) who showed that the label from [^{35}S]-cysteine was better incorporated into benzylpenicillin than label from [^{35}S]-sulphate. In competition experiments it was shown that L-cysteine or L-cystine were utilised in preference to [^{35}S]-sulphate, whereas D-cysteine and DL-penicillamine had no effect on the incorporation of label from sulphate into benzylpenicillin.

The biosynthetic fate of the α- and β-hydrogens of cysteine was studied by Arnstein and Crawhill (213) who synthesised DL-[2-^3H]- and DL-[3-^3H]-cystines. These were fed to cultures of P. chrysogenum and the resulting penicillin samples were isolated and degraded as in Fig. 3. Both precursors gave rise to radiolabelled benzylpenicillin and in both cases the label was found in the penicilloaldehyde degradation product. In the case of DL-[2-^3H]-cystine all the label was retained in the fomaldehyde dimedone indicating that it had been specifically incorporated at C-6. By inference the DL-[3-^3H]-cystine must have labelled the benzylpenicillin at C-5. The retention of the α-hydrogen of cysteine indicate that a 2,3-dehydrocysteinyl

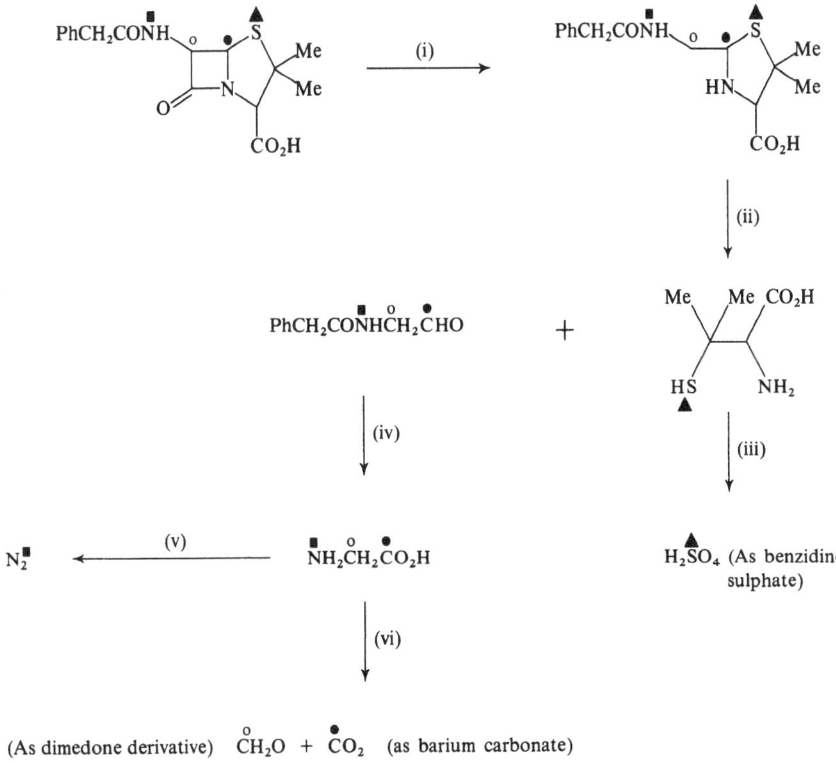

$(\bullet = {}^{14}\text{C}; \quad \blacktriangle = {}^{35}\text{S}; \quad \blacksquare = {}^{15}\text{N}; \quad \circ = {}^{3}\text{H})$

Reagents: (i) H⁺, Δ; (ii) HgCl₂, H₂S; (iii) Piries's reagent; (iv) Ag₂O, NH₃; (v) Kjeldal oxidation, NaOH, Br₂; (vi) Ninhydrin.

Fig. 3. Degradation of benzylpenicillin derived from L-[3-¹⁴C, ³⁵S, ¹⁵N] labelled cystine and DL-[2-³H]-cystine

residue was not involved in penicillin biosynthesis. This latter point was confirmed by BYCROFT *et al.* (*214*) who synthesised L-[2-³H]-cystine and fed it, together with L-[U-¹⁴C]-cystine, to a high producing strain of *P. chrysogenum*. They found that 84% of the tritium label was retained with respect to ¹⁴C-label in the resulting benzylpenicillin. The position of tritium label in the benzylpenicillin was identified by enzymically deacylating the penicillin to give 6APA (**44**) which was converted to 6-α-chloropenicillanic acid (**257**) *via* the diazo compound (**256**). Some 98% of the tritium label was lost during this conversion. These authors also studied the incorporation of L-[2-²H]-cystine into benzylpenicillin by ²H n.m.r. spectroscopy and confirmed that position 6 of the benzylpenicillin was labelled.

(44)   (256)   (257)

The origin of the C-5 hydrogen was investigated in more detail by Morecombe and Young (215) who synthesised $(2R,3R)$-[2,3-^{3}H$_2$]-cysteine and $(2R,3S)$-[3-^{3}H]-cysteine. These compounds and L-[3,3,3'3'-^{3}H$_4$]-cystine were individually mixed with L-[U-^{14}C]-cystine and fed to *P. chrysogenum* cultures. In duplicate experiments, the benzylpenicillin derived from L-[3,3,3'3'-^{3}H$_4$]-cystine showed 41% and 42% retention of tritium label, close to the theoretical maximum of 50%. These results were virtually identical with those of Adriens *et al.* (216) who performed an identical feeding experiment. The $(2R,3S)$-[3-^{3}H]-cysteine produced only 14% and 13% retention of tritium whereas $(2R,3R)$-[2,3-^{3}H$_2$]-cysteine produced 58% and 59% retention, obviously much better than the $(2R,3S)$-compound but somewhat lower than expected. However, when the benzylpenicillin derived from the labelled $(2R,3R)$-cysteine was chemically degraded (217), it was found that 75% of the retained label was located at position C-5, indicating that C-3 of cysteine is probably incorporated into C-5 of benzylpenicillin with retention of stereochemistry.

In a similar experiment Aberhart *et al.* (218) fed DL-cystines chirally labelled at carbon-3 with tritium to *P. chrysogenum* in combination with ^{14}C-cystine and found that with the $(3R)$ compound 76% of the tritium label was retained whereas with the $(3S)$ compound only 21% of the label was retained.

In summary, L-cysteine is the specific precursor of the β-lactam moiety of benzylpenicillin. The amino acid is incorporated with retention of configuration at the β-carbon and a 2,3-dehydro intermediate is not involved.

The incorporation of valine into benzylpenicillin was first demonstrated by Arnstein and Grant (210) and by Stevens *et al.* (220). The former authors fed DL-[4-^{14}C]-valine to a *P. chrysogenum* fermentation. The resulting labelled benzylpenicillin was degraded as shown in Fig. 3. Some 95% of the radioactivity of the benzylpenicillin was recovered in the acetone adduct (258) of the penicillamine, whereas the 2,4-dinitrophenylhydrazone of the benzylpenicilloaldehyde was found to be inactive. Stevens *et al.* (220) performed a similar experiment using DL-[1-^{14}C]-valine. Again, radioactivity was detected only in the penicillamine (259) degradation product. When the penicillamine was decarboxylated with ninhydrin all the activity was recovered in the liberated carbon dioxide, demonstrating that the carboxyl group of valine specifically labels the carboxyl group of benzyl-penicillin.

(258)   **(259)**

The α-centre of the valinyl moiety of penicillin has the D-configuration whereas natural valine has L-stereochemistry. Several attempts have been made to determine whether D- or L-valine is the precursor of the penicillins. STEVENS (*221*) added labelled valines to *P. chrysogenum* fermentations during the antibiotic production phase and found that L-[1-^{14}C]-valine was initially more rapidly incorporated into benzylpenicillin than D-[1-^{14}C]-valine. On this basis, it was concluded that L-valine was the precursor of penicillin. However, an alternative explanation would be that D-valine is the true precursor of penicillin but penetrates poorly into the mycelium whereas the L-isomer penetrates rapidly. Once inside, the L-isomer could be epimerised to the D-isomer. Indeed, ARNSTEIN and CLUBB (*222*) showed that washed mycelium of *P. chrysogenum* took up L-[1-^{14}C]-valine more rapidly than D-[1-^{14}C]-valine. More importantly, they also demonstrated that the relative rates of incorporation of the D- and L-isomers into benzylpenicillin were essentially the same as the relative rates of uptake into the mycelium, showing that it was the uptake that was the limiting factor. STEVENS and DE LONG (*223*) also concluded that D-[1-^{14}C]-valine was taken up by *P. chrysogenum* more slowly than L-[1-^{14}C]-valine. They also isolated mycelial protein and showed that the protein L-valine was labelled with equal efficiency by D- or L-[1-^{14}C]-valine indicating that D-valine can be readily epimerised by *P. chrysogenum*. Using washed mycelium from a different strain of *P. chrysogenum*, ARNSTEIN and MARGREITER (*224*) found that L-[1-^{14}C]-valine was incorporated into benzylpenicillin to a far greater extent than D-[1-^{14}C]-valine even after allowance was made for the poorer uptake of the D-isomer into the mycelium. They also found that DL-α-methylvaline inhibited the incorporation of D-[1-^{14}C]-valine into benzylpenicillin but had no effect on the incorporation of L-[1-^{14}C]-valine. This observation strongly suggests that the D-isomer is epimerised to the L-isomer before incorporation into penicillin.

WARREN *et al.* (*225*) achieved results similar to those above using a *Cephalosporium* sp. that produced penicillin N (**260**). They found that uptake of L-[1-^{14}C)-valine by washed mycelium was rapid compared to D-[1-^{14}C]-valine. The D-isomer was epimerised to the L-configuration inside the mycelium and D-[1-^{14}C]-valine labelled L-valine in the mycelial protein. No epimerisation of the L-isomer to D-valine was detected. These authors concluded that the labelling of (260) by D- and L-[1-^{14}C]-valine was consistent with the D-isomer being converted to the L-isomer before incorporation.

(260)

Further evidence for the epimerisation of D-valine was provided by BYCROFT et al. (226) who found that D-[1-^{14}C]- and L-[U-^{14}C]-valine were incorporated into benzylpenicillin to a similar extent by a high producing strain of P. chrysogenum even though the D-isomer was taken into the mycelium more slowly. However, when they fed D- or L-[U-^{14}C]-valine in combination with D- or L-[2-^{3}H]-valine they found that although some 37% of the ^{14}C-label was incorporated in each case, negligible amounts (ca. 0.2%) of tritium were retained from either isomer.

If this loss of tritium were due to epimerisation of valine by way of α-ketoisovaleric acid, substantial loss of label would be expected when [^{15}N]-valine is incorporated into penicillin. This does not appear to be the case. ARNSTEIN and CLUBB (222) fed DL-[2-^{14}C, ^{15}N]-valine to P. chrysogenum and isolated the penicillamine moeity of the resulting labelled benzylpenicil-lin. They found that 19% of the ^{15}N-label was retained with respect to the ^{14}C-label. In a second experiment they fed a mixture of L-[U-^{14}C]-valine and DL-[^{15}N]-valine and achieved 25% retention of ^{15}N-label. In both experiments the authors also isolated protein valine and found that retention of ^{15}N was 24% and 27%. In a similar experiment STEVENS and DE LONG (223) fed a mixture of L-[1-^{14}C]-valine and L-[^{15}N]-valine to a P. chrysogenum fermentation during the antibiotic production phase and isolated samples of labelled benzylpenicillin at short time intervals (3 and 6 hours) after addition of labelled precursor. The benzylpenicillin samples were degraded to penicillamines and the ^{14}C and ^{15}N content determined. Again, valine was also isolated from the mycelium. The retention of ^{15}N was 54% for the 3 hour sample (mycelial valine − 57%) and 44% for the 6 hour sample (mycelial valine − 42%). Clearly the above experiments de-monstrate that although substantial deamination of valine occurs in P. chrysogenum, the L-valine which is incorporated into benzylpenicillin is not deaminated to any greater extent than L-valine incorporated into protein. Therefore it seems unlikely that the epimerisation of L-valine to form the D-valinyl moiety of penicillin involves an α-keto intermediate. This was confirmed by BOOTH et al. (227) who fed a mixture of L-[U-^{14}C]-valine and L-[^{15}N]-valine to a high producing strain of P. chrysogenum. The ^{15}N content of the resulting benzylpenicillin was determined, without degradation, by ^{15}N-n.m.r. spectrometry which showed that 84% ($\pm 10\%$) of the nitrogen label was retained with respect to ^{14}C label. This demonstrated that L-valine is incorporated with little, if any, deamination.

Alternatively, the loss of the α-hydrogen atom of valine during incorporation could be due to a 2,3-dehydro-intermediate. Further evidence for such an intermediate was provided by ADRIENS et al. (216) who synthesised DL-[3-^{3}H]-valine and fed it with DL-[4-^{14}C]-valine, to a P. chrysogenum culture. The resulting phenoxymethylpenicillin was purified and combusted to separate the ^{14}C and ^{3}H labels as ^{14}CO$_2$ and ^{3}H$_2$O. It was found that only about 1% of the tritium label had been incorporated into the penicillin relative to ^{14}C incorporation. The possibility that the loss of the β-hydrogen of valine was due to a 3,4-dehydro intermediate was excluded by the work of ABERHART et al. (228). These authors synthesised DL-[4,4'-^{2}H$_6$]-valine which, when fed to P. chrysogenum, gave deuterated phenoxymethylpenicillin. ^{1}H-N.m.r. spectroscopy indicated that the gem-dimethyls were enriched to about 10 atom % with deuterium. This penicillin was esterified with diazomethane and examined by mass spectrometry in comparison with unlabelled methyl phenoxymethylpenicillinate. The major fragment produced by the latter was at m/z 174 due to the ion (261). In the

(261)

case of the labelled methyl phenoxymethylpenicillinate an additional ion at m/z 180 was observed indicating that all six deuterons had been incorporated from the methyl labelling valine. No significant ions between m/z 174 and m/z 180 were observed. KLUENDER et al. (229) similarly studied the incorporation of DL-[4,4'-^{2}H$_6$]-valine into penicillin N using washed mycelium of Cephalosporium acremonium. Again they found a peak at m/z 180 due to hexadeuterated (261), with no significant peaks between m/z 174 and m/z 180. These authors also synthesised (2S,3R)-[4-^{2}H$_3$]-valine and (2S,3S)-[4-^{2}H$_3$]-valine and predictably observed peaks at m/z 177 for the penicillins derived from each compound. It can be concluded therefore that a 3,4-dehydrovalinyl intermediate is not involved in penicillin biosynthesis.

The stereochemistry of incorporation of the methyl groups of valine into penicillin has been studied in several laboratories. $(2RS,3R)$-[4-^{13}C]-valine (**262**) synthesised by Baldwin *et al.* (*230*) was fed to a *P. chrysogenum* fermentation and the resulting phenoxymethylpenicillin (**263**) was examined by ^{13}C-n.m.r. spectrometry (*231*). Comparison of the n.m.r. spectrum with one of unlabelled phenoxymethylpenicillin indicated that the β-methyl group had been specifically enriched. Similarly Kluender *et al.* (*232*) synthesised $(2S,3S)$-[4-^{13}C]-valine and fed this compound to washed mycelium of *C. acremonium*. The resulting penicillin N was examined by ^{13}C-n.m.r. spectrometry and the α-methyl carbon was found to be specifically enriched. Further support for this finding came from Aberhart and Lin (*233*) who synthesised $(2RS,3S)$-[4-^{13}C]-valine (**264**) and found that this specifically labelled the α-methyl carbon of penicillin V (**263**).

(**262**)   (**263**)   (**264**)

In summary the D-valinyl moiety of the penicillin molecule is derived from L-valine. The epimerisation during incorporation does not involve deamination. A 2,3-dehydrovalinyl intermediate may be involved but a 3,4-dehydro intermediate is not. The *gem*-dimethyl groups of valine are incorporated into penicillin with retention of stereochemistry.

Because of the evidence that cysteine and valine were precursors of the penicillin nucleus Arnstein and Morris (*234*) investigated the incorporation of the dipeptides L-cystinyl-L-valine and L-cystinyl-D-valine into benzylpenicillin. They synthesised the two dipeptides labelled with ^{14}C in the valine carboxyl in both cases. They found that the LL-dipeptide was taken up by washed mycelium of *P. chrysogenum* at a faster rate than L-[U-^{14}C]-valine. Label from the LL-dipeptide was incorporated into mycelial protein indicating that extensive hydrolysis to cystine and L-[1-^{14}C]-valine had occurred. However when the ratio of specific radioactivities of penicillin to protein derived from labelled LL-dipeptide was compared with the same ratio obtained after feeding labelled valine it was found that the LL-dipeptide consistently gave the higher specific radioactivity indicating that at least some intact dipeptide was being incorporated into penicillin. As the specific activity of the penicillin was lower than that calculated for the intracellular dipeptide the authors concluded that the LL-dipeptide was not the exclusive precursor of penicillin and some other route must also be operating. The role of the LD-dipeptide unfortunately could not be ascertained as the authors found that this compound was not taken into the mycelium of *P. chrysogenum*.

In order to shed further light on the nature of the peptidyl intermediates in penicillin biosynthesis ARNSTEIN *et al.* (*235*) incubated *P. chrysogenum* mycelium with L-[U-^{14}C]-valine and then extracted the intracellular peptide and amino acid material into boiling aqueous ethanol. After applying various purification procedures the authors obtained a radioactive peptide with properties similar to a synthetic sample of γ-glutamylcysteinylvaline. ARNSTEIN and MORRIS (*236*) went on to sequence this peptide and concluded that it was δ-(α-aminoadipyl)cysteinylvaline (**265**). Due to lack of material the authors were unable to determine the stereochemistry of the tripeptide.

(**265**)

This compound was obviously implicated as a possible precursor of penicillin as penicillin N, produced by *C. acremonium,* possesses the D-δ-(α-aminoadipyl) side-chain (*237*). ARNSTEIN and MORRIS therefore proposed that their tripeptide was converted to penicillin N which in turn was transacylated to form the hydrophobic penicillins such as benzylpenicillin. Further weight was given to this hypothesis when FLYNN *et al.* (*238*) and COLE and BATCHELOR (*239*) isolated isopenicillin N (penicillin M) from culture filtrate and from the mycelium of *P. chrysogenum*. These authors demonstrated that isopenicillin N has the L-δ-(α-aminoadipyl) side-chain. WARREN *et al.* (*240*) found that both D- and L-α-amino-[6-^{14}C]-adipic acid labelled the side-chain of penicillin N produced by a *Cephalosporium* sp. but concluded, from the high isotope dilution, that the D-isomer was converted to the L-configuration before incorporation, even though the penicillin N side-chain has the D-configuration. Indeed, L-α-amino-[^{14}C]-adipic acid was found in the intracellular pool after feeding the ^{14}C-labelled D-isomer. These authors also noted that intracellular peptide material containing α-aminodipic acid, cysteine and valine was also labelled from L-α-amino-[^{14}C]-adipic acid (as well as by L-[1-^{14}C]-valine (*225*)). LODER and ABRAHAM (*241*) isolated this peptide material from the *Cephalosporium* sp. by homogenising the mycelium in trichloroacetic acid solution (to remove proteins) and precipitating it with cadmium chloride and bicarbonate. The precipitate was redissolved and thiol-containing material precipitated as the cuprous mercaptide. This was decomposed with H$_2$S and the free thiols oxidised to the S-sulphonates. Further purification revealed the presence of

S-sulphonylglutathione and three other peptides. Mass spectrometric methods indicated that one of these was δ-(α-aminodipyl)-cysteinylvaline. The optical rotatory dispersion spectrum of the intact peptide and the circular dichroism spectra of the constituent amino acids enabled the stereochemistry to be assigned as δ-(L-α-aminoadipyl)-L-cysteinyl-D-valine.

Later, the tripeptide from *P. chrysogenum* was also shown to have the LLD stereochemistry by CHAN *et al.* (*242*). These authors fed L-[U-^{14}C]-valine, DL-[1-^{14}C]-α-aminoadipic acid or L-[U-^{14}C]-cysteine to washed mycelium of *P. chrysogenum* and found that each labelled amino acid labelled the intracellular tripeptide. Each labelled tripeptide sample was isolated, oxidised to the sulphonic acid, further purified and then hydrolysed to its constituent amino acids. The amino acids were separated by two dimensional paper electrophoresis and chromatography, then reacted with D- and L-amino-acid oxidases to determine their stereochemistry. ADRIENS *et al.* (*243*) also concluded that the *P. chrysogenum* tripeptide possesses the LLD stereochemistry by passing the tripeptide through an amino acid analyser which could distinguish between the synthetic LLD and LLL-tripeptides. They could find no evidence of the LLL-tripeptide in *P. chrysogenum*. FAWCETT *et al.* (*244*) also isolated the tripeptide from *P. chrysogenum* and concluded that its stereochemistry was LLD, rather than LLL, from its electrophoretic mobility.

BAUER (*245*) reported that δ-(α-aminoadipyl)-cysteinylvaline was synthesised from its constituent amino acids by an extract of the mycelium of *P. chrysogenum,* and LODER and ABRAHAM (*246*) demonstrated that the LLD-tripeptide could be synthesised from δ-(L-α-aminoadipyl)-L-cysteine and DL-[1-^{14}C]-valine by a particulate extract of *Cephalosporium* sp. No synthesis of tripeptide was detected from δ-(D-α-aminoadipyl)-L-cysteine and valine, or from DL-[1-^{14}C]-α-aminoadipic acid and L-cysteinyl-L-valine or L-cysteinyl-D-valine. FAWCETT and ABRAHAM (*247*) reported that an enzyme from *Cephalosporium* sp. which coupled valine to δ-(L-α-aminoadipyl)-L-cysteine required an energy-generating system (ATP, phosphonenolpyruvate and pyruvate kinase) and was specific for L-valine, the D-isomer not being accepted. Clearly then, the epimerisation of L-valine to the D-configuration occurs at this stage, during its incorporation into the tripeptide. It will be recalled that the results of incorporating labelled valines into penicillins showed that a 2,3-dehydro-valinyl intermediate might be involved in penicillin biosynthesis. HUANG *et al.* (*248*) therefore fed L-[2,3-^{3}H]-valine to washed mycelium of *C. acremonium* and isolated the resulting labelled LLD-tripeptide. This was hydrolysed and the labelled D-valine isolated. When this was oxidised with a D-amino acid oxidase the resulting α-keto acid had the same tritium content as the D-valine, indicating that only the tritium on C-3 of the starting L-[2,3-^{3}H]-valine had been retained in the D-valine. Obviously the α-tritium had been lost during epimerisation but as the C-3 tritium had been retained a 2,3-dehydro-intermediate could not be involved. ADRIENS *et al.* (*249*) reached a similar conclusion using *P. chrysogenum*. These authors fed L-[1-^{14}C-2,3,4-^{3}H]-valine to washed mycelium and then isolated the LLD-tripeptide. By comparing the ^{14}C/^{3}H ratios of the labelled L-valine and the tripeptide they concluded that a 2,3-dehydrovalinyl intermediate was not involved in *P. chrysogenum* as only the α-tritium had been lost.

In order to demonstrate that δ-(α-aminoadipyl)-cysteinylvaline is the penicillin precursor, cell-free biosynthetic systems had to be employed as the tripeptide is not taken up by intact cells. FAWCETT *et al.* (*244, 250*) were the first to prepare such a cell-free system using enzymic digestion of the cell wall of *C. acremonium* in an isotonic medium to give protoplasts, which were then lysed by dilution. Again, an energy-generating system was added. A variety of labelled peptides were synthesised and added to the cell-free system. After incubation for 2 hours each reaction was divided into two. One half was oxidised so that any penicillin formed would be converted to penicillaminic acid (**266**) which was detected by two dimensional paper

electrophoresis and chromatography. The other half was examined for the presence of intact penicillin N by paper electrophoresis/chromatography, or for the presence of the corresponding penicilloic acid after treatment with penicillinase. The authors found that δ-(L-α-aminoadipyl)-L-cysteinyl-D-[4,4'-^{3}H]-valine was converted to a compound with the properties of penicillin N whereas the corresponding LLL-tripeptide was not. δ-(L-α-aminodipyl)-L-cysteinyl-D-(4,4'-^{3}H)-valine and δ-(L-α-aminoadipyl)-L-cysteinyl-D-[2-^{3}H]-valine both resulted in labelled penicillaminic acids, whereas tritiated LLL-tripeptide and DLD-tripeptide did not; neither did L-cysteinyl-D-[4,4'-^{3}H]-valine nor DL-[4,4'-^{3}H]-valine. The authors were uncertain whether the penicillin product of the LLD-tripeptide was penicillin N or iso-penicillin N as their analysis did not distinguish between them. Iso-penicillin N seemed the logical product as it also possesses the LLD-stereochemistry as opposed to the DLD-stereochemistry of peni-cillin N. However, at that time iso-penicillin N had not been found in *C. acremonium*. The retention of the valinyl α-tritium for labelled LLD-tripeptide indicated that a 2,3-dehydrovalinyl intermediate was not involved during the cyclisation of the tripeptide.

The nature of the cyclisation product of the LLD-tripeptide was examined by O'SULLIVAN et al. (251) and KONOMI et al. (252). The former authors synthesised δ-(L-α-amino-[4,5-³H]-adipyl)-L-cysteinyl-D-[4,4'-³H]-valine which was cyclised by a lysed protoplast preparation of C. acremonium. The resulting penicillin product was isolated and degraded with β-lactamase, followed by oxidation with performic acid or hydrolysis with 6NHCl. The resulting penicillaminic acid and free α-aminoadipic acid were purified by electrophoresis. Their specific radioactivities were found to be the same as in the starting tripeptide, indicating that no hydrolysis of the tripeptide had occurred prior to cyclisation. The α-aminoadipic acid residue from the penicillin product was oxidised by an L-amino acid oxidase proving that the penicillin was isopenicillin N. KONOMI et al. (252) also concluded that the cyclisation product was isopenicillin N on the basis of its antibacterial activity. They prepared penicillin N, isopenicillin N, LLL-tripeptide and LLD-tripeptide. They found that the LLD-tripeptide, when incubated in a cell-free system from C. acremonium, gave a product which possessed antibacterial activity against Staphylococcus aureus and Sarcina lutea, but not against Salmonella typhimurium or Pseudomonas aeruginosa. Isopenicillin N was found to have the same biological properties whereas penicillin N was active against the Salmonella and Pseudomonas spp. but not against Staphylococcus or Sarcina spp. The LLL-tripeptide did not give any biologically active compound in the cell-free system. It was found that the production of isopenicillin N from the LLD-tripeptide was increased by increased aeration.

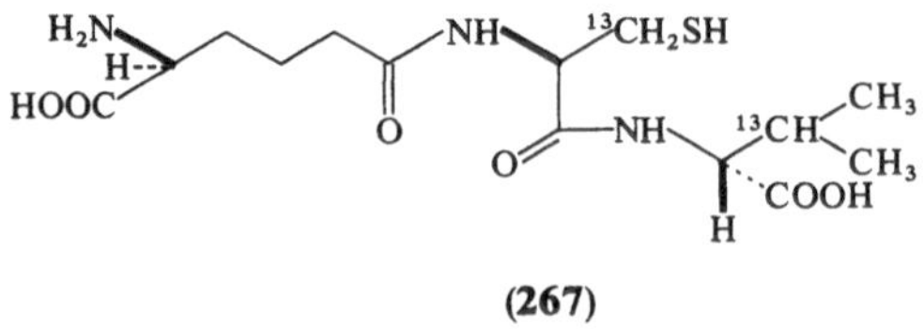

(267)

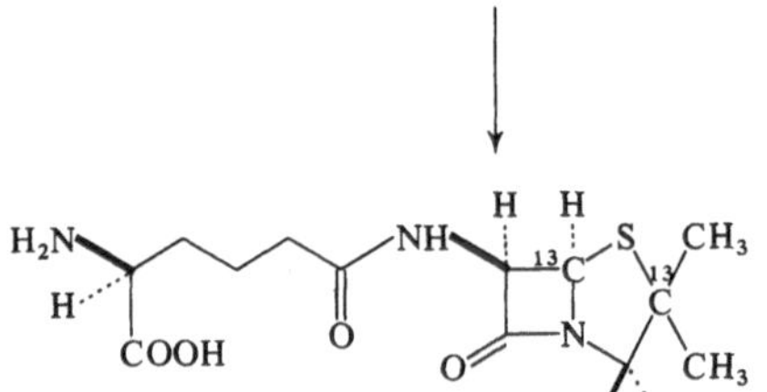

(268)

Further proof that the LLD-tripeptide is incorporated into isopenicillin N without prior hydrolysis was provided by BALDWIN et al. (253) who synthesised δ-(L-α-aminoadipyl)-L-[3-¹³C]-cysteinyl-D-[3-¹³C]-valine

(267). This compound was incubated with a cell-free preparation of *C. acremonium* in the probe of a ^{13}C-n.m.r. spectrometer. It was observed that as isopenicillin N was formed (measured by bioassay) the ^{13}C-signals of the substrate decreased and two new signals correspondingly increased in intensity. These new signals were assigned to C-5 and C-2 of isopenicillin N (268). Unfortunately, the hoped-for ^{13}C-^{13}C spin-spin coupling between C-5 and C-2 was not observed.

NEUSS *et al.* (254) confirmed that the product of the cell-free cyclisation is isopenicillin N by derivatising the side-chain amino group with 2,3,4,6-tetra-O-acetyl-β-D-glucopyranosyl isothiocyanate (GITC). The GITC-isopenicillin N derivative could be distinguished from the GITC-penicillin N derivative by reverse-phase h.p.l.c.

The biochemical conditions required for cyclisation of the LLD-tripeptide were further investigated by SAWADA *et al.* (255) who found that ferrous sulphate stimulated the cyclisation but that ascorbate, ATP or α-ketoglutarate were not required. Triton X-100 had a stimulatory effect on the reaction but the divalent cations Mn^{++}, Cu^{++} or Zn^{++} were all inhibitory, the last mentioned being particularly so. The disulphide dimer of the tripeptide could be used as a substrate. Cyclisation activity could be produced by ultrasonicating the mycelium of *C. acremonium* without first having to prepare protoplasts. The cyclising activity reached a maximum after growth of the organism had ceased.

The cyclising enzyme's requirement for oxygen, mentioned above, was studied by WHITE *et al.* (256), using two different methods. In the first experiment they conducted the cell-free cyclisation reaction in a closed vessel and monitored the consumption of dissolved oxygen in the reaction during isopenicillin N formation. Using two substrate concentrations they obtained figures of 1.33 and 1.14 mol of isopenicillin N formed per mole of dioxygen consumed. In the second experiment they followed the enzymic conversion of ^{13}C-labelled LLD-tripeptide to isopenicillin N by n.m.r. spectroscopy under oxygen-limited conditions. They calculated that the reaction mixture in the n.m.r. tube initially contained 0.46 μmol dioxygen/ml and noted that the conversion levelled-off when 0.49 μmol isopenicillin N/ml had been formed. From these experiments they concluded that one mol of dioxygen is required to convert one mol of LLD-tripeptide to isopenicillin N.

KUPKA *et al.* (257) increased the purity of the cyclising enzyme (isopenicillin N synthetase) from *C. acremonium* by ammonium sulphate precipitation followed by desalting on Sephadex G25. Further purification was achieved by ion-exchange chromatography on DEAE-Sepharose. The purified isopenicillin N synthetase was examined by SDS-PAGE which indicated that the molecular weight was approximately 31,000 daltons. It was found that the purified enzyme required the monomeric form of the

LLD-tripeptide substrate and would not cyclise the disulphide. The enzyme was stimulated by ferrous sulphate and by ascorbate. Various combinations of α-ketoglutarate, ATP, FAD, NAD and glutathione did not stimulate enzyme activity nor did Triton X-100. The Km value for LLD-tripeptide and isopenicillin N synthetase was determined as 0.3 mM. The pH optimum was at pH 7.5 – 8.0 and the temperature optimum 25 – 30° C. The enzyme converted the modified substrates (269), (270) and (271) to penicillins, but (272) was not cyclised. The product from the cyclisation of (271) was shown by h.p.l.c. not to be N-glycyl-isopenicillin N but probably isopenicillin N.

$$HOOC-CH_2(CH_2)_3CO-L-cys-D-val$$

(269)

$$\overset{DL}{HOOC-\underset{\underset{CH_3CONH}{|}}{C}H(CH_2)_3CO-L-cys-D-val}$$

(270)

$$HOOC-CH_2-\underset{\underset{NH_2}{|}}{N}HCOCH(CH_2)_3CO-L-cys-D-val$$

(271)

$$\overset{L}{HOOC-\underset{\underset{NH_2}{|}}{C}H(CH_2)_2CO-L-cys-D-val}$$

(272)

Bahadur *et al.* (*258*) also investigated the substrate specificity of isopenicillin N synthetase. They synthesised the tripeptides (273), (274) and (275), each modified in the D-valinyl residue. All three were converted to

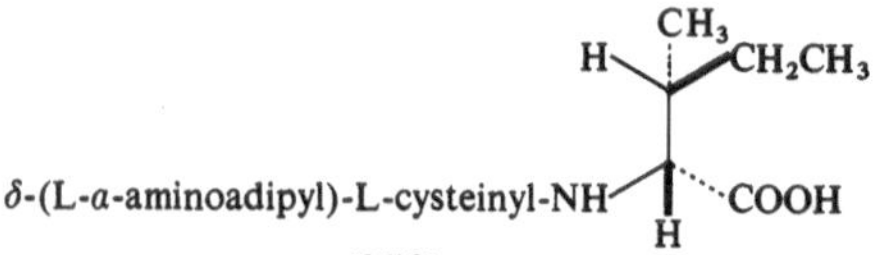

δ-(L-a-aminoadipyl)-L-cysteinyl-NH—C(CH₃)(CH₂CH₃)—COOH

(273)

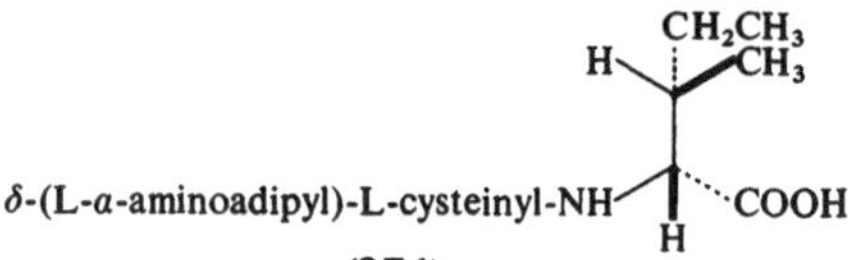

δ-(L-a-aminoadipyl)-L-cysteinyl-NH—C(CH₂CH₃)(CH₃)—COOH

(274)

δ-(L-a-aminoadipyl)-L-cysteinyl-NH—C(H)(CH₃)—COOH

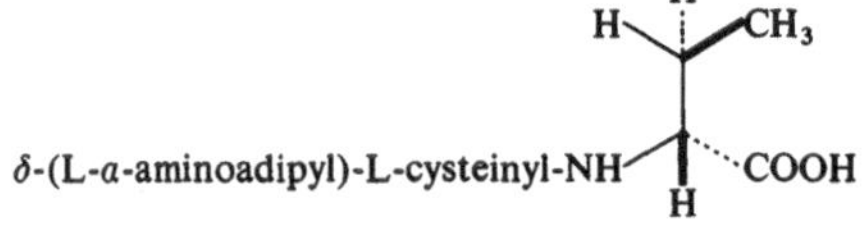

(275)

biologically active, penicillinase sensitive, compounds though not as efficiently as the natural substrate. The structures of the three cyclised products [(273), (274) and (275) gave (276), (277) and (278) respectively], were assigned by chemical degradations and comparison with reference compounds. Thus it was shown that when (273) and (274) were cyclised to (276) and (277) the stereochemistry of the terminal amino acid β-carbon

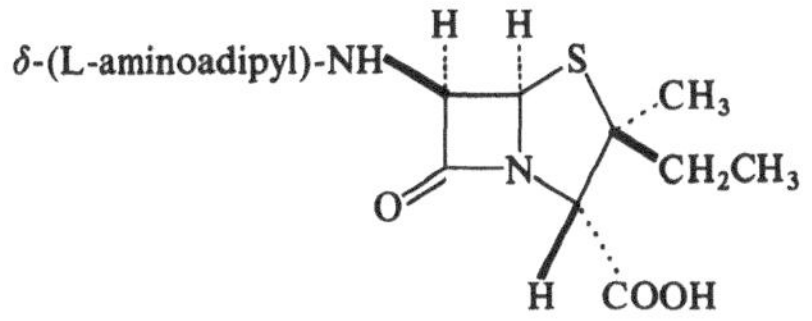

(276)

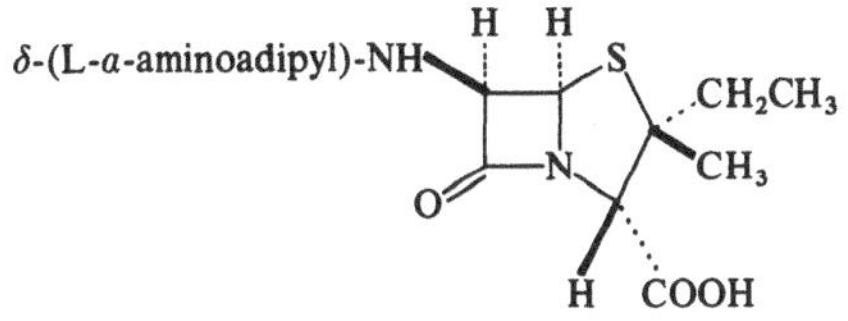

(277)

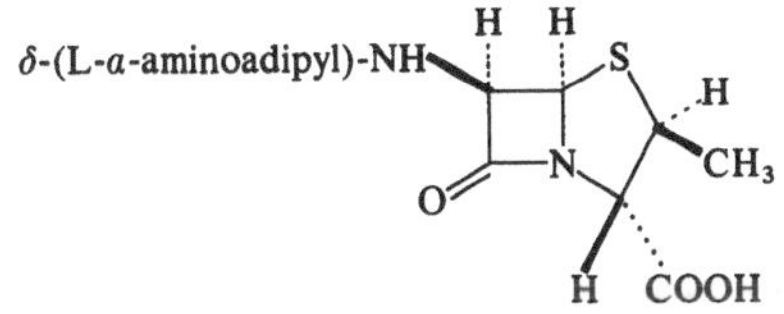

(278)

was retained in the product. In the case of (275), where the β-carbon of the valinyl residue was not chiral, the major product of cyclisation was (278). BAHADUR et al. (259) also demonstrated the conversion of (273) to (276) by carrying out the cell-free cyclisation in the n.m.r. probe. To confirm retention of stereochemistry at C-2 of (276) they oxidised (276) to the β-sulphoxide. Saturation of the C-2 methyl frequency at δ 1.35 produced an enhancement in the strength of the C-5 proton signal. This would only happen if the C-2 methyl group of (276) was in the α-orientation. A further modified tripeptide (279) was synthesised by BALDWIN et al. (260) following the discovery of the related (280) in the culture broth of *P. chrysogenum* by NEUSS et al. (261). No conversion of (279) to a penicillin by a cell-free extract of *C. acremonium* could be detected by bioassay, and ^{1}H-n.m.r. spectrometry indicated that (279) did not disappear. However, (279) was found to inhibit the conversion of the natural substrate to isopenicillin N.

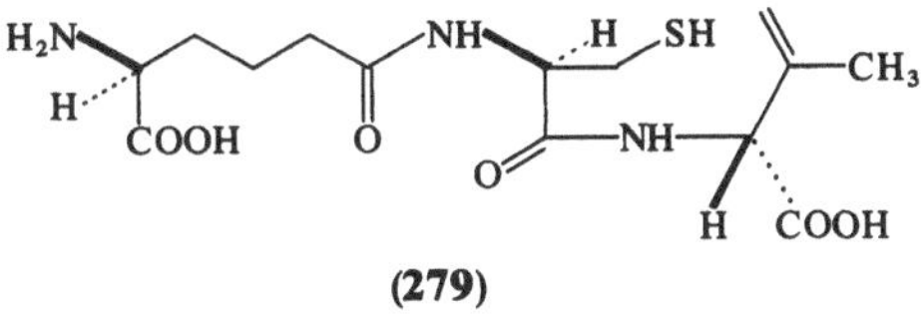

(279)

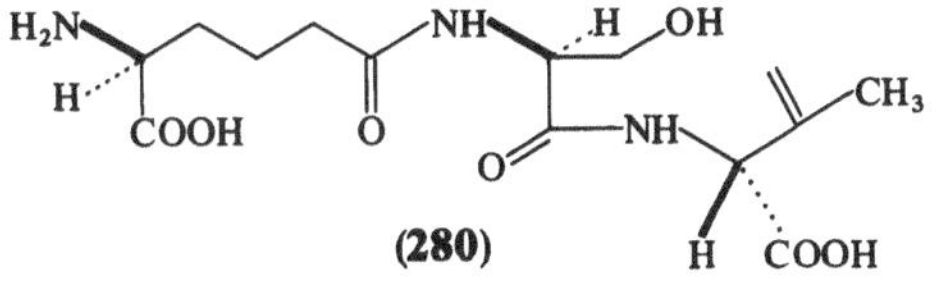

(280)

ABRAHAM et al. (262) also studied the substrate specificity of isopenicillin N synthetase from *C. acremonium*. They found that none of the following four tripeptides gave rise to a biologically active product when tested by bioassay:

δ-(L-α-aminoadipyl)-L-cysteinyl-L-valine,
γ-(L-glutamyl)-L-cysteinyl-D-valine,
δ-(L-α-aminoadipyl)-L-α-aminobutyryl-D-valine,
δ-(L-α-aminoadipyl)-L-alanyl-D-valine.

However, the first mentioned compound did inhibit the enzyme to some extent.

Isopenicillin N synthetase has also been studied in microorganisms other than *C. acremonium*. MEESSCHAERT *et al.* (*263*) examined the cyclisation of labelled LLD-tripeptide by lysed protoplasts of *P. chrysogenum*. Purification of the reaction mixture by a process which included analytical cation-exchange chromatography permitted isolation of a compound which they concluded was the penicilloic acid of isopenicillin N, formed by degradation of isopenicillin N on the ion-exchange column. When LLD-([2-^{3}H,1-^{14}C]-valine)-tripeptide was cyclised all of the tritium label was retained in the penicilloic acid indicating that a 2,3-dehydrovalinyl intermediate is not involved in *P. chrysogenum,* similar to the previously-mentioned result with *C. acremonium*. With LLD-([3,4-^{3}H,1-^{14}C]-valine)-tripeptide the analysis of the penicilloic acid indicated that only the C-3 tritium had been lost. With LLD-([3-^{3}H]-cysteinyl-[1-^{14}C]-valine)-tripeptide considerably more than the predicted 50% loss of tritium occurred presumably due to epimerisation of the penicilloic acid after ring opening. A second labelled compound was also detected in the cell-free reaction. This substance retained both the C-3 and C-4 tritium from the valine labelled peptide and 50% of the C-3 tritium from the cysteine labelled peptide. From these data the authors deduced that this compound was the monocyclic β-lactam (**281**). However, ABRAHAM *et al.* (*264*) synthesised (**281**), found it to be extremely unstable as the free thiol, and concluded that it was improbable that (**281**) could have been isolated from a cell-free reaction. These authors also synthesised the stable disulphide dimer of (**281**) and incubated this in cell-free systems from *C. acremonium* and *P. chrysogenum* but could not detect the formation of any bioactive material. The dimer did not inhibit the cyclisation of LLD-tripeptide in these systems. The involvement of free (**281**) in penicillin biosynthesis must therefore be doubtful.

(**281**)

JENSEN *et al.* (*265, 266*) studied tripeptide cyclisation in the procaryotic organism *Streptomyces clavuligerus*. Using an extract of sonicated cells they found that LLD-tripeptide was converted to a penicillinase sensitive product. On the basis of its antibacterial activity they concluded that it was a mixture of isopenicillin N and penicillin N, the latter presumably being produced by racemisation of isopenicillin N. Ascorbic acid and ferrous sulphate both stimulated the cyclising activity but ATP did not. Oxygen was essential for antibiotic production. The pH optimum was pH 7.0 – 8.0 (cf. enzyme from *C. acremonium*). The authors concluded that dithiothreitol was required for enzyme activity and for reduction of the disulphide dimer of the tripeptide substrate.

In addition to the monocyclic β-lactam (281) and the 2,3- or 3,4-dehydrovalinyl peptides discussed previously, other structures have been proposed as intermediates in the cyclisation of the LLD-tripeptide. BAHADUR *et al.* (267) synthesised (282), (283) and (284) and tested them as substrates in the cell-free systems from *C. acremonium*. None of these compounds was cyclised to isopenicillin N or any other biologically active compound. Chromatographic analysis indicated that the compounds remained unchanged in the cell-free system. However, (282) was a weak inhibitor of the cyclising enzyme.

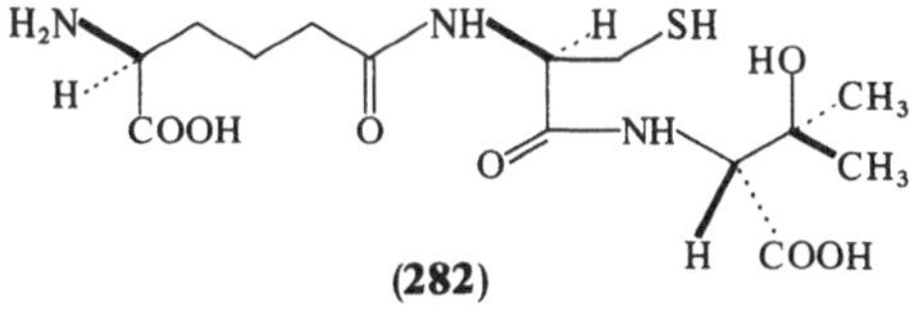

(282)

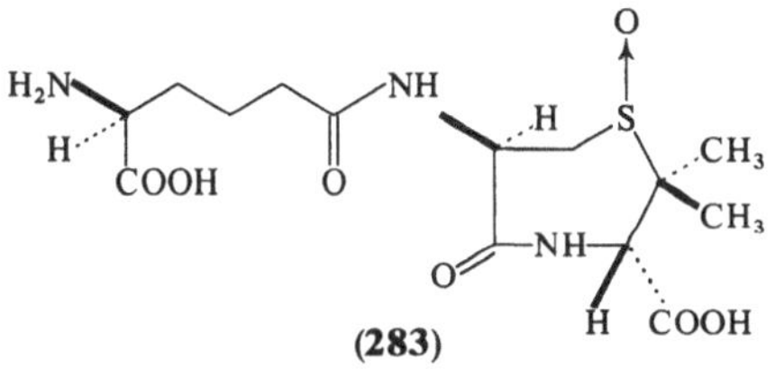

(283)

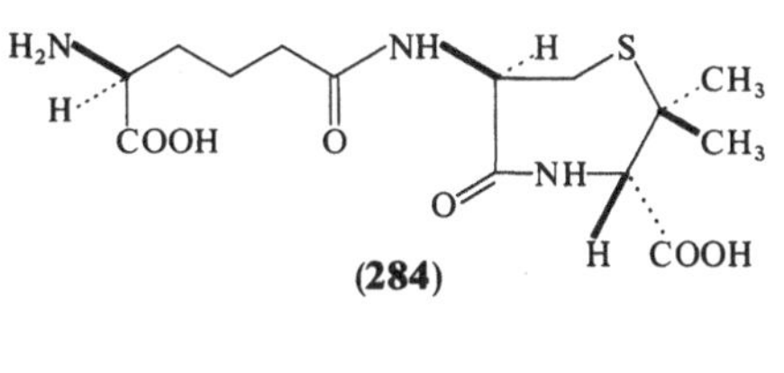

(284)

In order to establish whether intermediates such as (285) or (286) are involved in the cyclisation of LLD-tripeptide ADLINGTON *et al.* (268) synthesised the tripeptide containing L-α-aminoadipic acid labelled in all oxygens with $^{17}O/^{18}O$. This compound was cyclised in the *C. acremonium* cell-free system and the resulting isopenicillin N was derivatised and examined by chemical ionisation mass spectrometry. It was found that all the oxygen labels were retained, ruling out intermediates such as (285) and (286). Similarly, when unlabelled LLD-tripeptide was cyclised in a reaction mixture containing $H_2^{17}O$ and $H_2^{18}O$ no label was detected in the product, again ruling out (285) and (286) and also excluding such intermediates as enzyme bound thioesters, esters or amidines formed through any of the oxygen sites of the tripeptide.

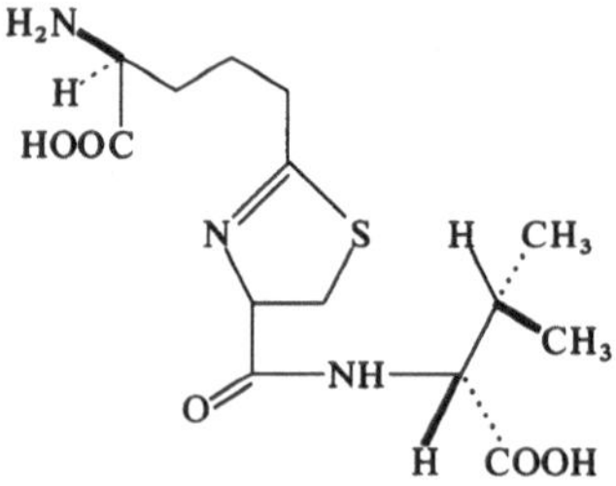

(285) 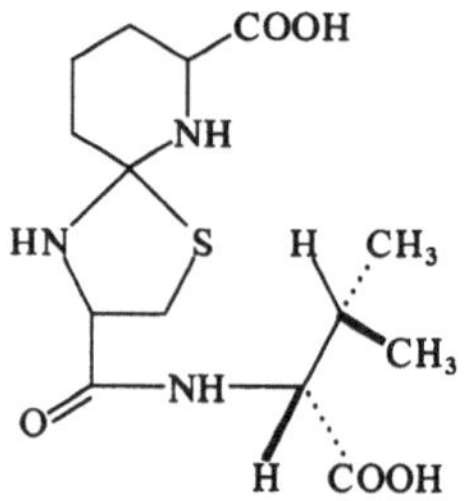   (286)

The metabolic fate of isopenicillin N differs depending on the organism. In *C. acremonium* Jayatilake *et al.* (*269*) demonstrated that lysed protoplasts could epimerise isopenicillin N to penicillin N, using a differential bioassay to detect the transformation. The epimerase was soluble, labile and intracellular. Intact protoplasts did not convert isopenicillin N to penicillin N. When penicillin N was added to lysed protoplasts generation of isopenicillin N could not be detected. Jensen *et al.* (*265, 266*) detected isopenicillin N epimerase activity in cell-free preparations of *S. clavuligerus* as discussed above.

Isopenicillin N is also the precursor of the "hydrophobic" penicillins, such as benzylpenicillin, produced by several strains of fungi including *P. chrysogenum*. When *P. chrysogenum* is grown in a medium containing a suitable mono-substituted acetic acid, such as phenylacetic acid or phenoxyacetic acid, then the corresponding penicillin, benzylpenicillin or phenoxymethylpenicillin is produced. A wide range of penicillins can be produced in this way from appropriate precursor acids (*270, 271*). Fawcett *et al.* (*272*) demonstrated that [2-β-methyl-³H]-isopenicillin N was converted to tritiated benzylpenicillin by cell-free extracts of *P. chrysogenum* in the presence of phenylacetyl-CoA. However, it was not clear whether this was due to transacylation or to deacylation to 6-aminopenicillanic acid [6-APA; (44)] and reacylation with phenylacetyl-CoA. Certainly, when *P. chrysogenum* is cultivated in the absence of a suitable side-chain precursor then 6-APA accumulates (*273*), presumably due to deacylation of isopenicillin N. However, there is no detailed report in the literature of an enzyme than can perform this deacylation. FAWCETT *et al.* (*272*) found that their cell-free system from *P. chrysogenum* could acylate 6-APA with phenylacetyl CoA. However penicillin N was not converted to benzylpenicillin in this system.

PRUESS and JOHNSON (*274*) partially purified an enzyme from *P. chrysogenum* which catalysed acyl-transfer from a number of [^{35}-S]-penicillins (penicillins V, G, K, X and dihydro-F) to 6-APA. The side-chains of penicillin N, methyl penicillin and phenylpenicillin were not transferred by this enzyme. The enzyme showed no penicillin amidase activity i. e. no 6-APA was generated when penicillin and enzyme were mixed in the absence of added 6-APA. Other compounds, such as esters of glycine, could act as acyl acceptors but not efficiently as 6-APA.

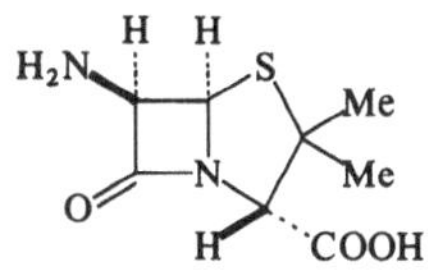

(44)

Penicillin amidase activity has been reported for many strains of fungi that are capable of producing hydrophobic penicillins. COLE (*275*) reported that five strains of *Penicillia*, *Aspergillus ochraceous*, *Trichophyton mentagrophytes*, *Epidermophyton floccosum* and *Cephalosporium* sp. were capable of deacylating phenoxymethylpenicillin. However, only *T. mentagrophytes* and *E. floccosum* produced any significant activity against benzylpenicillin. VANDERHAEGHE *et al.* (*276*) found that *Fusarium avenaceum* and *P. chrysogenum* could deacylate phenoxymethylpenicillin much faster than benzylpenicillin or penicillins with aliphatic side-chains. They found no evidence for deacylation of isopenicillin N or penicillin N. However when the side chain carboxyl group of these two penicillins was esterified formation of 6-APA was observed with the *P. chrysogenum* extract.

SPENCER and MAUNG (*277*) isolated a protein from *P. chrysogenum* which, after purification, appeared to be homogeneous and possessed acyltransferase activity, acylase and deacylase activity, and phenylacetyl-CoA hydrolase activity.

Numerous other reports exist of all these activities occurring separately or together in penicillin producing organisms and have been thoroughly reviewed by NEUSS and QUEENER (*278*). However, in spite of all of these publications the final stages in the production of the hydrophobic penicillins are still not well understood. Whether 6-APA is the key intermediate in hydrophobic penicillin biosynthesis or just a shunt metabolite is still to be determined.

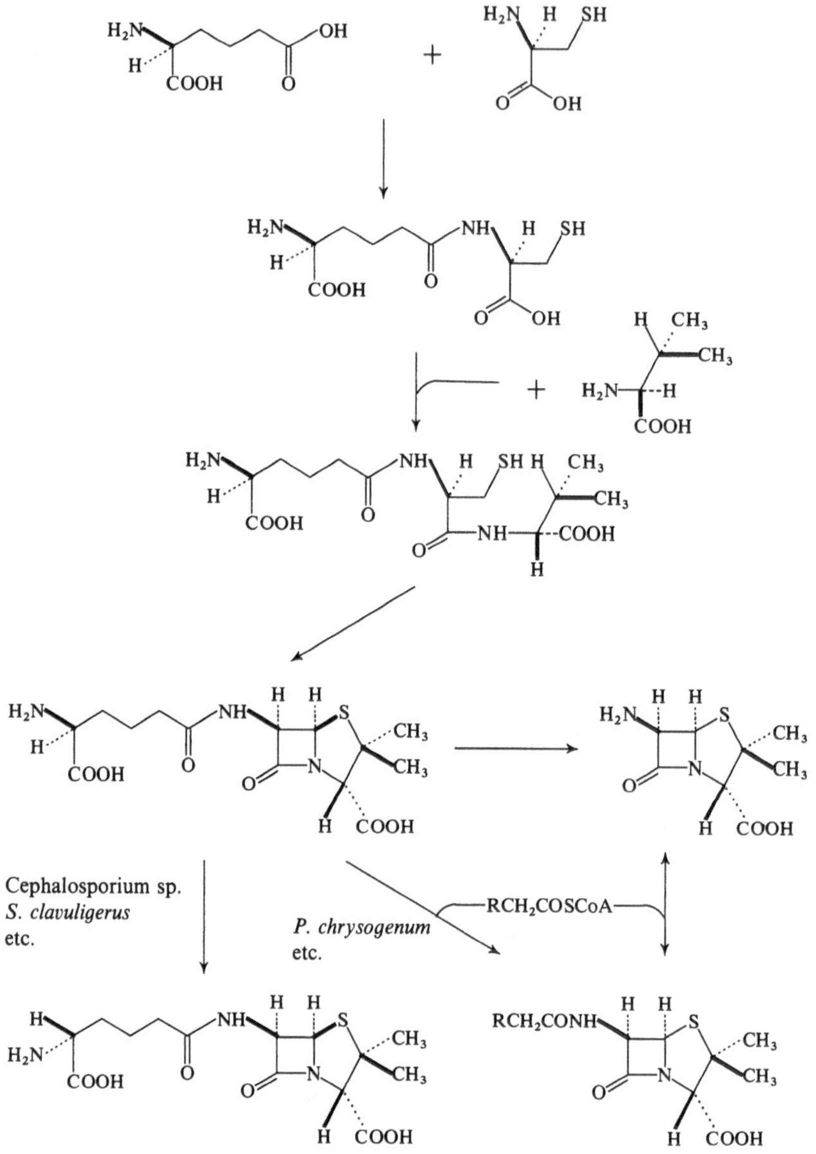

Fig. 4. Penicillin biosynthesis

In summary then, the penicillins are biosynthesised from the three amino acids L-α-aminoadipic acid, L-cysteine and L-valine. The dipeptide, δ-(L-α-aminoadipyl)-L-cysteine, is formed first which then combines with

L-valine to form δ-(L-α-aminoadipyl)-L-cysteinyl-D-valine. This is then cyclised to isopenicillin N. In some microorganisms, such as *Cephalosporium* sp. and *Streptomyces clavuligerus*, the δ-(L-α-aminoadipyl)-side-chain is epimerised to the D-configuration and the resulting penicillin N is excreted. In *Penicillium chrysogenum,* and other fungi that produce hydrophobic penicillins, isopenicillin N is converted to one or more penicillins with a monosubstituted acetyl side chain derived from the appropriate monosubstituted acetyl-CoA precursor. Whether this happens by transacylation of isopenicillin N, or deacylation to 6-APA and subsequent reacylation, or by both processes, is not clear. Penicillin biosynthesis is summarised schematically in Fig. 4.

4. Biosynthesis of Clavulanic Acid

In the first studies on clavulanic acid biosynthesis ELSON and OLIVER (*279*) fed [1-^{13}C]-acetate, [2-^{13}C]-acetate or [1,2-^{13}C]-acetate to cultures of *Streptomyces clavuligerus* during the clavulanic acid production phase. The resulting samples of ^{13}C-labelled clavulanic acid were isolated as the benzyl ester (**287**) and examined by ^{13}C-n.m.r. spectroscopy. The labelling patterns

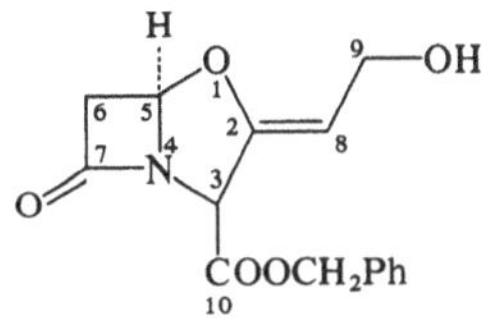

(**287**)

indicated that the five-carbon skeleton (C-10, C-3, C-2, C-8 and C-9) was derived from acetate (**288**) *via* the tricarboxylic acid (TCA) cycle and that α-ketoglutarate (**289**) and glutamate (**290**) were probably involved as precursors. Further metabolism of the labelled α-ketoglutarate in the TCA cycle to malate or oxalacetate (**291**), followed by decarboxylation (say to phosphoenolpyruvate) would generate a suitable three carbon precursor for the β-lactam ring skeleton and explain why the β-lactam carbons were labelled by acetate. Further metabolism of the labelled oxalacetate round the TCA cycle resulted in labelling also of C-2, C-3 and C-10 of clavulanic acid, making interpretation of the results difficult.

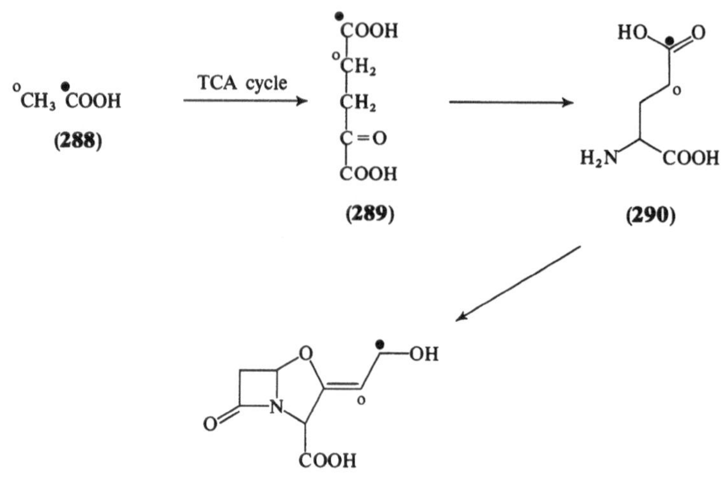

$^{\circ}CH_3\ ^{\bullet}COOH$

(288)

$\xrightarrow{\text{TCA cycle}}$

$^{\bullet}COOH$
$^{\circ}CH_2$
CH_2
$C=O$
$COOH$

(289)

$\longrightarrow$

(290)

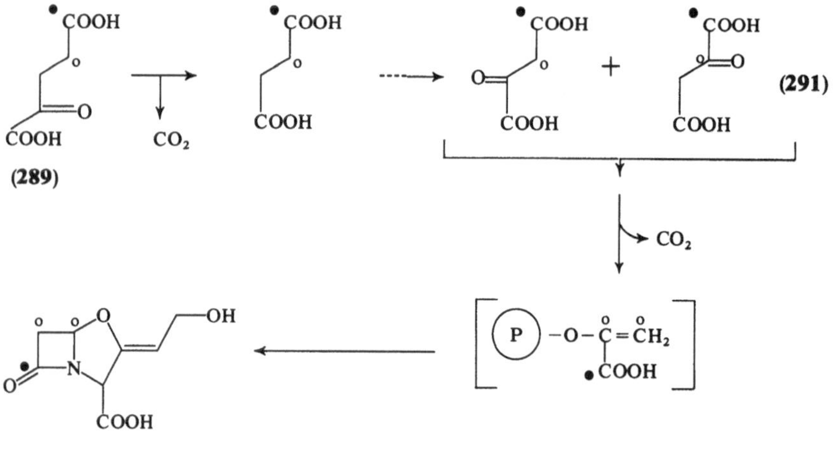

$^{\bullet}COOH$
CO_2
(289)

$^{\bullet}COOH$
$COOH$

$\dashrightarrow$

$^{\bullet}COOH$
$COOH$

$+$

$^{\bullet}COOH$
$COOH$

(291)

$\xrightarrow{\ \ } CO_2$

$\left[\begin{array}{c} \text{P} -O-\overset{\circ}{C}=\overset{\circ}{C}H_2 \\ ^{\bullet}COOH \end{array} \right]$

The labelling of the β-lactam carbons by acetate indicated that gluconeogenesis was playing a role in clavulanic acid biosynthesis. Therefore, compounds which can be metabolised in the gluconeogenesis-glycolysis sequence should label the β-lactam carbons. This was found to be the case. When the authors fed [1,3-$^{13}C_2$]-glycerol (292) to *S. clavuligerus* the resulting clavulanic acid was labelled specifically at C-5 and C-7. In the

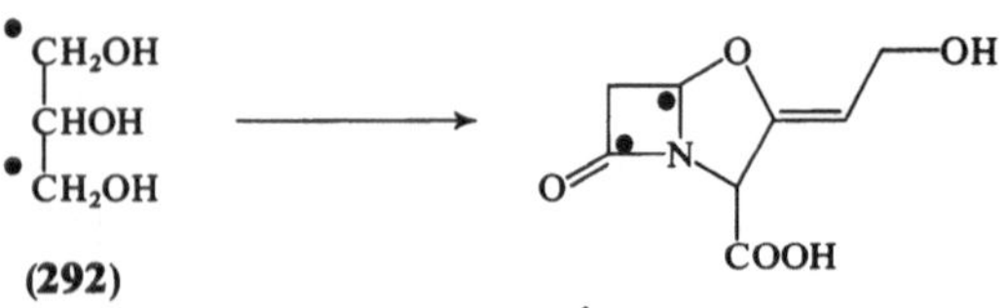

$^{\bullet}CH_2OH$
$CHOH$
$^{\bullet}CH_2OH$

(292)

$\longrightarrow$

References, pp. 89—106

n.m.r. spectrum, a long-range ^{13}C-^{13}C spin-spin coupling was observed between these two carbons indicating that the carbon skeleton of glycerol had been incorporated intact.

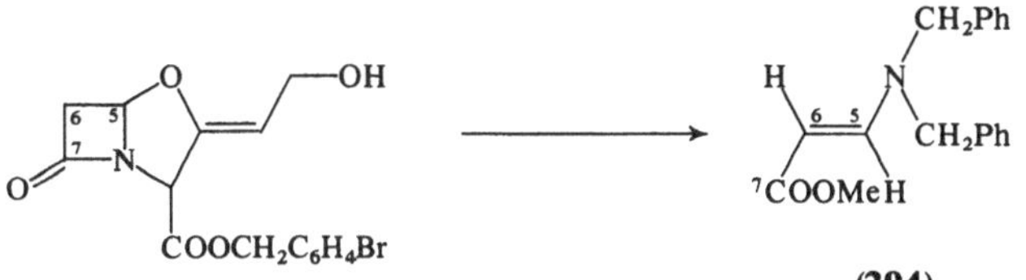

(293)   (294)

This result was confirmed using ^{14}C-labelled glycerol and a degradation technique (57). [U-^{14}C]-Glycerol was fed to a *S. clavuligerus* fermentation and the resulting ^{14}C-labelled clavulanic acid was isolated as the crystalline *p*-bromobenzyl ester (293). This was degraded with dibenzylamine in methanol to yield the three β-lactam carbon atoms in the form of the crystalline aminoacrylate (294). When the specific radioactivities of the starting material and degradation product were compared it was found that ~95% of the radioactivity was retained in the degradation product. [1-^{13}C]-Propionate and [3-^{13}C]-propionate did not specifically label the β-lactam ring, labelling C-2, C-3 and C-10 as well. Neither compound labelled C-8 or C-9 indicating that propionate was not being converted to acetate. The authors concluded that propionate was being metabolised by the TCA cycle *via* methylmalonyl CoA.

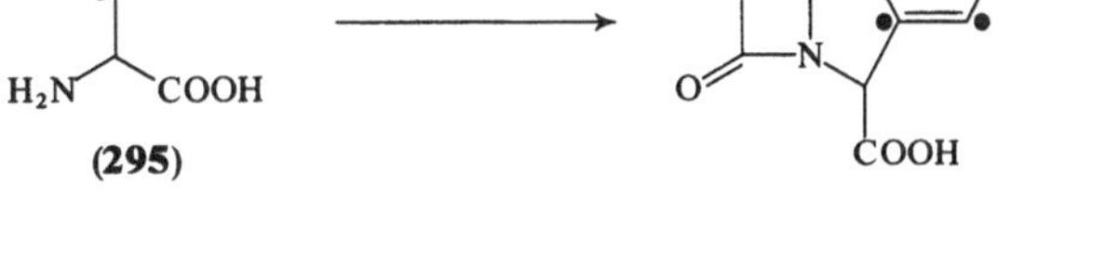

(295)

Subsequently ELSON *et al.* (280) studied the incorporation of labelled glutamate. When DL-[3,4-$^{13}C_2$]-glutamate (295) was fed to a *S. clavuligerus* culture the resulting clavulanic acid was predominantly labelled at C-2 and C-8. The ^{13}C-n.m.r. spectrum showed a spin-spin coupling between these carbons showing that the bond had not been broken prior to incorporation. Additional lesser ^{13}C-enrichments and couplings were observed which were consistent with some of the glutamate being deaminated to α-ketoglutarate and hence being metabolised *via* the TCA cycle.

The biosynthesis of clavulanic acid is summarised in Fig. 5.

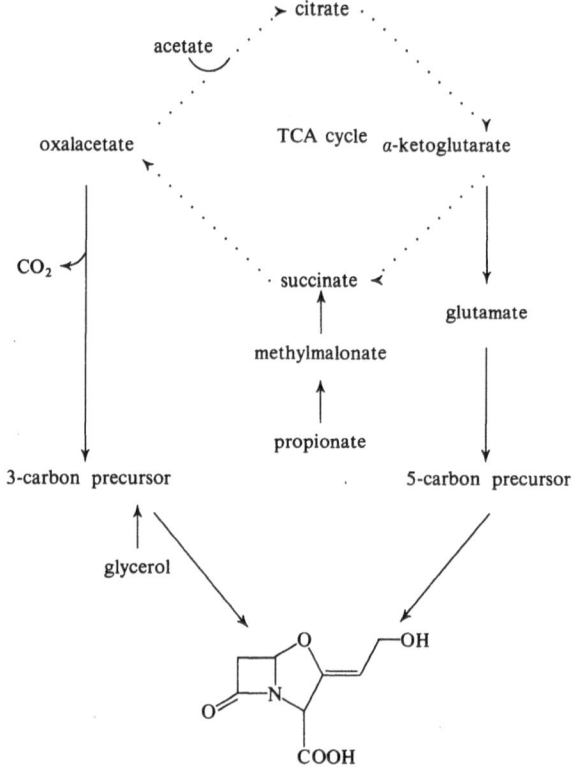

Fig. 5. Clavulanic acid biosynthesis

5. Biosynthesis of Carbapenem Antibiotics

Little has been published on the biosynthesis of the carbapenem antibiotics. ALBERS-SCHÖNBERG *et al.* (*281*) reported that the carbapenem nucleus in thienamycin is derived from glutamate and acetate as shown in Fig. 6, but so far no detailed account of this work has been published. Box

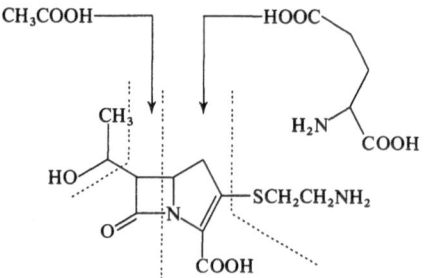

Fig. 6. Biosynthesis of thienamycin

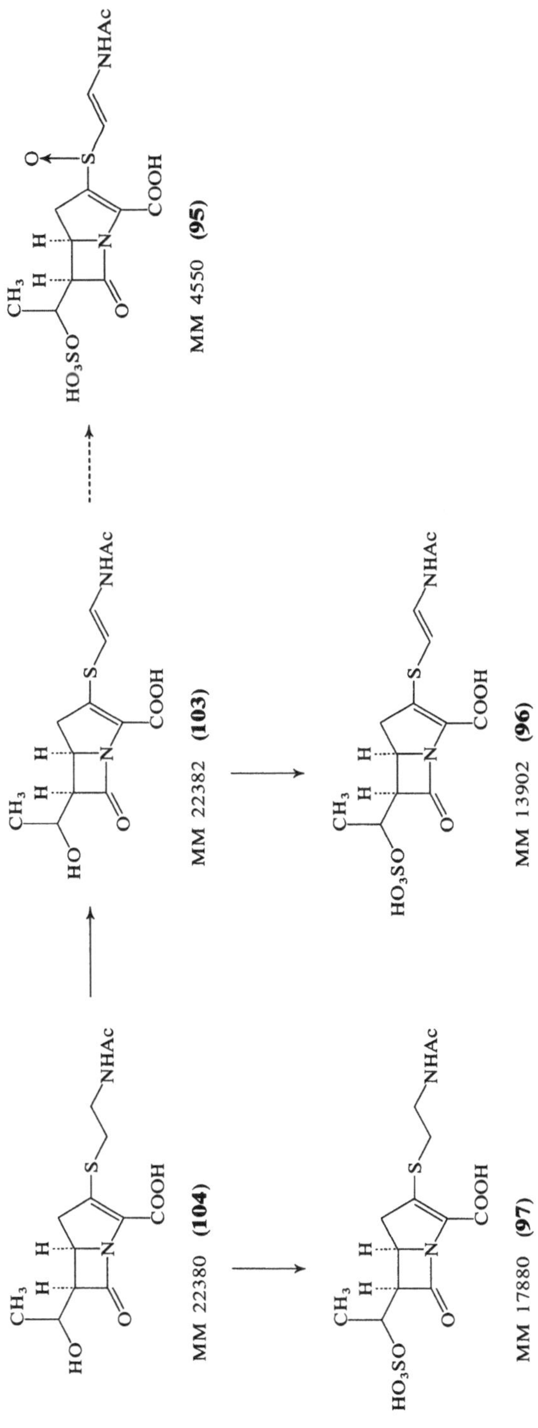

Fig. 7. Biosynthetic conversions of *cis*-olivanic acids

et al. (73) studied the biosynthetic modification of the side-chains of the cis-olivanic acids. They found that when MM 22380 (**104**) and MM 22382 (**103**) were incubated with a blocked mutant of Streptomyces olivaceus they were transformed into related olivanic acids MM 4550 (**95**), MM 13902 (**96**) and MM 17880 (**97**). They proposed the order of biosynthetic events shown in Fig. 7.

Fukagawa et al. (282) presented evidence that the OA-6129 group of carbapenems, which possess the pantothenyl side-chain, could be early precursors of the carbapenem antibiotics. These compounds were produced by a blocked mutant of Streptomyces fulvoviridis which lacked an acylase. This enzyme (A933 acylase) was isolated from the parent culture and was found to be able to remove the pantothenyl side-chain from the OA-6129 antibiotics:

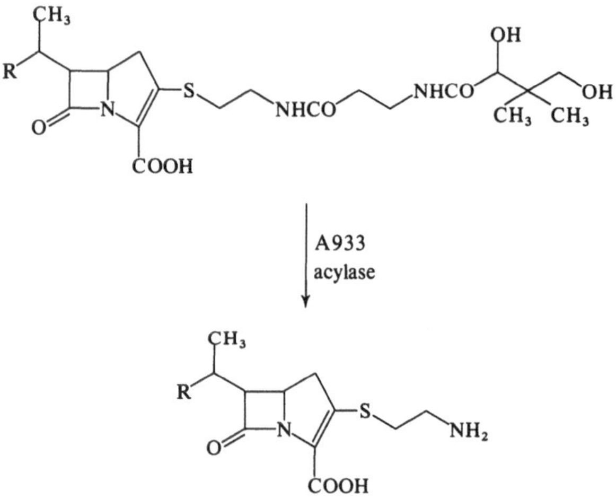

The A933 acylase could also transacylate the OA-6129 antibiotics with acyl CoA derivatives and acylate the free amino compounds.

6. Biosynthesis of the Cephalosporins

Cephalosporin C (**179**), like penicillin N, is derived from the amino acids α-aminodipic acid, cysteine and valine as shown below. The acetoxy side chain is derived from acetate. The incorporation of the three amino acids was first demonstrated by Trown et al. (285) who fed DL-α-amino-[2-^{14}C]-adipic acid, DL-[1-^{14}C]-valine or a mixture of DL- and meso-[3-^{14}C]-cystine to fermentations of a Cephalosporium sp. In each case the resulting cephalosporin C was found to be radioactive. The cephalosporin C samples

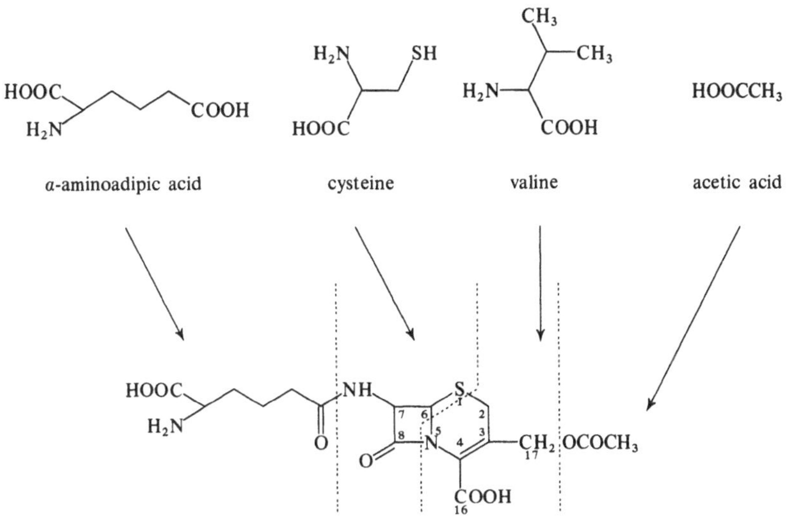

(179)

were then chemically degraded by acid hydrolysis or by hydrogenolysis followed by acid hydrolysis. The hydrolysis mixtures were examined by paper electrophoresis and chromatography and radioautography. It was found that when labelled α-aminoadipic acid was the precursor, 99% of the radioactivity of the cephalosporin C was recovered as α-aminoadipic acid in the hydrolysis mixture, showing that this precursor was being exclusively incorporated into the C-7 acylamino side chain. When valine was the precursor 90% of the radioactivity of the resulting cephalosporin C was found in the hydrolysis products derived from the dehydrovalinyl moiety, again indicating a specific incorporation. However in the case of cystine only 59% of the radioactivity of cephalosporin C was recovered as glycine (derived from C-6 and C-7) in the hydrolysis mixture, significant amounts of radioactivity also appearing in the α-aminoadipyl (13%) and valinyl (11.5%) residues. Using similar methodology it was also shown (286) that the C-3 acetoxy group was labelled when [1-^{14}C]-acetate was fed.

The stereochemistry of cysteine incorporation was investigated by HUDDLESTON et al. (287) who synthesised (2R,3R)-[U-^{14}C,2,3-^{3}H$_2$]-cysteine and (2R,3S)-[U-^{14}C, 3-^{3}H]-cysteine. These compounds were incorporated into cephalosporin C using washed mycelium of *Cephalosporium acremonium*. With (2R,3S)-[U-^{14}C,3-^{3}H]-cysteine 18% of the tritium label was retained with respect to ^{14}C-label, whereas with (2R,3R)-[U-^{14}C,2,3-^{3}H$_2$]-cysteine 50% of the tritium label was retained. As this latter precursor was tritium labelled on both carbons 2 and 3, the resulting cephalosporin C was degraded by acid hydrolysis and oxidation to glycine to determine the amount of tritium incorporated at positions 6 and

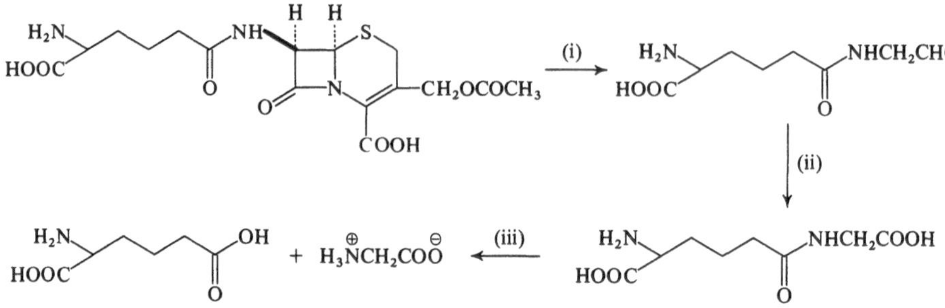

Reagents: (i) H⁺; (ii) Ag₂O; (iii) H⁺.

7. From the $^{14}C/^3H$ ratio of the glycine the retention of tritium at C-6 was calculated as approximately 65% whereas at C-7 it was only 35%. The greater retention of tritium from the $3R$ compared with the $3S$ molecule indicated a marked stereo chemical selectivity and suggested that cysteine was incorporated without inversion of the C-3 centre. The authors noted that these results were very similar to those found for the incorporation of chirally labelled cysteines into penicillin. The reason for this similarity will become apparent later in this text.

The incorporation of valine into cephalosporin C has been further studied by several groups. WARREN et al. (225) noted that L-[1-^{14}C]-valine labelled cephalosporin C, penicillin N and interacellular α-aminoadipylcysteinylvaline when fed to a Cephalosporium sp. fermentation. D-[1-^{14}C]-valine also labelled the two antibiotics but after epimerisation to the L-form. NEUSS et al. (231) fed (2RS,3R)-[4-^{13}C]-valine to a C. acremonium fermentation and examined the resulting ^{13}C-cephalosporin C by ^{13}C-n.m.r. spectroscopy. The cephalosporin C was specifically enriched at C-2, no other ^{13}C-enrichments being detected. Similarly the side chain methylene (C-17) of cephalosporin C was shown to be specifically enriched by (2S,3S)-[4-^{13}C]-valine by KLUENDER et al. (232). Both of the hydrogens at C-2 and C-17 were shown to be derived from (2S,3R)-[4,4,4-$^2H_3$]-valine and (2S,3S)-[4,4,4-$^2H_3$]-valine respectively (229), the deuterated cephalosporin C samples being examined by mass and n.m.r. spectrometry. The incorporation of (2S,3S)-[^{15}N,4,4,4-$^2H_3$]-valine into cephalosporin C and penicillin N was reported by HUANG et al. (248) who examined the methyl esters of these antibiotics by mass spectrometry, monitoring the valinyl fragments (296) and (261). Peaks at m/z 233 and

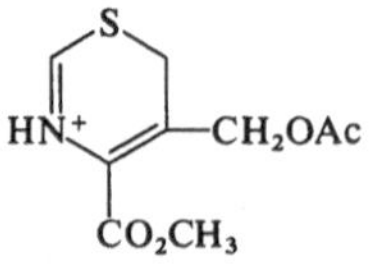

(296) m/z 230

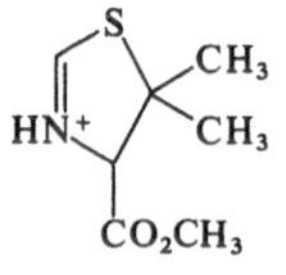

(261) m/z 174

m/z 178 indicated that some intact valine had been incorporated into both antibiotics. However there were much more intense peaks at m/z 232 and m/z 177 indicating that extensive deamination of the 15N-labelled valine had occurred. The isotopic ratios of the valinyl fragments of cephalosporin C and penicillin N were very similar and the authors concluded that both antibiotics were derived from the same precursor tripeptide i.e. α-aminoadipylcysteinylvaline.

The cephalosporin produced by *Streptomyces clavuligerus,* cephamycin C (**203**), has also been shown to be derived from the amino acids α-aminoadipic acid, cysteine and valine (*288*). DL-[1-^{14}C]-Cysteine, DL-[3-^{14}C]-cysteine, DL-[1-^{14}C]-valine and DL-α-amino-[1-^{14}C]-adipic acid all labelled cephamycin C. Acid hydrolysis of cephamycin C, followed by separation of the hydrolysis mixture using an amino acid analyser, indicated that DL-α-amino-[1-^{14}C]-adipic acid was specifically incorporated into the acylamino side chain. The 7-methoxy group was shown to be derived from L-[*methyl*-^{14}C]-methionine.

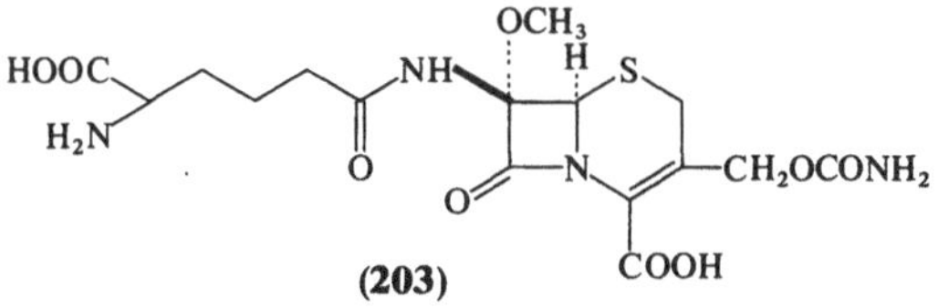

(**203**)

The accumulation of results presented so far suggested that the biosynthesis of the cephalosporins and penicillins was essentially the same in many respects and that, like the penicillins, the cephalosporins were probably derived from α-aminoadipylcysteinylvaline. The cephalosporin producing organisms that have been most studied to date *(C. acremonium* and *S. clavuligerus)* both also produce penicillin N. There are, therefore, two possible routes from the tripeptide to the cephem products, one by direct cyclisation of the tripeptide, the other by enzymic ring expansion of penicillin N. The first evidence that penicillin N could be converted enzymically to a cephalosporin was provided by KOHSAKA and DEMAIN (*289*) who demonstrated that lysed protoplasts of *C. acremonium* could convert penicillin N to another antibiotic or antibiotics which were destroyed by cephalosporinase but not by penicillinase. The product of this cell-free enzymic reaction was characterised as deacetoxycephalosporin C (**297**) by YOSHIDA *et al. (290)*. These authors also showed that cell-free

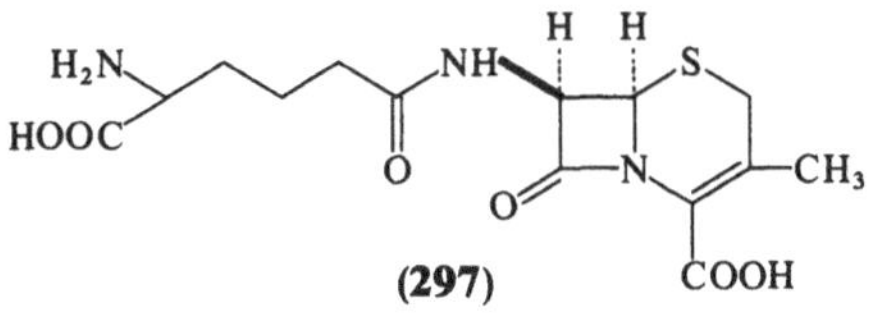

(**297**)

extracts of mutants of *C. acremonium* which could not produce penicillin N or cephalosporin C in fermentation could convert penicillin N to de-acetoxycephalosporin C, whereas extracts of mutants which could produce penicillin N but not cephalosporin C could not perform this reaction. Further evidence for this conversion was provided by Baldwin et al. (*291, 292*) who synthesised [6-^{3}H]-penicillin N and [10-^{14}C,6-^{3}H]-penicillin N (**298**). These compounds were added to cell-free extracts of *C. acremonium*

(298)

* = ^{14}C

and the resulting deacetoxycephalosporin C samples were purified by Amberlite XAD-2 chromatography, h.p.l.c., dilution with authentic de-acetoxycephalosporin C, and conversion into crystalline derivatives. After recrystallisation to constant radioactivity it was found (after correction for dilution) that the specific radioactivity was the same as the starting penicillin N and, in the case of the double labelled material, the isotope ratio was the same as the starting penicillin N. The tritium label was shown to be located at C-7 of deacetoxycephalosporin C by chemical conversion to the 7-methoxy derivative by the method of Koppel and Koehler (*185*) whereupon 95% of the radioactivity was lost.

Hook et al. (*293*) found that the enzymic ring expansion of penicillin N by *C. acremonium* extracts was markedly stimulated by aeration, ferrous ions and ascorbate, a result which was confirmed by Sawada et al. (*294*) who found that the optimum concentrations of these co-factors were 0.04 m*M* and 0.67 m*M* respectively. These latter authors also showed that the optimal pH was pH 7.2, the optimal temperature 25° C and the optimal penicillin N concentration 0.07 m*M*. Adenosine triphosphate was required at 0.83 m*M* but phospho*enol*pyruvate and pyruvate kinase (used in the earlier experiments) were not found to be necessary. Magnesium sulphate and potassium chloride slightly stimulated the reaction whereas cupric sulphate and zinc sulphate were inhibitory. Sawada et al. (*295*) showed that active cell-free extracts could be prepared by ultrasonication of the mycelium of *C. acremonium* without the need to first form protoplasts. Felix et al. (*296*) using ether permeablised mycelium of *C. acremonium* confirmed the earlier findings (*293*) that α-ketoglutarate had a stimulating effect on the ring expanding enzyme. Kupka et al. (*257, 297*) partially purified the enzyme and showed it to be soluble, to have a molecular weight of 31,000 daltons and a Km of 0.03 mM for penicillin N. The enzyme would not accept isopenicillin N,6-aminopenicillanic acid, adipylpenicillin, benzylpenicillin or ampicillin as substrate. The last three antibiotics mentioned acted as competitive inhibitors. MILLER *et al.* (*298*) showed that the synthetic cepham exomethylenecephalosporin C (**299**), was not converted to deacetoxycephalosporin C in the *C. acremonium* cell-free system but, again, acted as a weak inhibitor. The cepham (**300**), which was isolated from a *C. acremonium* culture broth, also did not form deacetoxy-cephalosporin C in the ring expanding system.

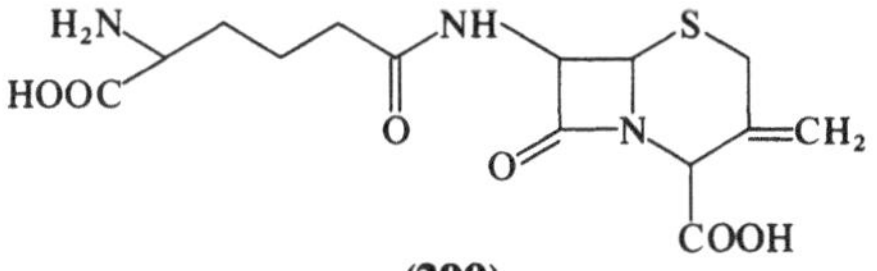

(299)

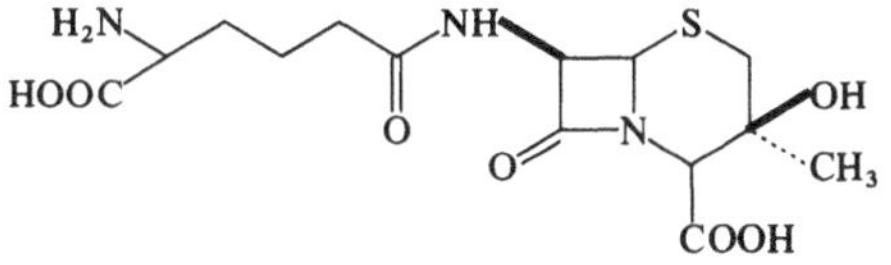

(300)

JENSEN *et al.* (*299, 300*) studied the ring expansion of penicillin N to deacetoxycephalosporin C by cell-free extracts of *S. clavuligerus*. As with the *Cephalosporium* enzyme they found a requirement for α-ketoglutarate but found that a range of other α-keto acids would not serve as co-substrates. Adenosine triphosphate was not required. The enzymic activity was not precipitated by centrifugation at 100,000 xg. 6-Aminopenicillanic acid, or penicillins V, G, K or dihydro-F were not accepted as substrates.

In both *C. acremonium* and *S. clavuligerus* deacetoxycephalosporin C is oxidised to deacetylcephalosporin C (**301**). LIERSCH *et al.* (*301*) observed that cell-free extracts of *C. acremonium* could convert, deacetoxycephalo-sporin C to cephalosporin C and that the enzymic system required molecular oxygen, manganese ions and NADH (or NADPH) as co-factors. From this they concluded that an oxygenase was involved and postulated (**301**) as the intermediate in cephalosporin C production. STEVENS *et al.*

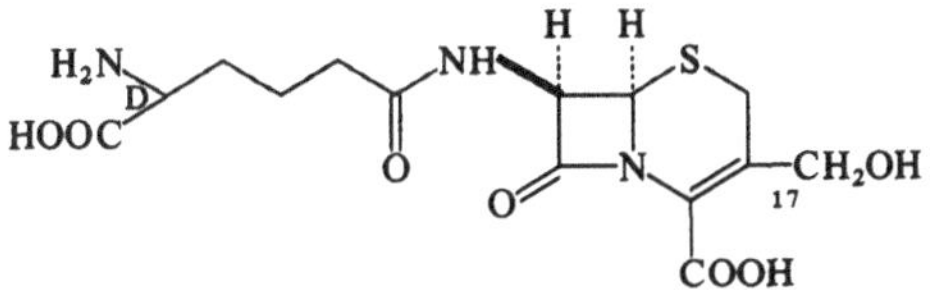

(301)

(*302*) also came to this conclusion after shaking *C. acremonium* mycelium in the presence of $^{18}O_2$. The resulting cephalosporin C was purified, derivatised and examined by mass spectrometry. From analysis of the molecular ion and fragment ions they concluded that one atom of ^{18}O had been incorporated which was located at C-17. Using similar methodology O'Sullivan *et al.* (*303*) demonstrated that the oxygen at C-17 of cephamycin C (**203**) produced by *S. clavuligerus* is also derived from molecular oxygen.

Fujisawa *et al.* (*304*) provided further evidence that deacetoxycephalosporin C was converted to deacetylcephalosporin C by isolating the latter compound from a cell-free system of *C. acremonium* to which deacetoxycephalosporin C had been added. After purification by chromatography and electrophoresis the product was examined by t.l.c. and u.v. and i.r. spectroscopy. It was found to be indentical with authentic deacetylcephalosporin C.

Turner *et al.* (*305*) isolated the deacetoxycephalosporin C oxygenase from both *C. acremonium* and *S. clavuligerus* by DEAE-cellulose column chromatography. They concluded that the enzymes were dioxygenases as they both required α-ketoglutarate, ferrous ions, ascorbate and dithiothreitol for full activity. The *C. acremonium* enzyme was markedly activated by preincubation with ferrous ions. Both enzymes where inhibited by high concentrations of nucleotides. The pH optimum of the *S. clavuligerus* enzyme was about pH 7.0, whereas that of *C. acremonium* was about pH 6.0. Neither enzyme would accept the deacetoxycephalosporins (**302**) − (**307**) as substrates.

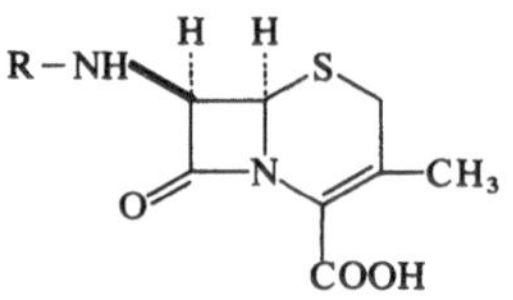

(**302**) R=H
(**303**) R=HCO
(**304**) R=$C_6H_5CH_2CO$
(**305**) R=$C_6H_5OCH_2CO$
(**306**) R=$HO_2C(CH_2)_3CO$

(**307**) R=

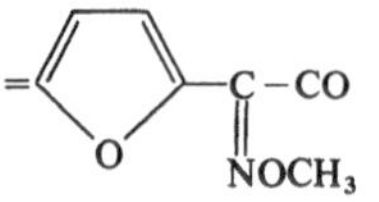

Liersch *et al.* (*301*) and Fujisawa and Kanzaki (*306*) demonstrated that cell-free extracts of *C. acremonium* could transacetylate deacetylcephalosporin C with [*acetyl*-1-^{14}C]-acetyl CoA. The transacetylase could not utilise acetate plus CoASH nor could it transacetylate serine or homoserine. However it could transacetylate deacetylcephaloram, deacetylcephacetrile and 7-aminodeacetylcephalosporanic acid. It required divalent cations for activity, Mg^{++} being the most stimulatory. The pH optimum was pH 7.0 – 7.5. FUJISAWA and KANZAKI also showed that mutants of *C. acremonium* which produced deacetylcephalosporin C but not cephalosporin C lacked the transacetylase activity.

Deacetylcephalosporin C is acetylated to give cephalosporin C (**179**) in *C. acremonium*. In *Streptomyces clavuligerus*, however, deacetylcephalosporin C is converted to the carbamoyl derivative (**308**). BREWER *et al.* (*307, 308*) purified the O-carbamoyltransferase from this organism forty-fold by batch adsorption onto DEAE-cellulose followed by hydroxyapatite. The enzyme required carbamoylphosphate and ATP for activity and was further stimulated by Mg^{++} and Mn^{++} ions. Other deacetylcephalosporins (**309**) – (**315**) were also accepted as substrates. The enzyme activity was optimum at pH 6.75.

(308)

(309) R = H R¹ = H

(310) $R = C_6H_5CH_2CO$ R¹ = H

(311) R = R¹ = H

(312) R = R¹ = H

(313) R = R¹ = H

(314) R = R¹ = H

(315) R = $R^1 = OCH_3$

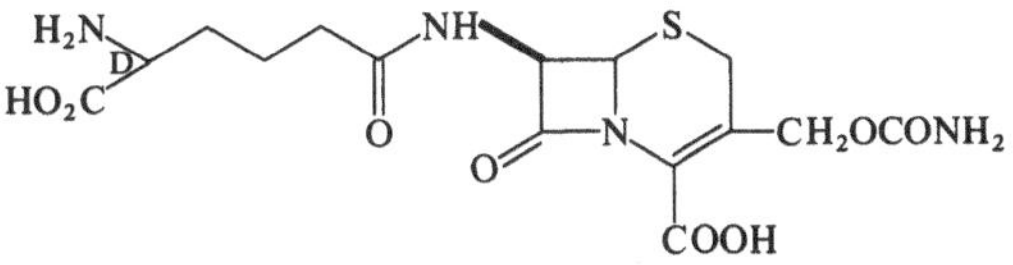

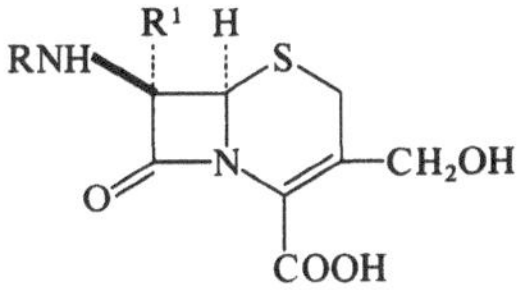

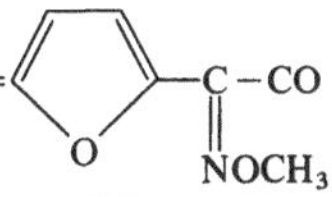

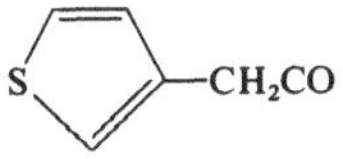

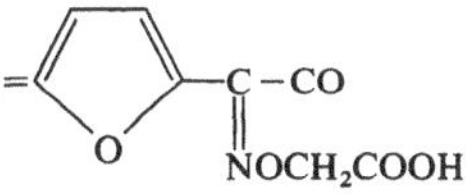

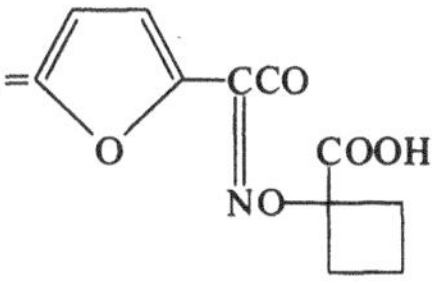

The terminal stage of cephalosporin biosynthesis in *S. clavuligerus* is the introduction of the 7α-methoxy group to form cephamycin C (**203**). The process almost certainly occurs in two stages: initially an oxidation with

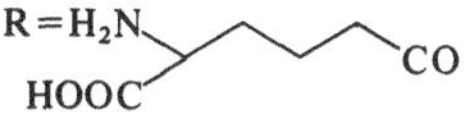

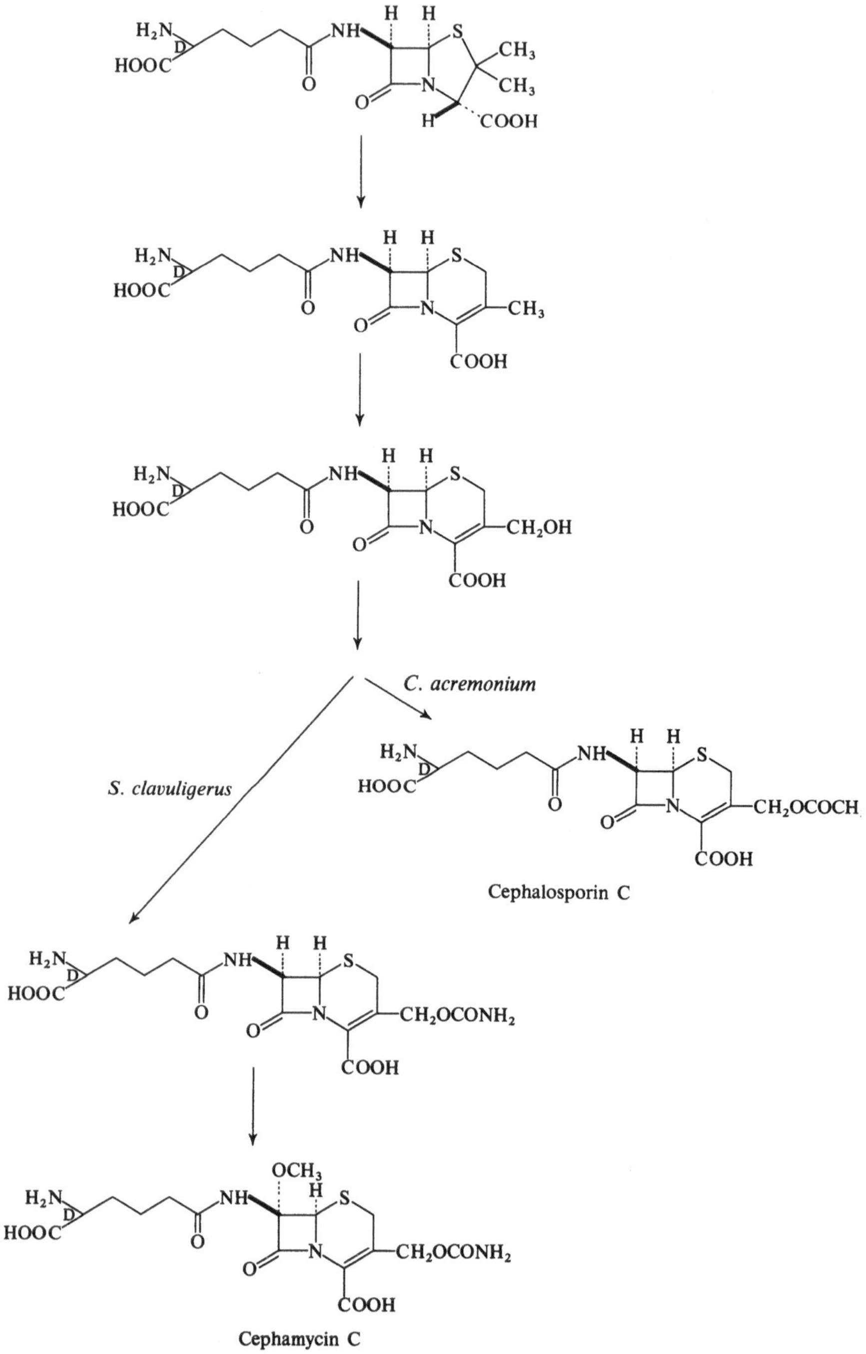

Fig. 8. Biosynthesis of the cephalosporins

molecular oxygen to form the 7α-hydroxy cephalosporin followed by O-methylation. The incorporation of molecular oxygen at C-7 of cephamycin C was demonstrated by O'SULLIVAN *et al.* (*303*) who shook washed mycelium of *S. clavuligerus* under an atmosphere of $^{18}O_2$. The resulting cephamycin C was purified, derivatised and examined by chemical ionisation mass spectrometry. Analysis of the fragment ions confirmed incorporation of ^{18}O at C-7. Later, O'SULLIVAN and ABRAHAM (*309*) showed that cell-free extracts of *S. clavuligerus* could convert (**308**) to [^{3}H]-cephamycin C in the presence of S-adenosyl-L-[*methyl*-^{3}H]-methionine. The product was identical with authentic cephamycin C when examined by paper electrophoresis followed by chromatography. Ferrous ions and α-ketoglutarate were required for the methoxylation to occur. Cephalosporin C was also methoxylated in this system, even though *S. clavuligerus* does not produce this antibiotic, and deacetoxycephalosporin C was methoxylated to a lesser extent. Deacetylcephalosporin C was not converted to 7α-methoxydeacetylcephalosporin C and therefore it is unlikely that the latter is a precursor of cephamycin C. There was no detectable methoxylation of cephalothin, cefuroxime, penicillin N, clavulanic acid or nocardicin A. HOOD *et al.* (*310*) confirmed that the methoxylation of cephalosporins occurs in two stages. By adding cephalosporin C to an aerated cell-free system from *S. clavuligerus* containing α-ketoglurate and ferrous ions, but *not* S-adenosyl methionine, they detected (by h.p.l.c.) a new cephalosporinase-labile substance. When examined by fast atom bombardment mass spectrometry, n.m.r. and u.v. spectroscopy, and h.p.l.c., this was found to be identical with synthetic 7α-hydroxy-cephalosporin C. Addition of synthetic 7α-hydroxycephalosporin C and S-adenosylmethionine to the cell-free system caused the production of cephalosporin C.

To summarise, cephalosporins are produced by ring expansion of penicillin N to deacetoxycephalosporin C which is then oxidised to deacetylcephalosporin C. In *C. acremonium*, this last compound is acetylated to cephalosporin C. In *S. clavuligerus* deacetylcephalosporin C is transcarbamoylated then converted to the 7α-methoxy derivative cephamycin C, probably *via* a 7α-hydroxy intermediate. Cephalosporin biosynthesis is shown diagramatically in Fig. 8.

References

1. KIKUCHI, T., and S. UYEO: Pachysandra Alkaloids VIII. Structures of Pachystermine-A and -B, Novel Type Alkaloids Having a β-Lactam Ring. Chem. Pharm. Bull. **15**, 549 (1967).

2. STEWART, W. W.: Isolation and Proof of Structure of Wildfire Toxin. Nature **229**, 174 (1971).

3. SCANNELL, J. P., D. L. PRUESS, J. F. BLOUNT, H. A. AX, M. KELLETT, F. WEISS, T. C. DEMNY, T. H. WILLIAMS, and A. STEMPEL: Antimetabolites Produced By Microorganisms. XII. (S)-Alanyl-3-[α-(S)-Chloro-3-(S)-hydroxy-2-oxo-3-azetidinyl-methyl]-(S)-Alanine, A New β-Lactam Containing Natural Product. J. Antibiotics **28**, 1 (1975).

4. TAKITA, T., Y. MURAOKA, T. YOSHIOKA, A. FUJII, K. MAEDA, and Y. UMEZAWA: The Chemistry of Bleomycin. IX. The Structures of Bleomycin and Phleomycin. J. Antibiotics **25**, 755 (1972).

5. TAKITA, T., Y. MURAOKA, T. NAKATANI, A. FUJII, Y. UMEZAWA, H. NAGANAWA, and H. URREZAWA: Chemistry of Bleomycin. XIX. Revised Structures of Bleomycin and Phleomycin. J. Antibiotics **31**, 801 (1975).

6. FLYNN, E. H., Ed.: Cephalosporins and Penicillins. New York and London: Academic Press. 1972.

7. AOKI, H., H. SAKAI, M. KOHSAKA, T. KONAMI, J. HOSODA, Y. KUBOCHI, E. IGUCHI, and H. IMANAKA: Nocardicin A, A New Monocyclic β-Lactam Antibiotic. I. Discovery, Isolation and Characterisation. J. Antibiotics **29**, 890 (1976).

8. HASHIMOTO, M., T. KOMORI, and T. KAMIYA: Nocardicin A, A New Monocyclic β-Lactam Antibiotic. II. Structure Determination of Norcardicins A and B. J. Antibiotics **29**, 890 (1976).

9. HASHIMOTO, M., T. KOMORI, and T. KAMIYA: Nocardicin A and B, Novel Mono-cyclic β-Lactam Antibiotics from a *Nocardia* species. J. Amer. Chem. Soc. **98**, 3023 (1976).

10. KAMORI, T., K. KUNIGITA, K. NAKAHARA, H. AOKI, and H. IMANAKA: Production of 3-Aminonocardicinic Acid from Nocardicin C by Microbial Enzymes. Agric. Biol. Chem. **42**, 1439 (1978).

11. KAMIYA, T.: Studies on the New Monocyclic β-Lactam Antibiotics, Nocardicins. Recent Advances in the Chemistry of β-Lactam Antibiotics (J. ELKS, Ed.). Special Publication Number 28, p. 281. The Chemical Society, 1977.

12. SCHAFFNER-SABBA, K., B. W. MULLER, R. SCARTAZZINI, and H. WEHRLI: Ein einfacher Zugang zu 3-Amino-nocardicinsäure. Helv. Chim. Acta **63**, 321 (1980).

13. FOGLIO, M., G. FRANCESHI, P. LOMBARDI, C. SCARAFILE, and F. ARCAMONE: From the Penicillin to the Nocardicin Skeleton: An Alternative Route. J. C. S. Chem. Commun. **1978**, 1101.

14. KAMIYA, T., M. HASHIMOTO, O. NAKAGUCHI, and T. OKU: Total Synthesis of Monocyclic β-Lactam Antibiotics, Nocardicin A and D. Tetrahedron **35**, 323 (1979).

15. KAMIYA, T., T. OKU, O. NAKAGUCHI, H. TAKENO, and M. HASHIMOTO: A Novel Synthesis of Nocardicins and their Analogues. Tetrahedron Letters **1978**, 5119.

16. CURRAN, W. V., M. L. SASSIVER, A. S. ROSS, T. L. FIELDS, and J. H. BOOTHE: The Total Synthesis of Nocardicin A. J. Antibiotics **35**, 329 (1982).

17. KOPPEL, G. A., L. MCSHANE, F. JOSE, and R. D. G. COOPER: Total Synthesis of Nocardicin A. Synthesis of 3-ANA and Nocardicin A. J. Amer. Chem. Soc. **100**, 3933 (1978).

18. MATTINGLY, P. G., and M. J. MILLER: Synthesis of 2-Azetidinones from Serine-hydroxamates: Approaches to the Synthesis of 3-Aminonocardicinic Acid. J. Org. Chem. **46**, 1557 (1981).

19. WASSERMANN, H. H., D. J. HLASTA, A. W. TREMPER, and J. S. WU: Applications of New β-Lactam Syntheses to the Preparation of (±)-3-Aminonocardicinic Acid. J. Org. Chem. **46**, 2999 (1981).

20. WASSERMANN, H. H., and D. J. HLASTA: A Synthesis of (±)-3-Aminonocardicinic Acid (3-ANA). J. Amer. Chem. Soc. **100**, 6780 (1978).

21. WASSERMANN, H. H., A. W. TREMPER, and J. S. WU: β-Lactams from Azetidine Carboxylates: Tetrahedron Letters **1979**, 1089.

22. CHIBA, K., M. MORI, and Y. BAN: A Novel Synthesis of (±)-3-Aminonocardicinic Acid. J. C. S. Chem. Commun. **1980**, 770.

23. IMADA, A., K. KITANO, K. KINTAKA, M. MUROI, and M. ASAI: Sulfazecin and Isosulfazecin, Novel β-Lactam Antibiotics of Bacterial Origin. Nature **289**, 590 (1981).

24. ASAI, I., K. HAIBARA, M. MUROI, K. KINTAKA, and T. KISHI: Sulfazecin, A Novel β-Lactam Antibiotic of Bacterial Origin. Isolation and Chemical Characterisation. J. Antibiotics **34**, 621 (1981).

25. KINTAKA, K., K. HAIBARA, M. ASAI, and A. IMADA: Isosulfazecin, A New β-Lactam Antibiotic Produced by An Acidophilic Pseudomonad. J. Antibiotics **34**, 1081 (1981).

26. SYKES, R. B., C. M. CIMARUSTI, D. P. BONNER, K. BUSH, D. M. FLOYD, N. H. GEORGOPAPADAKOU, W. H. KOSTER, W. LIU, W. L. PARKER, P. A. PRINCIPE, M. L. RATHNUM, W. A. SLUSARCHYK, W. H. TREJO, and J. S. WELLS: Monocyclic β-Lactam Antibiotics Produced by Bacteria. Nature **291**, 489 (1981).

27. PARKER, W. L., W. H. KOSTER, C. M. CIMARUSTI, D. M. FLOYD, W. LIU, and M. L. RATHNUM: SQ26, 180, A Novel Monobactam. II. Isolation, Structure Determination and Synthesis. J. Antibiotics **35**, 189 (1982).

28. PARKER, W. L., and M. L. RATHNUM: EM 5400, A Family of Monobactam Antibiotics Produced by Agrobacterium Radiobacter; II. Isolation and Structure Determination. J. Antibiotics **35**, 300 (1982).

29. CIMARUSTI, C. M., H. E. APPLEGATE, H. W. CHANG, D. M. FLOYD, W. M. KOSTER, W. A. SLUSARCHYK, and M. G. YOUNG: Monobactams. The Conversion of 6-APA to (S)-3-Amino-2-oxoazetidine-1-sulfonic Acid and Its 3-(RS)-Methoxy Derivative. J. Org. Chem. **47**, 179 (1982).

30. FLOYD, D. M., A. W. FRITZ, and C. M. CIMARUSTI: Monobactams. Stereospecific Synthesis of (S)-3-Amino-2-oxoazetidine-1-sulfonic Acid. J. Org. Chem. **47**, 176 (1982).

31. BREUER, H., C. M. CIMARUSTI, TH. DENZEL, W. H. KOSTER, W. A. SLUSARCHYK, and H. D. TREUNER: Monobactams-Structure-Activity Relationships Leading to SQ26, 776: J. Antimicrobial Chemotherapy **8**, 21–28 Supp. E (1981).

32. CLARKE, H. T., J. R. JOHNSON, and Sir R. ROBINSON (Eds.): The Chemistry of Penicillin. Princeton University Press. 1949.

33. DOYLE, F. P., and J. H. C. NAYLER: Penicillins and Related Structures. Advances in Drug Research, Vol. 1 (HARPER, N. J., and A. B. SIMMONDS, Eds.), p. 1–69. New York: Academic Press. 1964.

34. NAYLER, J. H. C.: Advances in Penicillin Research. Advances in Drug Research **7**, 1–105 (1973).

35. KACZKA, E., and K. FOLKERS: Desthiobenzylpenicillin and Other Hydrogenolysis Products of Benzylpenicillin, reference 32, p. 243–268.

36. CROWFOOT, D., C. W. BUNN, B. W. ROGERS-LOW, and A. TURNER-JONES: The X-Ray Crystallographic Investigation of the Structure of Penicillin, reference 32, p. 310–366.

37. BATCHELOR, F. R., F. P. DOYLE, J. H. C. NAYLER, G. N. ROLINSON: Synthesis of Penicillin: 6-Aminopenicillanic Acid in Penicillin Fermentations: Nature **183**, 257 (1959).

38. ROLINSON, G. N., F. R. BATCHELOR, D. BUTTERWORTH, J. CAMERON-WOOD, M. COLE, G. C. EUSTACE, M. V. HART, M. RICHARDS, and E. B. CHAIN: Formation of 6-Aminopenicillanic Acid from Penicillin by Enzymatic Hydrolysis: Nature **187**, 236 (1960).

39. WEISSENBURGER, H. W. O., and M. G. VAN DER HOEVEN: An Efficient Nonenzymatic Conversion of Benzylpenicillin to 6-Aminopenicillanic Acid: Rec. Trav. Chim. Pays-Bas Belg. **89**, 1081 (1970).

40. DU VIGNEAUD, V., J. L. WOOD, and M. E. WRIGHT: The Condensation of Oxazolones and D-Penicillamine and the Resultant Antibiotic Activity, reference 32, p. 892–920.

41. SHEEHAN, J. C., K. R. HENERY-LOGAN, and D. A. JOHNSON: The Synthesis of Substituted

Penicillins and Simpler Structural Analogs. VII. The Cyclisation of a Penicilloate Derivative to Methyl Phthalimidopenicillante. J. Amer. Chem. Soc. **75**, 3292 (1953).

42. Sheehan, J. C., and K. R. Henery-Logan: The Total Synthesis of Penicillin V. J. Amer. Chem. Soc. **79**, 1262 (1957); **81**, 3089 (1959).

43. — — A General Synthesis of Penicillins. J. Amer. Chem. Soc. **81**, 5836 (1959); **84**, 2983 (1962).

44. Bose, A. K., G. Spiegelman, and M. S. Manhas: Studies on Lactams. X. Total Synthesis of 5,6-*trans*-Penicillin V Methyl Ester. J. Amer. Chem. Soc. **90**, 4506 (1968).

45. Firestone, R. A., N. S. Maciejewicz, R. W. Ratcliffe, and B. G. Christensen: Total Synthesis of β-Lactam Antibiotics. IV. Epimerization of 6(7)-Aminopenicillins and -cephalosporins from α to β. J. Org. Chem. **39**, 437 (1974).

46. Sammes, P. G. (Ed.): Topics in Antibiotic Chemistry Vol. 4. The Chemistry and Antimicrobial Activity of New Synthetic β-Lactam Antibiotics. Chichester: Ellis Horwood Ltd. 1980.

47. Baldwin, J. E., M. A. Christie, S. B. Haber, and L. I. Kruse: Stereospecific Synthesis of Penicillins. Conversion from a Peptide Precursor. J. Amer. Chem. Soc. **98**, 3045 (1976).

48. Baldwin, J. E., and M. A. Christie: Stereospecific Synthesis of Penicillins. Stereoelectronic Control in the Conversion of a Peptide into a Penicillin. J. Amer. Chem. Soc. **100**, 4597 (1978).

49. Brown, A. G., D. Butterworth, M. Cole, G. Hanscombe, J. D. Hood, C. Reading, and G. N. Rolinson: Naturally-Occurring β-Lactamase Inhibitors with Antibacterial Activity. J. Antibiotics **29**, 668 (1976).

50. Howarth, T. T., A. G. Brown, and T. J. King: Clavulanic Acid, a Novel β-Lactam isolated from *Streptomyces clavuligerus;* X-Ray Crystal Structure Analysis. J. C. S. Chem. Commun. **1976**, 267.

51. Brown, A. G., J. Goodacre, J. B. Harbridge, T. T. Howarth, R. J. Ponsford, I. Stirling, and T. J. King: Clavulanic Acid; a Novel Fused β-Lactam isolated from *Streptomyces clavuligerus.* Recent Advances in the Chemistry of β-Lactam Antibiotics (J. Elks, Ed.), pp. 295 – 298, Special Publication No. 28. London: The Chemical Society. 1977; J. Chem. Soc. Perkin I, 1984, 635.

52. Davies, J. S., and T. T. Howarth: Clavulanic Acid. Rearrangement to 3,4-Disubstituted Pyrroles. Tetrahedron Letters **23**, 3109 (1982).

53. Cooper, R. D. G.: Clavulanic Acid and Derivatives. Topics in Antibiotic Chemistry, Vol. 3 (P. Sammes, Ed.), pp. 57 – 73. Chichester: Ellis Horwood. 1980.

54. Cherry, P. C., and C. E. Newall: Clavulanic Acid. Chemistry and Biology of β-Lactam Antibiotics, Vol. 2 (R. Morin and M. Gorman, Eds.), pp. 361 – 402. New York: Academic Press. 1982.

55. Brown, A. G.: New Naturally Occurring β-Lactam Antibiotics and Related Compounds. J. Antimicrobial Chemotherapy **7**, 15 – 48 (1981).

56. Brown, A. G., T. T. Howarth, I. Stirling, and T. J. King: The Formation and Crystal Structure Analysis of Isoclavulanic Acid. Tetrahedron Letters **1976**, 4203.

57. Stirling, I., and S. W. Elson: Studies on the Biosynthesis of Clavulanic Acid II. Chemical Degradation of ¹⁴C-Labelled Clavulanic Acid. J. Antibiotics **32**, 1125 (1979).

58. Corbett, D. F., T. T. Howarth, and I. Stirling: Oxidation of Clavulanic Acid and a Ready Synthesis of the 7-Oxo-4-oxa-1-azabicyclo-[3.2.0]-hept-2-ene Ring System. J. C. S. Chem. Commun. **1977**, 808.

59. Cherry, P. C., G. I. Gregory, C. E. Newall, P. Ward, and N. S. Watson: Reaction of Sulphur Nucleophiles with Activated Derivatives of Clavulanic Acid. J. C. S. Chem. Commun. **1978**, 467.

60. Hunt, E.: Decarboxylation of Clavulanic Acid and its 9-Methyl Ether. J. Chem. Research (S) **1981**, 64.

61. CHERRY, P. C., C. E. NEWALL, and N. S. WATSON: Synthesis of Antibacterial Pen-2-em-3-carboxylic Acids from Clavulanic Acid. J. C. S. Chem. Commun. **1979**, 663.
62. GILPIN, M. L., J. B. HARBRIDGE, T. T. HOWARTH, and T. J. KING: Wittig Reactions with β-Lactam Carbonyls: A Convenient Means of Protection. X-Ray Crystal Structure of p-Nitrobenzyl-(2R,5R)-Z-7-Methoxycarbonylmethylene-Z-3-(β-phthalimido-ethylidene)-4-oxa-1-azabicyclo-[3.2.0]-heptane-2-carboxylate. J. C. S. Chem. Commun. **1981**, 929.
63. READING, C.: U. K. Patent 1, 547, 222 (1979).
64. BROWN, D., J. R. EVANS, and R. A. FLETTON: Structures of Three Novel β-Lactams from *Streptomyces Clavuligerus*. J. C. S. Chem. Commun. **1979**, 282.
65. MÜLLER, J. C., V. TOOME, D. L. PRUESS, J. F. BLOUNT, and M. WEIGELE: Ro 22-5417, A New Clavam Antibiotic From *Streptomyces Clavuligerus* III. Absolute Stereochemistry. J. Antibiotics **36**, 217 (1983).
66. WANNING, M., H. ZÄHNER, B. KRONE, and A. ZEECH: Ein neues antifungisches β-Lactam-Antibiotikum der Clavam-Reihe. Tetrahedron Letters **22**, 27 (1981).
67. BENTLEY, P. H., P. D. BERRY, G. BROOKS, M. L. GILPIN, E. HUNT, and I. ZOMAYA: Total Synthesis of (±)-Clavulanic Acid. J. C. S. Chem. Commun. **1977**, 748.
68. BENTLEY, P. H., G. BROOKS, M. L. GILPIN, and E. HUNT: A New Total Synthesis of (±)-Clavulanic Acid. Tetrahedron Letters **1979**, 1889.
69. BROWN, A. G., D. F. CORBETT, A. J. EGLINGTON, and T. T. HOWARTH: Structures of Olivanic Acid Derivatives MM 4550 and MM 13902; Two New, Fused β-Lactams isolated from *Streptomyces olivaceus*. J. C. S. Chem. Commun. **1977**, 523.
70. CORBETT, D. F., A. J. EGLINGTON, and T. T. HOWARTH: Structure Elucidation of MM 17880, a New Fused β-Lactam Antibiotic isolated from *Streptomyces olivaceus;* a Mild β-Lactam Degradation Reaction. J. C. S. Chem. Commun. **1977**, 953.
71. BUTTERWORTH, D., M. COLE, G. HANSCOMB, and G. N. ROLINSON: Olivanic Acids, A Family of β-Lactam Antibiotics with β-Lactamase Inhibitory Properties Produced by *Streptomyces* species. I. Detection, Properties and Fermentation Studies. J. Antibiotics **32**, 287 (1979).
72. HOOD, J. D., S. J. BOX, and M. S. VERRALL: ibid. II. Isolation and Characterisation of the Olivanic Acids MM 4550, MM 13902 and MM 17880 from *Streptomyces olivaceus*. J. Antibiotics **32**, 295 (1979).
73. BOX, S. J., J. D. HOOD, and S. R. SPEAR: Four Further Antibiotics Related to Olivanic Acid Produced by *Streptomyces olivaceus*: Fermentation, Isolation, Characterisation and Biosynthetic Studies. J. Antibiotics **32**, 1239 (1979).
74. BROWN, A. G., D. F. CORBETT, A. J. EGLINGTON, and T. T. HOWARTH: Structures of Olivanic Acid Derivatives MM 22380, MM 22381, MM 22382 and MM 22383; Four New Antibiotics Isolated from *Streptomyces olivaceus*. J. Antibiotics **32**, 961 (1979).
75. — — — — Some Aspects of the Chemistry of the Olivanic Acids. Recent Advances in the Chemistry of β-Lactam Antibiotics, No. 38 (G. I. GREGORY, Ed.), pp. 255 – 268, Special Publication. London: The Chemical Society. 1980; Tetrahedron **39**, 2551 (1983).
76. MAEDA, K., S. TAKAHASHI, M. SEZAKI, I. IINUMA, H. NAGANAWA, S. KONDO, M. OHNO, and H. UMEZAWA: Isolation and Structure of a β-Lactamase Inhibitor from Streptomyces. J. Antibiotics **30**, 770 (1977).
77. CASSIDY, P. J., G. ALBERS-SCHÖNBERG, T. T. GOEGELMAN, T. MILLER, B. H. ARISON, E. O. STAPLEY, and J. BIRNBAUM: Epithienamycins II. Isolation and Structure Assignment. J. Antibiotics **34**, 637 (1981).
78. BOX, S. J., D. F. CORBETT, K. G. ROBINS, S. R. SPEAR, and M. J. VERRALL: A New Olivanic Acid Derivative Produced by *Streptomyces olivaceus:* Isolation and Structural Studies. J. Antibiotics **35**, 1394 (1982).
79. KYOWA HAKKO KOGYO, K. K.: Antibiotic 8U-207 Production by Cultivating *Streptomyces* Organism. J5 7, 002, 693 (1980).

80. Albers-Schönberg, G., B. H. Arison, O. T. Hensens, J. Hirshfield, K. Hoogsteen, E. A. Kaczka, R. E. Rhodes, J. S. Kahan, R. W. Ratcliffe, E. Walton, L. J. Ruswinkle, R. B. Morin, and B. G. Christensen: Structure and Absolute Configuration of Thienamycin. J. Amer. Chem. Soc. **100**, 6491 (1978).

81. Kahan, J. S., F. M. Kahan, R. T. Goegelman, S. A. Currie, M. Jackson, E. O. Stapley, T. W. Miller, A. K. Miller, D. Hendlin, S. Mochales, S. Hernandez, H. B. Woodruff, and J. Birnbaum: Thienamycin, A New β-Lactam Antibiotic. I. Discovery, Taxonomy, Isolation and Physical Properties. J. Antibiotics **32**, 1 (1979).

82. Kahan, J. S., F. M. Kahan, R. T. Goegelman, E. O. Stapley, and S. Hernandez: N-Acetyl Thienamycin. U.S. Pat. 4,165.379 (1979).

83. Kahan, J. S.: Antibiotic N-Acetyl-Dehydro-Thienamycin. U.S. Pat. 4,162,323 (1979).

84. Kempf, A. J., and K. E. Wilson: Fermentation Process for 6-Hydroxymethyl-2-(2-Aminoethylthio)-1-Carbadethiapen-2-em-3-carboxylic acid. U.S. Pat. 4,247,640 (1981).

85. Okamura, K., S. Hirata, Y. Okumura, Y. Fukagawa, Y. Shimauchi, K. Kouno, T. Ishikura, and J. Lein: PS-5, A New β-Lactam Antibiotic from *Streptomyces*. J. Antibiotics **31**, 480 (1978).

86. Yamamoto, K., T. Yoshioka, Y. Kato, N. Shibamoto, K. Okamura, Y. Shimauchi, and T. Ishikura: Structure and Stereochemistry of Antibiotic PS-5. J. Antibiotics **32**, 796 (1980).

87. Shibamoto, N., A. Koki, M. Nishino, K. Nakamura, K. Kiyoshima, K. Okamura, M. Okabe, R. Okamoto, Y. Fukagawa, Y. Shimauchi, and T. Ishikura: PS-6 and PS-7, New β-Lactam Antibiotics Isolation, Physicochemical Properties and Structures. J. Antibiotics **32**, 1128 (1980).

88. Shibamoto, N., M. Nishino, K. Okamura, Y. Fukagawa, and T. Ishikura: PS-8, A Minor Carbapenem Antibiotic. J. Antibiotics **35**, 763 (1982).

89. Rosi, D., M. L. Drozd, M. F. Kuhrt, L. Terminiello, P. E. Came, and S. J. Daum: Mutants of *Streptomyces cattleya* Producing N-Acetyl and Deshydroxy Carbapenems Related to Thienamycin. J. Antibiotics **34**, 341 (1981).

90. Nakayama, M., S. Kimura, S. Tanabe, T. Mizoguchi, I. Watanabe, T. Mori, K. Miyahara, and T. Kawasaki: Structures and Absolute Configurations of Carpetimycins A and B. J. Antibiotics **34**, 818 (1981).

91. Harada, S., S. Shinagaua, Y. Nazaki, M. Asai, and T. Kishi: C-19393 S_2 and H_2, New Carbapenem Antibiotics. II-Isolation and Structures. J. Antibiotics **33**, 1425 (1980).

92. Tsuji, N., K. Nagashima, M. Kobayashi, J. Shoji, T. Kato, Y. Terui, H. Nakai, and M. Shiro: Asparenomycins A, B and C, New Carbapenem Antibiotics. III-Structures. J. Antibiotics **35**, 24 (1982).

93. Ito, T., N. Ezaki, K. Ohba, S. Amano, Y. Kondo, S. Miyadoh, T. Shomura, M. Sezaki, T. Niwa, M. Kojima, S. Inouye, Y. Yamada, and T. Niida: A Novel β-Lactamase Inhibitor, SF-2103A Produced by a *Streptomyces*. J. Antibiotics **35**, 533 (1982).

94. Tsuji, N., K. Nagashima, M. Kobayashi, Y. Terui, K. Matsumoto, and E. Kondo: The Structures of Pluracidomycins, New Carbapenem Antibiotics. J. Antibiotics **35**, 536 (1982).

95. Okabe, M. S., I. Azuma, I. Kojima, K. Kouno, R. Okamoto, Y. Fukagawa, and T. Ishikura: Studies on the OA-6129 Group of Antibiotics, New Carbapenem Compounds. I. Taxonomy, Isolation and Physical Properties. J. Antibiotics **35**, 1255 (1982).

96. Yoshioka, T., I. Kojima, K. Isshiki, A. Watanabe, Y. Shimauchi, M. Okabe, Y. Fukagawa, and T. Ishikura: Structures of OA-6129A, B_1, B_2 and C, New Carbapenem Antibiotics. Tetrahedron Letters **23**, 5177 (1982).

97. Parker, W. L., M. L. Rathnum, J. S. Wells, W. H. Trejo, P. A. Principe, and R. B. Sykes: SQ 27,860, A Simple Carbapenem Produced by Species of *Serratia* and *Erwinia*. J. Antibiotics **35**, 653 (1982).

98. WILSON, K., and J. KEMPF: US Patent 4,335,212.

99. HARADA, S., N. YUKIMASA, S. SHINAGAWA, and K. KITANO: C-19393 E₅, A New Carbapenem Antibiotic. Fermentation, Isolation and Structure. J. Antibiotics 35, 957 (1982).

100. SHIONOGI, Co.: Japanese Patent Application Publication No. J5 7,102,890.

101. Kowa Company Ltd.: Novel Antibiotics (KA-6643 series): E.P. Publication No. 0,050,961.

102. TANABE, S., M. OKUCHI, M. NAKAYAMA, S. KIMURA, A. IWASAKI, T. MIZOGUCHI, A. MURAKAMI, H. ITOH, and T. MORI: A New Carbapenem-Antibiotic, 6643-X. J. Antibiotics 35, 1237 (1982).

103. RATCLIFFE, R. W., and G. ALBERS-SCHÖNBERG: The Chemistry of Thienamycin and Other Carbapenem Antibiotics. Chemistry and Biology of β-Lactam Antibiotics, Volume 2 (R. B. MORIN and M. GORMAN, eds.), pp. 227 – 313. Academic Press, 1982.

104. JOHNSTON, D. B. R., S. M. SCHMITT, F. A. BOUFFORD, and B. G. CHRISTENSEN: Total Synthesis of (±) Thienamycin. J. Amer. Chem. Soc. 100, 313 (1978).

105. SCHMITT, S. M., D. B. R. JOHNSTON, and B. G. CHRISTENSEN: Thienamycin Total Synthesis 3. Total Synthesis of (±) Thienamycin and (±)-8-Epithienamycin. J. Org. Chem. 45, 1142 (1980).

106. CAMA, L. D., B. G. CHRISTENSEN: Total Synthesis of Thienamycin Analogues 1. Synthesis of the Thienamycin Nucleus and dl-Descysteaminylthienamycin. J. Amer. Chem. Soc. 100, 8006 (1978).

107. BAXTER, A. J. G., K. H. DICKINSON, P. M. ROBERTS, T. C. SMALE, and R. SOUTHGATE: Synthesis of 7-Oxo-1-azabicyclo[3.2.0]hept-2-ene-2-carboxylates: the Olivanic Acid Ring System. J. Chem. Soc. Chem. Commun. 1979, 236.

108. BATESON, J. H., A. J. C. BAXTER, K. H. DICKINSON, R. I. HICKLING, R. J. PONSFORD, P. M. ROBERTS, T. C. SMALE, and R. SOUTHGATE: Total Synthesis of Olivanic Acid Analogues and Related β-Lactam Antibiotics: Recent Advances in the Chemistry of β-Lactam Antibiotics (G. I. GREGORY, ed.). RSC Special Publication No. 38, 1981.

109. PFAENDLER, H. R., J. GOSTELI, R. B. WOODWARD, and G. RIHS: Structure, Reactivity, and Biological Activity of Strained Bicyclic β-Lactams. J. Amer. Chem. Soc. 103, 4526 (1981).

110. BATESON, J. H., A. J. G. BAXTER, P. M. ROBERTS, T. C. SMALE, and R. SOUTHGATE: Olivanic Acid Analogues. Part 1. Total Synthesis of the 7-Oxo-1-azabicyclo[3.2.0]hept-2-ene-2-carboxylate System and Some Related β-Lactams. J. Chem. Soc. Perkin 1 1981, 3242.

111. PONSFORD, R. J., and R. SOUTHGATE: Total Synthesis of Olivanic Acids and Related Compounds: Preparation of (±)-MM 22383 and (±) N-Acetyldehydrothienamycin. J. Chem. Soc. Chem. Commun. 1980, 1085.

112. SALZMANN, T. N., R. W. RATCLIFFE, B. G. CHRISTENSEN, and F. A. BOUFFORD: A Stereocontrolled Synthesis of (+)-Thienamycin. J. Amer. Chem. Soc. 102, 6161 (1980).

113. MELILLO, D. G., I. SHINKAI, T. LIU, K. RYAN, and M. SLETZINGER: A Practical Synthesis of (±)-Thienamycin. Tetrahedron Letters 1980, 2783.

114. CORBETT, D. F., S. COULTON, and R. SOUTHGATE: Inversion of Configuration at C-8 in the Olivanic Acids: Conversion to the Thienamycins and Other Novel Derivatives. J. Chem. Soc. Perk. Trans. 1, 1982, 3011.

115. KARADY, S., J. S. AMATO, R. A. REAMER, and L. M. WEINSTOCK: Stereospecific Conversion of Penicillin to Thienamycin. J. Amer. Chem. Soc. 103, 6765 (1981).

116. REIDER, P. J., and E. J. J. GRABOWSKI: Total Synthesis of Thienamycin: A New Approach From Aspartic Acid. Tetrahedron Letters 23, 2293 (1982).

117. SHINKAI, I., R. A. REAMER, F. W. HARTNER, T. LIU, and M. SLETZINGER: A Direct Transformation of Bicyclic Keto Esters to N-Formimidoyl Thienamycin. Tetrahedron Letters 23, 4903 (1982).

118. Tufariello, J. J., G. E. Lee, P. A. Senaratue, and M. A. Nuri: Thienamycin, A Solution of the Stereochemical Problem. Tetrahedron Letters **1979**, 4359.

119. Kametani, T., S. P. Huang, S. Yokohama, Y. Suzuki, and M. Ihara: Studies on the Syntheses of Heterocyclic Compounds. 800. A Formal Total Synthesis of (±)-Thienamycin and a (±)-Decysteaminylthienamycin Derivative. J. Amer. Chem. Soc. **102**, 2060 (1980).

120. Kametani, T., S. P. Huang, A. Nakayama, and T. Hondo: Further Studies on the Synthesis of Thienamycin: a Facile and Stereoselective Synthesis of a Bicyclic β-Keto Ester by 1,3-Dipolar Cycloaddition. J. Org. Chem. **47**, 2328 (1982).

121. Kametani, T.: Synthesis of Carbapenem Antibiotics. Heterocycles **17**, 463 (1982).

122. Shiozaki, M., and T. Hiraoka: A Stereocontrolled Formal Total Synthesis of (±)-Thienamycin. Tetrahedron **38**, 3457 (1982).

123. Miyashita, M., N. Chida, and A. Yoshikoshi: Synthesis of the Precursor of (+)-Thienamycin utilising D-Glucosamine. J. Chem. Soc. Chem. Commun. **1982**, 1354.

124. Shibasaki, M., A. Nishida, and S. Ikegami: A Simple Preparation of (+)-4-Phenylthioazetidin-2-one and an Asymmetric Synthesis of (+)-Thienamycin. J. Chem. Soc. Chem. Commun. **1982**, 1324.

125. Ikota, N., O. Yoshino, and K. Koga: Synthetic Studies on Optically Active β-Lactams. Stereocontrolled Synthesis of Chiral Thienamycin Intermediates from D-Glucose. Chem. Pharm. Bull. **30**, 1929 (1982).

126. Hanessian, S., D. Desilets, G. Rancourt, and R. Fortin: The Total, Stereocontrolled Synthesis of a Chemical Precursor to (+)-Thienamycin. A Formal Synthesis of the Antibiotic. Can. J. Chem. **60**, 2292 (1982).

127. Shibasaki, M., A. Nishida, and S. Ikegami: A Mild Method for The Conversion Of Propiolic Esters to β-Keto Esters. Application to the Formal Total Synthesis of (+)-Thienamycin. Tetrahedron Letters **23**, 2875 (1982).

128. Shiozaki, M., N. Ishida, T. Hiraoka, and H. Yanagisawa: Stereocontrolled Synthesis of Chiral Intermediates of Thienamycin from Threonines. Tetrahedron Letters **22**, 5205 (1981).

129. Kametani, T., S. P. Huang, T. Nagahara, S. Yokohama, and M. Ihara: Studies on the Synthesis of Heterocyclic Compounds. Part 877. An Alternative Synthesis of Protected (±)-Thienamycin and a Related Compound. J. Chem. Soc. Perkin 1 **1981**, 964.

130. Bateson, J. H., R. I. Hickling, P. M. Roberts, T. C. Smale, and R. Southgate: Olivanic Acids and Related Compounds: Total Synthesis of (±)-PS-5 and (±)-6-Epi-PS-5. J. Chem. Soc. Chem. Commun. **1980**, 1084.

131. Kametani, T., T. Honda, A. Nakayama, Y. Sasakai, T. Mochizuki, and K. Fukumoto: A Short and Stereoselective Synthesis of the Carbapenem Antibiotic PS-5. J. Chem. Soc. Perkin Trans. 1 **1981**, 2228.

132. Favara, D., A. Omodei-Salè, P. Consonni, and A. Depaoli: A Facile Synthesis of Trans-(+)-4-Carboxymethyl-3-Ethylazetidin-2-one and Its Conversion into Natural PS-5. Tetrahedron Letters **23**, 3105 (1982).

133. Corbett, D. F., and A. J. Eglington: Conversion of the Olivanic Acids into Antibiotics of the PS-5 Type: Use of a New Carboxy Protecting Group. J. Chem. Soc. Chem. Commun. **1980**, 1083.

134. Natsugari, H., Y. Matsushita, N. Tamura, K. Yoshioka, and M. Ochiai: Synthesis of 5,6-*cis*-Carbapenem Related to C-19393H$_2$. J. Chem. Soc. Perkin Trans. 1 **1983**, 403.

135. Iimori, T., Y. Takahashi, T. Izawa, S. Kobayashi, and M. Ohno: Stereocontrolled Synthesis of a *cis*-Carbapenem Antibiotic (−)-Carpetimycin A. J. Amer. Chem. Soc. **105**, 1659 (1983).

136. Newton, G. G. F., and E. P. Abraham: Cephalosporin C, a New Antibiotic containing Sulphur and D-α-Aminoadipic Acid. Nature **175**, 548 (1955).

137. ABRAHAM, E. P., and G. G. F. NEWTON: The Structure of Cephalosporin C. Biochem. J. **79**, 377 (1961).
138. HODGKIN, D. C., and E. N. MASLEN: The X-ray Analysis of the Structure of Cephalosporin C. Biochem. J. **79**, 393 (1961).
139. HALE, W. C., G. G. F. NEWTON, and E. P. ABRAHAM: Derivatives of Cephalosporin C formed with certain Heterocyclic Tertiary Bases. Biochem. J. **79**, 403 (1961).
140. JEFFERY, D. J., E. P. ABRAHAM, and G. G. F. NEWTON: Deacetylcephalosporin C. Biochem. J. **81**, 591 (1961).
141. HUBER, F. M., R. H. BALTZ, and P. G. CALTRIDER: Formation of Desacetylcephalosporin C in Cephalosporin C Fermentation. Applied Microbiology **16**, 1011 (1968).
142. FUJISAWA, Y., H. SHIRAFUJI, M. KIDA, K. NARA, M. YONEDA, and T. KANZAKI: New Findings on Cephalosporin C Biosynthesis. Nature New Biology **246**, 154 (1973).
143. HIGGINS, C. E., R. L. HAMILL, T. H. SANDS, M. M. HOEHN, N. E. DAVIS, R. NAGARAJAN, and L. D. BOECK: The Occurrence of Deacetoxy-Cephalosporin C in Fungi and Streptomyces. J. Antibiotics **27**, 298 (1974).
144. NAGARAJAN, R., L. D. BOECK, R. L. HAMILL, C. E. HIGGENS, and K. S. YANG: Deacetoxycephalosporin C from Streptomyces and Fungi. J. C. S. Chem. Commun. **1974**, 321.
145. KANZAKI, T., T. FUKITA, H. SHIRAFUJI, and Y. FUJISANA: Occurrence of a 3-Methylthiomethylcephem Derivative in a Culture Broth of Cephalosporium Mutant. J. Antibiotics **27**, 361 (1974).
146. KANZAKI, T., T. FUKITA, K. KITANO, K. KATAMOTO, K. NARA, and Y. NAKAO: Occurrence of a Novel Cephalosporin Compound in the Culture Broth of a *Cephalosporium acremonium* Mutant. J. Ferment. Technol. **54**, 720 (1976).
147. KITANO, K., Y. FUJISAWA, K. KATAMOTO, K. NARA, and Y. NAKAO: Occurrence of 7β-(4-Carboxybutanamido)-cephalosporin Compounds in the Culture Broth of Some Strains of the Genus *Cephalosporium*. J. Ferment. Technol. **54**, 712 (1976).
148. TRAXLER, P., H. J. TREICHLER, and J. NÜESCH: Synthesis of *N*-Acetyldeacetoxy Cephalosporin C by a Mutant of *Cephalosporium acremonium*. J. Antibiotics **38**, 605 (1975).
149. NAGARAJAN, R., L. D. BOECK, M. GORMAN, R. L. HAMILL, C. E. HIGGENS, M. M. HOEHN, W. M. STARK, and J. G. WHITNEY: β-Lactam Antibiotics from *Streptomyces*. J. Amer. Chem. Soc. **93**, 2308 (1971).
150. SHOJI, J., R. SAKAZAKI, K. MATSUMOTO, T. TANIMOTO, Y. TERUI, S. KOZUKI, and E. KONDO: Isolation of 7β-(5-Hydroxy-5-carboxyvaleramido)-3-hydroxymethyl-3-cephem-4-carboxylic acid from *Streptomyces* sp. J. Antibiotics **36**, 167 (1983).
151. LODER, B., G. G. F. NEWTON, and E. P. ABRAHAM: The Cephalosporin C Nucleus (7-Aminocephalosporanic Acid) and some of its Derivatives. Biochem. J. **79**, 408 (1961).
152. MORIN, R. B., B. G. JACKSON, E. H. FLYNN, R. W. ROESKE, and S. L. ANDREWS: Chemistry of Cephalosporin Antibiotics XIV. The Reaction of Cephalosporin C with Nitrosyl Chloride. J. Amer. Chem. Soc. **91**, 1396 (1969).
153. FECHTIG, B., H. PETER, H. BICKEL, and E. VISCHER: Über die Darstellung von 7-Aminocephalosporansäure. Helv. Chim. Acta **51**, 1108 (1968).
154. HATFIELD, L. D., W. H. W. LUNN, B. G. JACKSON, L. R. PETERS, L. C. BLASZCZAK, J. W. FISHER, J. P. GARDNER, and J. M. DUNIGAN: Application of Phosphorus-Halogen Compounds in Cleavage of the 7-Amide Group of Cephalosporins. Recent Advances in the Chemistry of β-Lactam Antibiotics (G. I. GREGORY, Ed.), Special Publication No. 38, pp. 109 – 124. London: The Chemical Society. 1980.
155. HEUSLER, K.: Total Synthesis of Penicillins and Cephalosporins; in reference 6, pp. 255 – 279.
156. HOLDEN, K. G.: Total Synthesis of Penicillins, Cephalosporins, and their Nuclear

Analogs: Chemistry and Biology of β-Lactam Antibiotics, Vol. 2 (R. B. Morin and M. Gorman, Eds.), pp. 100 – 164. New York: Academic Press. 1982.

157. Heymes, R., G. Amiard, and G. Nominé: Accès par synthèse totale aux analogues de la céphalosporine C. II. Lactone de la désacétylcéphalothine. Bull. Soc. Chim. Fr. 1974, 563.

158. Dolfini, J. E., J. Schwartz, and F. Weisenborn: Synthesis of Dihydrothiazines Related to Deacetylcephalosporin Lactones. An Alternative Total Synthesis of Deacetylcephalosporin Lactone. J. Org. Chem. 34, 1582 (1969).

159. Neidleman, S. L., S. C. Pan, L. A. Last, and J. E. Dolfini: Chemical Conversion of Desacetylcephalothin Lactone into Desacetylcephalothin. The Final Link in a Total Synthesis of Cephalosporanic Acid Derivatives. J. Med. Chem. 13, 386 (1970).

160. Edwards, J. A., A. Guzman, R. Johnson, P. J. Beeby, and J. H. Fried: A New Total Synthesis of (±)-Desacetylcephalothin Lactone. A Synthesis of Novel Furo-[3,4-C]-cephams. Tetrahedron Letters 1974, 2031.

161. Morin, R. B., B. G. Jackson, R. A. Mueller, E. R. Lavagnino, W. B. Scanlon, and S. L. Andrews: Chemistry of Cephalosporin Antibiotics. XV. Transformations of Penicillin Sulphoxides. A Synthesis of Cephalosporin Compounds. J. Amer. Chem. Soc. 91, 1401 (1969).

162. Cooper, R. D. G., and D. O. Spry: Rearrangements of Cephalosporins and Penicillins, in reference 6, pp. 184 – 254.

163. Cooper, R. D. G., and G. A. Koppel: The Chemistry of Penicillin Sulphoxide: Chemistry and Biology of β-Lactam Antibiotics, Vol. 1 (R. B. Morin and M. Gorman, Eds.), pp. 1 – 92. New York: Academic Press. 1982.

164. Woodward, R. B.: Recent Advances in the Chemistry of Natural Products. Science 153, 487 (1966).

165. Woodward, R. B., K. Heusler, J. Gosteli, P. Naegeli, W. Oppolzer, R. Ramage, S. Ranganathan, and H. Vorbrüggen: The Total Synthesis of Cephalosporin C. J. Amer. Chem. Soc. 88, 852 (1966).

166. Stapley, E. O., M. Jackson, S. Hernandez, S. B. Zimmerman, S. A. Currie, S. Mochales, J. M. Mata, H. B. Woodruff, and D. Hendlin: Cephamycins, a New Family of β-Lactam Antibiotics. I. Production by Actinomycetes. Antimicrob. Ag. Chemother. 2, 122 (1972).

167. Miller, T. W., R. T. Goegelman, R. G. Weston, I. Putter, and F. J. Wolf: Cephamycins, a New Family of β-Lactam Antibiotics. II. Isolation and Chemical Characterization. Antimicrob. Ag. Chemother. 2, 132 (1972).

168. Albers-Schönberg, G., B. H. Arison, and J. L. Smith: New β-Lactam Antibiotics: Structure Determination of Cephamycins A and B. Tetrahedron Letters 1972, 2911.

169. Fukase, H., T. Hasegawa, K. Hatano, H. Iwasaki, and M. Yoneda: C-2801X, A New Cephamycin-Type Antibiotic. II. Isolation and Characterization. J. Antibiotics 29, 113 (1976).

170. Supplement to Index of Antibiotics from Actinomycetes. J. Antibiotics 29, 43 (1976).

171. Supplement to Index of Antibiotics from Actinomycetes. J. Antibiotics 30, 88 (1977).

172. Gushima, H., S. Watanabe, T. Saito, T. Sasaki, H. Eiki, Y. Oka, and T. Osono: Oganomycin A, A New Cephamycin-Type Antibiotic Produced by Streptomyces oganensis and its Derivatives, Oganomycins B, GA and GB. J. Antibiotics 34, 1507 (1981).

173. Inouye, S., M. Kojima, T. Shomura, K. Iwamatsu, T. Niwa, Y. Kondo, T. Niida, Y. Ogawa, and K. Kusama: Discovery, Isolation and Structure of Novel Cephamycins of Streptomyces chartreusis. J. Antibiotics 36, 115 (1983).

174. Gordon, E. M., and R. B. Sykes: Cephamycin Antibiotics. Chemistry and Biology of β-Lactam Antibiotics, Vol. 1 (R. B. Morin and M. Gorman, Eds.), pp. 199 – 370. New York: Academic Press. 1982.

175. SLOCOMBE, B., M. J. BASKER, P. H. BENTLEY, J. P. CLAYTON, M. COLE, K. R. COMBER, R. A. DIXON, R. A. EDMONDSON, D. JACKSON, D. J. MERRIKIN, and R. SUTHERLAND: BRL 17421, a Novel β-Lactam Antibiotic, Highly Resistant to β-Lactamases, Giving High and Prolonged Serum Levels in Humans. Antimicrob. Ag. Chemother. **20**, 38 (1981).
176. KARADY, S., S. H. PINES, L. M. WEINSTOCK, F. E. ROBERTS, G. S. BRENNER, A. M. HOINOWSKI, T. Y. CHENG, and M. SLETZINGER: Semisynthetic Cephalosporins via a Novel Acyl Exchange Reaction. J. Amer. Chem. Soc. **94**, 1410 (1972).
177. WEINSTOCK, L. M., S. KARADY, F. E. ROBERTS, A. M. HOINOWSKI, G. S. BRENNER, T. B. K. LEE, W. C. LUMA, and M. SLETZINGER: The Chemistry of Cephamycins. IV. Acylation of Amides in the Presence of Neutral Acid Scavengers. Tetrahedron Letters **1975**, 3979.
178. CAMA, L. D., and B. G. CHRISTENSEN: Substituted Penicillins and Cephalosporins. VII. A Stereospecific Introduction of the C-6(7)-α-Methoxy Group. Tetrahedron Letters **1973**, 3505.
179. LUNN, W. H. W., R. W. BURCHFIELD, T. K. ELZEY, and E. V. MASON: Cleavage of 7-Methoxycephalosporin C Derivatives with Phosphorus Pentachloride. Tetrahedron Letters **1974**, 1307.
180. KARADY, S., J. S. AMATO, L. M. WEINSTOCK, and M. SLETZINGER: The Chemistry of Cephamycins. VI. Cleavage of the 7-Amido Group. Tetrahedron Letters **1978**, 407.
181. APPLEGATE, H. E., C. M. CIMARUSTI, and W. A. SLUSARCHYK: Deacylation of Amides: Removal of the Acyl Side-chain from Cephamycin Derivatives. J. C. S. Chem. Commun. **1980**, 293.
182. SHIOZAKI, M., N. ISHIDA, K. IINO, and T. HIRAOKA: Cleavage and Some Modifications of the 7-Amide Group of the Cephamycins. Tetrahedron **36**, 2735 (1980).
183. CAMA, L. D., W. J. LEANZA, T. R. BEATTIE, and B. G. CHRISTENSEN: Substituted Penicillin and Cephalosporin Derivatives. Stereospecific Introduction of the C-6(7)-Methoxy Group. J. Amer. Chem. Soc. **94**, 1408 (1972).
184. BALDWIN, J. E., F. J. URBAN, R. D. G. COOPER, and F. L. JOSE: Direct 6-Methoxylation of Penicillin Derivatives. A Convenient Pathway to Substituted β-Lactam Antibiotics. J. Amer. Chem. Soc. **95**, 2401, 1973.
185. KOPPEL, G. A., and R. E. KOEHLER: Functionalization of C-6(7) of Penicillins and Cephalosporins. A One-Step Stereoselective Synthesis of 7-α-Methoxycephalosporin C. J. Amer. Chem. Soc. **95**, 2403 (1973).
186. GORDON, E. M., H. W. CHANG, and C. M. CIMARUSTI: Sulfenyl Transfer Rearrangement of Thiooximes. A Novel Conversion of Cephalosporins to 7α-Methoxycephalosporins. J. Amer. Chem. Soc. **99**, 5504 (1977).
187. KOBAYASHI, T., K. IINO, and T. HIRASKA: A Novel Route to 7α-Methoxycephalosporins. J. Amer. Chem. Soc. **99**, 5505 (1977).
188. GORDON, E. W., H. W. CHANG, C. M. CIMARUSTI, B. TOEPLITZ, and J. Z. GOUGOUTAS: Sulfenyl Transfer Rearrangements of Sulfenimines (Thiooximes). A Novel Synthesis of 7α-Methoxycephalosporins and 6α-Methoxypenicillins. J. Amer. Chem. Soc. **102**, 1690 (1980).
189. SUGIMURA, Y., K. IINO, Y. IWANO, T. SAITO, and T. HIRAOKA: A Novel Synthesis of 7-Methoxycephalosporins and 6-Methoxypenicillins. Tetrahedron Letters **1976**, 1310.
190. SAITO, T., Y. SUGIMURA, Y. IWANO, K. IINO, and T. HIRAOKA: A New Synthetic Route to 7α-Methoxycephalosporins. J. C. S. Chem. Commun. **1976**, 516.
191. TAYLOR, A. W., and G. BURTON: Formation and 6α-Substitution of 6β-(2-Carboxy)-Ketenimino Penicillins. Tetrahedron Letters **1977**, 3831.
192. YANAGISAWA, H., M. FUKUSHIMA, A. ANDO, and H. NAKAO: A Novel General Method for Synthesising 7α-Methoxycephalosporins. Tetrahedron Letters **1975**, 2705.
193. SLUSARCHYK, W. A., H. E. APPLEGATE, P. FUNKE, W. H. KOSTER, M. S. PUAR, M. YOUNG, and J. E. DOLFINI: Synthesis of 6-Methylthiopenicillins and 7-Heteroatom-Substituted Cephalosporins. J. Org. Chem. **38**, 943 (1973).

194. Applegate, H. E., J. E. Dolfini, M. S. Puar, W. A. Slusarchyk, and B. Toeplitz: Synthesis of 7α-Methoxycephalosporins. J. Org. Chem. **39**, 2794 (1974).

195. Applegate, H. E., C. M. Cimarusti, J. E. Dolfini, P. T. Funke, W. H. Koster, M. S. Puar, W. A. Slusarchyk, and M. G. Young: Synthesis of 2-, 4-, and 7-Methylthio-Substituted Cephalosporins. J. Org. Chem. **44**, 811 (1979).

196. Jen, T., T. Frazee, and J. R. E. Hoover: A Stereospecific Synthesis of C-6(7)-Methoxypenicillin and -cephalosporin Derivatives. J. Org. Chem. **38**, 2857 (1973).

197. Spitzer, W. A., and T. Goodson: The Synthesis of S-Methyl and O-Methyl β-Lactam Antibiotics. Tetrahedron Letters **1973**, 273.

198. Cama, L. D., and B. G. Christensen: Substituted Penicillins and Cephalosporins. VII. A Stereospecific Introduction of the C-6(7)-α-Methoxy Group. Tetrahedron Letters **1973**, 3505.

199. Ratcliffe, R. W., and B. G. Christensen: Total Synthesis of β-Lactam Antibiotics II. (±)-Cephalothin. Tetrahedron Letters **1973**, 4649.

200. — — Total Synthesis of β-Lactam Antibiotics III. (±)-Cefoxitin. Tetrahedron Letters **1973**, 4653.

201. Nakatsuka, S., H. Tanino, and Y. Kishi: Biogenetic-Type Synthesis of Penicillin-Cephalosporin Antibiotics. I. A Stereocontrolled Synthesis of the Penam- and Cephem-Ring Systems from an Acyclic Tripeptide Equivalent. J. Amer. Chem. Soc. **97**, 5008 (1975).

202. Kishi, Y.: Synthetic Studies in the Field of Natural Products. Pure and Appl. Chem. **43**, 423 (1975).

203. Cooper, R. D. G.: Structural Studies on Penicillin Derivatives. VIII. A Possible Model Biosynthetic Route to Penams and Cephems. J. Amer. Chem. Soc. **94**, 1018 (1972).

204. Otsuka, H., W. Nagata, M. Toshioka, M. Narisada, T. Yoshida, Y. Harada, and H. Yamada: Discovery and Development of Moxalactam (6059-5): The Chemistry and Biology of 1-Oxacephems. Medicinal Research Reviews **1**, 217–248 (1981). John Wiley & Sons Inc.

205. Nakayama, M., S. Kimura, T. Mizoguchi, S. Tanabe, A. Iwasaki, A. Murakami, M. Okuchi, H. Itoh, and T. Mori: New β-Lactam Antibiotics, Carpetimycins C and D. J. Antibiotics **36**, 943 (1983).

206. Hosoda, J., N. Tani, T. Konomi, S. Ohsawa, H. Oaki, and H. Imanaka: Incorporation of ^{14}C-Amino Acids into Nocardicin A by Growing Cells. Agric. Biol. Chem. **41**, (10), 2007–2012 (1977).

207. Townsend, C. A., and A. M. Brown: Nocardicin A: Biosynthetic Experiments with Amino Acid Precursors. J. Amer. Chem. Soc. **105**, 913–918 (1983).

208. Townsend, C. A., A. M. Brown, and L. T. Nguyen: Nocardicin A: Stereochemical and Biomimetic Studies of Monocyclic β-Lactam Formation. J. Amer. Chem. Soc. **105**, 919–927 (1983).

209. O'Sullivan, J., A. M. Gillum, C. A. Aklouis, M. L. Souser, and R. B. Sykes: Biosynthesis of Monobactam Compounds: Origin of the Carbon Atoms in the β-Lactam Ring. Antimicrob. Ag. Chemoth. **21**, 558 (1982).

210. Arnstein, H. R. V., and P. T. Grant: The Biosynthesis of Penicillin. 1. The Incorporation of Some Amino Acids into Penicillin. Biochem. J. **57**, 353 (1954).

211. — — The Biosynthesis of Penicillin. 2. The Incorporation of Cystine into Penicillin. Biochem. J. **57**, 360 (1954).

212. Stevens, C. M., P. Vohra, E. Inamine, and O. A. Roholt: Utilisation of Sulphur Compounds for the Biosynthesis of Penicillins. J. Biol. Chem. **204**, 1001 (1953).

213. Arnstein, H. R. V., and J. C. Crawhill: The Biosynthesis of Penicillin. 6. A Study of the Mechanism of the Formation of the Thiazolidine-β-Lactam Rings Using Tritium-Labelled Cystine. Biochem. J. **67**, 180 (1957).

214. Bycroft, B. W., C. M. Wels, K. Corbett, and D. A. Lowe: Incorporation of [α-^{2}H]- and

[α-³H]-L-Cystine into Penicillin G and the Location of the Label Using Isotope Exchange and ²H-Nuclear Magnetic Resonance. J. C. S. Chem. Commun. **1975**, 123.

215. MORECOMBE, D. J., and D. W. YOUNG: Synthesis of Chirally Labelled Cysteines and the Steric Origin of C(5) in Penicillin Biosynthesis. J. C. S. Chem. Commun. **1975**, 198.

216. ADRIENS, P., H. VANDERHAEGHE, B. MEESSCHAERT, and H. EYSSEN: Incorporation of Double Labelled L-Cystine and DL-Valine in Penicillin: Antimicrob. Ag. Chemoth. **8**, 15 (1975).

217. YOUNG, D. W., D. J. MORECOMBE, and P. K. SEN: The Stereochemistry of β-Lactam Formation in Penicillin Biosynthesis. Eur. J. Biochem. **75**, 133 (1977).

218. ABERHART, D. J., L. J. LIN, and J. Y.-R. CHU: Studies on the Biosynthesis of β-Lactam Antibiotics. II. Synthesis and Incorporation into Penicillin G of (2RS,2′RS,3R,3′R)-[3,3′-³H₂]-Cystine and (2RS,2′RS,3S,3′S)-[3,3′-³H₂]-Cystine. J. C. S. Perkin 1 **1975**, 2517.

219. BALDWIN, J. E., P. D. BAILEY, G. GALLACHER, K. A. SINGLETON, and (in part) P. H. WALLACE: Stereospecific Synthesis of Tabtoxin. J. Chem. Soc. Chem. Commun. **1983**, 1049.

220. STEVENS, C. M., P. VOHRA, and C. W. DE LONG: Utilisation of Valine in the Biosynthesis of Penicillins. J. Biol. Chem. **211**, 297 (1954).

221. STEVENS, C. M., E. INAMINE, and C. W. DE LONG: The Rates of Incorporation of L-Cystine and D- and L-Valine in Penicillin Biosynthesis. J. Biol. Chem. **219**, 405 (1956).

222. ARNSTEIN, H. R. V., and M. E. CLUBB: The Biosynthesis of Penicillin. 5. Comparison of Valine and Hydroxyvaline as Penicillin Precursors. Biochem. J. **65**, 618 (1957).

223. STEVENS, C. M., and C. W. DE LONG: Valine Metabolism and Penicillin Biosynthesis. J. Biol. Chem. **230**, 991 (1958).

224. ARNSTEIN, H. R. V., and H. MARGREITER: The Biosynthesis of Penicillin. 7. Further Experiments on the Utilisation of L- and D-Valine and the Effect of Cysteine and Valine Analogues on Penicillin Biosynthesis. Biochem. J. **68**, 339 (1958).

225. WARREN, S. C., G. G. F. NEWTON, and E. P. ABRAHAM: The Role of Valine in the Biosynthesis of Penicillin N and Cephalosporin C by a *Cephalosporium* sp. Biochem. J. **103**, 902 (1967).

226. BYCROFT, B. W., C. M. WELS, K. CORBETT, A. P. MALONEY, and D. A. LOWE: Biosynthesis of Penicillin G from D- and L-[¹⁴C]- and (α-³H]-Valine. J. C. S. Chem. Commun. **1975**, 923.

227. BOOTH, H., B. W. BYCROFT, C. M. WELS, K. CORBETT, and A. P. MALONEY: Application of ¹⁵N Pulsed Fourier Transform Nuclear Magnetic Resonance Spectroscopy to Biosynthesis Studies; incorporation of L-[¹⁵N]-Valine in Penicillin G. J. C. S. Chem. Commun. **1976**, 110.

228. ABERHART, D. J., J. Y.-R. CHU, N. NEUSS, C. H. NASH, J. L. OCCOLOWITZ, L. L. HUCKSTEP, and N. DE LA HIGUERA: Retention of Valine Methyl Hydrogens in Penicillin Biosynthesis. J. C. S. Chem. Commun. **1974**, 564.

229. KLUENDER, H., F.-C. HUANG, A. FRITZBERG, H. K. SCHNOES, C. J. SIH, P. A. FAWCETT, and E. P. ABRAHAM: Studies on the Incorporation of (2S,3R)-[4,4,4-²H₃]-Valine and (2S,3S)-[4,4,4-²H₃]-Valine into β-Lactam Antibiotics. J. Am. Chem. Soc. **96**, 4054 (1974).

230. BALDWIN, J. E., J. LÖLIGER, W. RASTETTER, N. NEUSS, L. L. HUCKSTEP, and N. DE LA HIGUERA: Use of Chiral Isopropyl Groups in Biosynthesis. Synthesis of (2RS,3R)-[4-¹³C]-Valine. J. Am. Chem. Soc. **95**, 3796 (1973), see also p. 6511 (correction).

231. NEUSS, N., C. H. NASH, J. E. BALDWIN, P. A. LEMKE, and J. B. GRUTZNER: Incorporation of (2RS,3R)-[4-¹³C]-Valine into Cephalosporin C. J. Am. Chem. Soc. **95**, 3797 (1973); see also p. 6511 (correction).

232. KLUENDER, H., C. H. BRADLEY, C. J. SIH, P. A. FAWCETT, and E. P. ABRAHAM: Synthesis and Incorporation of (2S,3S)-[4-¹³C]-Valine into β-Lactam Antibiotics. J. Am. Chem. Soc. **95**, 6149 (1973).

233. ABERHART, D. J., and L. J. LIN: Studies on the Biosynthesis of β-Lactam Antibiotics.

Part 1. Stereospecific synthesis of (2RS,3S)-[4,4,4-^{2}H$_3$]-, (2RS,3S)-[4-^{3}H]-, (2RS,3R)-[4-^{3}H]- and (2RS,3S)-[4-^{13}C]-Valine. Incorporation of (2RS,3S)-[4-^{13}C]-Valine into Penicillin V. J. C. S. Perkin I **1974**, 2320.

234. Arnstein, H. R. V., and D. Morris: The Utilisation of L-Cystinyl-L-Valine for Penicillin Biosynthesis. Biochem. J. **76**, 323 (1960).

235. Arnstein, H. R. V., M. Artman, D. Morris, and E. J. Toms: Sulphur Containing Amino Acids and Peptides in the Mycelium of *Penicillium chrysogenum*. Biochem. J. **76**, 353 (1960).

236. Arnstein, H. R. V., and D. Morris: The Structure of a Peptide Containing α-Aminoadipic Acid, Cystine and Valine, Present in the Mycelium of *Penicillium chrysogenum*. Biochem. J. **76**, 357 (1960).

237. Abraham, E. P., G. G. F. Newton, and C. W. Hale: Purification and Some Properties of Cephalosporin N, a New Penicillin. Biochem. J. **58**, 94 (1954).

238. Flynn, E. H., M. H. McCormick, M. C. Stamper, H. de Valera, and C. W. Godzeski: A New Natural Penicillin from *Penicillium chrysogenum*. Nature **84**, 4594 (1962).

239. Cole, M., and F. R. Batchelor: Aminoadipylpenicillin in Penicillin Fermentations. Nature **198**, 383 (1963).

240. Warren, S. C., G. G. F. Newton, and E. P. Abraham: Use of α-Aminoadipic Acid for the Biosynthesis of Penicillin N and Cephalosporin C by a *Cephalosporium* sp. Biochem. J. **103**, 891 (1967).

241. Loder, P. B., and E. P. Abraham: Isolation and Nature of Intracellular Peptide from a Cephalosporin C-Producing *Cephalosporium* sp. Biochem. J. **123**, 471 (1971).

242. Chan, J. A., F.-C. Huang, and C. J. Sih: The Absolute Configuration of the Amino Acids in δ-(α-Aminoadipyl)cysteinylvaline from *Penicillium chrysogenum*. Biochemistry **15**, 177 (1976).

243. Adriens, P., B. Meesschaert, W. Wuyts, H. Vanderhaeghe, and H. Eyssen: Presence of δ-(L-α-Aminoadipyl)-L-cysteinyl-D-valine in Fermentations of *Penicillium chrysogenum*. Antimicrob. Ag. Chemoth. **8**, 638 (1975).

244. Fawcett, P. A., J. J. Usher, J. A. Huddleston, R. C. Bleaney, J. J. Nisbet, and E. P. Abraham: Synthesis of δ-(α-Aminoadipyl)cysteinylvaline and its Role in Penicillin Biosynthesis. Biochem. J. **157**, 651 (1976).

245. Bauer, K.: Zur Biosynthese der Penicilline: Bildung von δ-(α-Aminoadipyl)cysteinylvalin in Extracten von *Penicillium chrysogenum*. Z. Naturforsch. B, **25**, 1125 (1970).

246. Loder, P. B., and E. P. Abraham: Biosynthesis of Peptides Containing α-Aminoadipic Acid and Cysteine in Extracts of a *Cephalosporium* sp. Biochem. J. **123**, 477 (1971).

247. Fawcett, P. A., and E. P. Abraham: δ-(α-Aminoadipyl)cysteinylvaline Synthetase. In: Methods in Enzymology, Vol. 43 (J. H. Hash, ed.), p. 471. New York: Academic Press. 1971.

248. Huang, F.-C., J. A. Chan, C. J. Sih, P. A. Fawcett, and E. P. Abraham: The Nonparticipation of α,β-Dehydrovalinyl Intermediates in the Formation of δ-(L-α-Aminoadipyl)-L-cysteinyl-D-valine. J. Am. Chem. Soc. **97**, 3858 (1975).

249. Adriens, P., B. Meesschaert, H. Vanderhaeghe, and H. Eyssen: Incorporation of Double-Labelled Valine into δ-(L-α-Aminoadipyl)-L-cysteinyl-D-valine by *P. chrysogenum*. Arch. Int. Physiol. Biochim. **84**, 767 (1976).

250. Fawcett, P. A., P. B. Loder, M. J. Duncan, T. J. Beesley, and E. P. Abraham: Formation and Properties of Protoplasts from Antibiotic-Producing Strains of *Penicillium chrysogenum* and *Cephalosporium acremonium*. J. Gen. Microbiol. **79**, 293 (1973).

251. O'Sullivan, J., R. C. Bleaney, J. A. Huddleston, and E. P. Abraham: Incorporation of ^{3}H from δ-(L-α-Amino[4,5-^{3}H]adipyl)-L-cysteinyl-D-[4,4-^{3}H]-valine into Isopenicillin N. Biochem. J. **184**, 421 (1979).

252. Konomi, T., S. Herchen, J. E. Baldwin, M. Yoshida, N. A. Hunt, and A. L. Demain:

Cell-Free Conversion of δ-(L-α-Aminoadipyl)-L-cysteinyl-D-valine into an Antibiotic with the Properties of Isopenicillin N in *Cephalosporium acremonium*. Biochem. J. **184**, 427 (1979).

253. BALDWIN, J. E., B. L. JOHNSON, J. J. USHER, E. P. ABRAHAM, J. A. HUDDLESTON, and R. L. WHITE: Direct N.M.R. Observation of Cell-Free Conversion of (L-α-Amino-δ-adipyl)-L-cysteinyl-D-valine into Isopenicillin N. J. C. S. Chem. Commun. **1980**, 1271.

254. NEUSS, N., D. M. BERRY, J. KUPKA, A. L. DEMAIN, S. W. QUEENER, D. C. DUCKWORTH, and L. L. HUCKSTEP: High Performance Liquid Chromatography (HPLC) of Natural Products V: The Use of HPLC in the Cell-Free Biosynthetic Conversion of α-Aminoadipyl-cysteinyl-valine (LLD) into Isopenicillin N. J. Antibiotics **35**, 580 (1982).

255. SAWADA, Y., J. E. BALDWIN, P. D. SINGH, N. A. SOLOMON, and A. L. DEMAIN: Cell-Free Cyclisation of δ-(L-α-Aminoadipyl)-L-cysteinyl-D-valine to Isopenicillin N. Antimicrob. Ag. Chemoth. **18**, 465 (1980).

256. WHITE, R. L., E.-M. M. JOHN, J. E. BALDWIN, and E. P. ABRAHAM: Stoichiometry of Oxygen Consumption in the Biosynthesis of Isopenicillin N from a Tripeptide. Biochem. J. **203**, 791 (1982).

257. KUPKA, J., Y.-Q. SHEN, S. WOLFE, and A. L. DEMAIN: Studies on the Ring-Cyclisation and Ring Expanding Enzymes of β-Lactam Biosynthesis in *C. acremonium*. Can. J. Microbiol. **29**, 488 (1983).

258. BAHADUR, G. A., J. E. BALDWIN, J. J. USHER, E. P. ABRAHAM, G. S. JAYATILAKE, and R. L. WHITE: Cell-Free Biosynthesis of Penicillins. Conversion of Peptides into New β-Lactam Antibiotics. J. Am. Chem. Soc. **103**, 7650 (1981).

259. BAHADUR, G., J. E. BALDWIN, L. D. FIELD, E.-M. M. LEHTONAN, J. J. USHER, C. A. VALLEJO, E. P. ABRAHAM, and R. L. WHITE: Direct [1]H-N.M.R. Observation of the Cell-Free Conversion of δ-(α-Aminoadipyl)-L-cysteinyl-D-valine and δ-(L-α-Aminoadipyl)-L-cysteinyl-D-(−)-isoleucine into Penicillins. J. C. S. Chem. Commun. **1981**, 917.

260. BALDWIN, J. E., B. CHAKRAVARTI, L. D. FIELD, J. A. MURPHY, K. R. WHITTEN, E. P. ABRAHAM, and G. JAYATILAKE: The Synthesis of L-α-Aminoadipyl-L-cysteinyl-D-3,4-didehydrovaline, a Potent Inhibitor of Isopenicillin N Synthetase. Tetrahedron **38**, 2773 (1982).

261. NEUSS, N., R. D. MILLER, C. A. AFFOLDER, W. NAKATSUKASA, J. A. MABE, L. L. HUCKSTEP, N. DE LA HIGUERA, A. H. HUNT, J. L. OCCOLOWITZ, and J. H. GILLIAM: High Performance Liquid Chromatography (HPLC) of Natural Products. III. Isolation of New Tripeptides from the Fermentation Broth of *P. chrysogenum*. Helv. Chim. Acta **63**, 1119 (1980).

262. ABRAHAM, E. P., J. A. HUDDLESTON, G. S. JAYATILAKE, J. O'SULLIVAN, and R. L. WHITE: Conversion of δ-(L-α-Aminoadipyl)-L-cysteinyl-D-valine to Isopenicillin N in Cell-Free Extracts of *Cephalosporium acremonium*. In: Recent Advances in the Chemistry of β-Lactam Antibiotics (G. I. GREGORY, ed.), p. 125. London: Royal Society of Chemistry. 1980.

263. MEESSCHAERT, B., P. ADRIENS, and H. EYSSEN: Studies on the Biosynthesis of Isopenicillin N with a Cell-Free Preparation of *Penicillium chrysogenum*. J. Antibiotics **33**, 722 (1980).

264. ABRAHAM, E. P., R. M. ADLINGTON, J. E. BALDWIN, M. J. CRIMMIN, L. D. FIELD, G. S. JAYATILAKE, and R. L. WHITE: Monocyclic β-Lactam Tripeptide, 1-(D-Carboxy-2-methylpropyl)-3-L-(δ-L-2-aminoadipamido)-4-L-mercaptoazetidin-2-one, a Putative Intermediate in Penicillin Biosynthesis. J. C. S. Chem. Commun. **1982**, 1130.

265. JENSEN, S. E., D. W. S. WESTLAKE, and S. WOLFE: Cyclisation of δ-(L-α-Aminoadipyl)-L-cysteinyl-D-valine to Penicillins by Cell-Free Extracts of *Streptomyces clavuligerus*. J. Antibiotics **35**, 483 (1982).

266. — — — High Performance Liquid Chromatographic Assay of Cyclisation Activity in Cell-Free Systems from *Streptomyces clavuligerus*. J. Antibiotics **35**, 1026 (1982).

267. BAHADUR, G. A., J. E. BALDWIN, T. WAN, M. JUNG, E. P. ABRAHAM, J. A. HUDDLESTON, and R. L. WHITE: On the Proposed Intermediacy of β-Hydroxyvaline- and Thiazepinone-Containing Peptides in Penicillin Biosynthesis. J. C. S. Chem. Commun. **1981**, 1146.

268. ADLINGTON, R. M., R. T. APLIN, J. E. BALDWIN, L. D. FIELD, E.-M. M. JOHN, E. P. ABRAHAM, and R. L. WHITE: Conversion of $^{17}O/^{18}O$-Labelled δ-(α-Aminoadipyl)-L-cysteinyl-D-valine into $^{17}O/^{18}O$-Labelled Isopenicillin N in a Cell-Free Extract of *C. acremonium*. J. C. S. Chem. Commun. **1982**, 137.

269. JAYATILAKE, G. S., J. A. HUDDLESTON, and E. P. ABRAHAM: Conversion of Isopenicillin N into Penicillin N in Cell-Free Extracts of *Cephalosporium acremonium*. Biochem. J. **194**, 645 (1981).

270. BEHRENS, O. K., J. CORSE, R. G. JONES, E. C. KLEIDERER, Q. F. SOPER, F. R. VAN ABEELE, L. M. LARSON, J. C. SYLVESTER, W. J. HAINES, and H. E. CARTER: Biosynthesis of Penicillins. II. Utilisation of Deuterophenylacetyl-^{15}N-DL-valine in Penicillin Biosynthesis. J. Biol. Chem. **175**, 765 (1948).

271. BEHRENS, O. K., J. CORSE, D. E. HUFF, R. G. JONES, Q. F. SOPER, and C. W. WHITEHEAD: Biosynthesis of Penicillins. III. Preparation and Evaluation of Precursors for New Penicillins. J. Biol. Chem. **175**, 771 (1948).

272. FAWCETT, P. A., J. J. USHER, and E. P. ABRAHAM: Behaviour of Tritium Labelled Isopenicillin N and 6-Aminopenicillanic Acid as Potential Penicillin Precursors in an Extract of *Penicillium chrysogenum*. Biochem. J. **151**, 741 (1975).

273. BATCHELOR, F. R., E. B. CHAIN, and G. N. ROLINSON: 6-Aminopenicillanic Acid. I. 6-Aminopenicillanic Acid in Penicillin Fermentations. Proc. Roy. Soc. B, **154**, 478 (1961).

274. PRUESS, D. L., and M. J. JOHNSON: Penicillin Acyltransferase in *Penicillium chrysogenum*. J. Bact. **94**, 1502 (1967).

275. COLE, M.: Formation of 6-Aminopenicillanic Acid, Penicillins and Penicillin Acylase by Various Fungi. Applied Microbiol. **14**, 98 (1966).

276. VANDERHAEGHE, H., M. CLAESEN, A. VLIETUICK, and G. PARMENTIER: Specificity of Penicillin Acylase of *Fusarium* and of *Penicillium chrysogenum*. Applied Microbiol. **16**, 1557 (1968).

277. SPENCER, B., and C. MAUNG: Multiple Activities of Penicillin Acyltransferase of *Penicillium chrysogenum*. Biochem. J. **118**, 29 P (1970).

278. NEUSS, N., and S. W. QUEENER: β-Lactam Antibiotics: Chemistry and Biology, Vol. II (R. B. MORIN and M. GORMAN, eds.). New York: Academic Press. 1982.

279. ELSON, S. W., and R. S. OLIVER: Studies on the Biosynthesis of Clavulanic Acid. I. – Incorporation of ^{13}C-Labelled Precursors. J. Antibiotics **31**, 586 (1978).

280. ELSON, S. W., R. S. OLIVER, B. W. BYCROFT, and E. A. FARUK: Studies on the Biosynthesis of Clavulanic Acid. III. Incorporation of DL-[3,4-^{13}C$_2$] Glutamic Acid. J. Antibiotics **35**, 81 (1982).

281. ALBERS-SCHÖNBERG, G., B. H. ARISON, E. KACZKA, F. M. KAHAN, J. S. KAHAN, B. LAGO, W. M. MAIESE, R. E. RHODES, and J. L. SMITH: Abstracts of the Sixteenth Interscience Conference on Antimicrobial Agents and Chemotherapy (1976).

282. FUKAGAWA, Y., K. KUBO, K. OKAMURA, and T. ISHIKAWA: Biosynthesis of Carbapenem Antibiotics. In: Trends in Antibiotic Research (H. UMEZAWA, A. DEMAIN, T. HATA, and C. HUTCHINSON, eds.). Japanese Antibiotics Research Association, Tokyo. 1982.

283. SINGH, P. D., J. H. JOHNSON, P. C. WARD, J. SCOTT WELLS, W. H. TREJO, and R. B. SYKES: A New Monobactam Produced by a *Flexibacter* sp. J. Antibiotics **36**, 1245 (1983).

284. COOPER, R., K. BUSH, P. A. PRINCIPE, W. H. TREJO, J. SCOTT WELLS, and R. B. SYKES: Two New Monobactam Antibiotics Produced by a *Flexibacter* sp. J. Antibiotics **36**, 1252 (1983).

285. TROWN, P. W., B. SMITH, and E. P. ABRAHAM: Biosynthesis of Cephalosporin C from Amino Acids. Biochem. J. **86**, 284 (1963).

286. TROWN, P. W., E. P. ABRAHAM, G. G. F. NEWTON, C. W. HALE, and G. A. MILLER: Incorporation of Acetate into Cephalosporin C. Biochem. J. **84**, 157 (1962).
287. HUDDLESTON, J. A., E. P. ABRAHAM, D. W. YOUNG, D. J. MORECOMBE, and P. K. SEN: The Stereochemistry of β-Lactam Formation in Cephalosporin Biosynthesis. Biochem. J. **169**, 705 (1978).
288. WHITNEY, J. G., D. R. BRANNON, J. A. MABE, and K. J. WICKER: Incorporation of Labelled Precursors into A 16886B, a Novel β-Lactam Antibiotic Produced by *Streptomyces clavuligerus*. Antimicrob. Ag. Chemother. **1**, 247 (1972).
289. KOHSAKA, M., and A. L. DEMAIN: Conversion of Penicillin N to Cephalosporin(s) by Cell-Free Extracts of *Cephalosporium acremonium*. Biochem. Biophys. Res. Commun. **70**, 465 (1976).
290. YOSHIDA, M., T. KONOMI, M. KOHSAKA, J. E. BALDWIN, S. HERCHEN, P. D. SINGH, N. A. HUNT, and A. L. DEMAIN: Cell-Free Ring Expansion of Penicillin N to Deacetoxy-cephalosporin C by *Cephalosporium acremonium* CW-19 and its Mutants. Proc. Natl. Acad. Sci. USA **75**, 6253 (1978).
291. BALDWIN, J. E., S. R. HERCHEN, and P. D. SINGH: Syntheses of Penicillin N, [6α-^{3}H] Penicillin N and [10-^{14}C, 6α-^{3}H] Penicillin N. Biochem. J. **186**, 881 (1980).
292. BALDWIN, J. E., P. D. SINGH, M. YOSHIDA, Y. SAWADA, and A. L. DEMAIN: Incorporation of ^{3}H and ^{14}C from [6α-^{3}H] Penicillin N and [10-^{14}C, 6α-^{3}H] Penicillin N into Deacetoxycephalosporin C. Biochem. J. **186**, 889 (1980).
293. HOOK, D. J., L. T. CHANG, R. P. ELANDER, and R. B. MORIN: Stimulation of the Conversion of Penicillin N to Cephalosporin by Ascorbic Acid, α-Ketoglutarate and Ferrous Ions in Cell-Free Extracts of Strain of *Cephalosporium acremonium*. Biochem. Biophys. Res. Commun. **87**, 258 (1979).
294. SAWADA, Y., N. A. HUNT, and A. L. DEMAIN: Further Studies on Microbiological Ring-Expansion of Penicillin N. J. Antibiotics **32**, 1303 (1979).
295. SAWADA, Y., N. A. SOLOMON, and A. L. DEMAIN: Stimulation of Cell-Free Ring Expansion of Penicillin N by Sonication and Triton X-100. Biotech. Lett. **2**, 43 (1980).
296. FELIX, H. R., H. H. PETER, and H. J. TREICHLER: Microbiological Ring Expansion of Penicillin N. J. Antibiotics **34**, 567 (1981).
297. KUPKA, J., Y.-Q. SHEN, S. WOLFE, and A. L. DEMAIN: Partial Purification and Properties of the α-Ketoglutarate-Linked Ring − Expansion Enzyme of β-Lactam Biosynthesis of *Cephalosporium acremonium*. FEMS Microbiol. Lett. **16**, 1 (1982).
298. MILLER, R. D., L. L. HUCKSTEP, J. P. McDERMOTT, S. W. QUEENER, S. KUKOLJA, D. O. SPRY, T. K. ELZEY, S. M. LAWRENCE, and N. NEUSS: High Performance Liquid Chromatography (HPLC) of Natural Products. IV. The Use of HPLC in Biosynthetic Studies of Cephalosporin C in the Cell-Free System. J. Antibiotics **34**, 984 (1981).
299. JENSEN, S. E., D. W. S. WESTLAKE, R. J. BOWERS, and S. WOLFE: Cephalosporin Formation by Cell-Free Extracts from *Streptomyces clavuligerus*. J. Antibiotics **35**, 1351 (1982).
300. JENSEN, S. E., D. W. S. WESTLAKE, and S. WOLFE: Analysis of Penicillin N Ring Expansion Activity from *Streptomyces clavuligerus* by Ion-Pair High-Pressure Liquid Chromatography. Antimicrob. Ag. Chemoth. **24**, 307 (1983).
301. LIERSCH, M., J. NÜESCH, and H. J. TREICHLER: Final Steps in the Biosynthesis of Cephalosporin C. In: Second International Symposium on the Genetics of Industrial Micro-organisms (K. D. MACDONALD, ed.), p. 179. London: Academic Press. 1976.
302. STEVENS, C. M., E. P. ABRAHAM, F.-C. HUANG, and C. J. SIH: Incorporation of Molecular Oxygen at C-17 of Cephalosporin C During its Biosynthesis. Fed. Proc. **34**, 625 (1975).
303. O'SULLIVAN, J., R. T. APLIN, C. M. STEVENS, and E. P. ABRAHAM: Biosynthesis of a 7-α-Methoxycephalosporin. Incorporation of Molecular Oxygen. Biochem. J. **179**, 47 (1979).
304. FUJISAWA, Y., M. KIKUCHI, and T. KANZAKI: Deacetylcephalosporin C Synthesis by Cell-Free Extracts of *Cephalosporium acremonium*. J. Antibiotics **30**, 775 (1977).

305. Turner, M. K., J. E. Farthing, and S. J. Brewer: The Oxygenation of [3-Methyl-^{3}H] desacetoxycephalosporin C [7β-(5-D-Aminoadipamido)-3-methylceph-3-em-4-carboxylic acid] to [3-Hydroxymethyl-^{3}H] deacetylcephalosporin C by 2-Oxoglutarate-Linked Dioxygenases from *Acremonium chrysogenum* and *Streptomyces clavuligerus*. Biochem. J. **173**, 839 (1978).

306. Fujisawa, Y., and T. Kanzaki: Role of Acetyl-CoA: Deacetylcephalosporin C Acetyltransferase in Cephalosporin C Biosynthesis by *Cephalosporium acremonium*. Agr. Biol. Chem. **39**, 2043 (1975).

307. Brewer, S. J., T. T. Boyle, and M. K. Turner: The Carbamoylation of the 3-Hydroxymethyl Group of 7α-Methoxy-7β-(5-D-aminoadipamido)-3-hydroxymethyl-ceph-3-em-4-carboxylic acid (Desacetyl-7α-methoxycephalosporin C) by Homogenates of *Streptomyces clavuligerus*. Biochem. Soc. Trans. **5**, 1026 (1977).

308. Brewer, S. J., P. M. Taylor, and M. K. Turner: An Adenosine Triphosphate Dependant Carbamoyl-phosphate-3-hydroxymethylcephem O-carbamoyl-transferase from *Streptomyces clavuligerus*. Biochem. J. **185**, 555 (1980).

309. O'Sullivan, J., and E. P. Abraham: The Conversion of Cephalosporins to 7α-Methoxycephalosporins by Cell-Free Extracts of *Streptomyces clavuligerus*. Biochem. J. **186**, 613 (1980).

310. Hood, J. D., A. L. Elson, M. L. Gilpin, and A. G. Brown: Identification of 7α-Hydroxycephalosporin C as an Intermediate in the Methoxylation of Cephalosporin C by a Cell-Free Extract of *Streptomyces clavuligerus*. J. C. S. Chem. Commun. **1983**, 1187.

(Received April 6, 1984)

New Techniques for the Mass Spectrometry
of Natural Products

By I. HOWE and M. JARMAN, Drug Metabolism Team, Section of Drug Development, Cancer Research Campaign Laboratory, Institute of Cancer Research, Sutton, Surrey, U.K.

With 7 Figures

Contents

I. Introduction

The power and utility of mass spectrometry (MS) in solving problems in natural product chemistry has been demonstrated on numerous occasions over the past 25 years. Although only electron-impact (EI) ionization was available in the early days of natural-product MS, notable successes were achieved in correlating spectra with structure, particularly with natural steroids, terpenes and alkaloids.

Since these initial studies, MS has diversified to such an extent that it is employed for structural investigations on a wide range of natural product classes. Broadly, three areas of technological developments have led to these advances, namely: (i) the progress in ionization and mass separation techniques so that a wide range of compounds (in terms of volatility and mass) can be handled, (ii) the coupling of mass spectrometers to instruments (i.e. gas- and liquid-chromatographs) which effect prior physical separation of compounds, so that complex mixtures can be investigated, and (iii) the advancement in computer techniques, enabling operation of the mass spectrometer, data storage and structural identification to become much more streamlined.

The scope of this review is limited largely to the newer techniques employed to characterize natural products by mass spectrometry. However, in order to maintain a reasonable perspective on the subject the older, more established techniques are introduced in those instances where their application is preferred for a particular compound class. The definition of natural products is taken to include chemical components of, and products from, plant and animal systems.

Numerous reviews have concentrated on the mass spectra of particular compound classes and these will be referenced as they arise in the subsequent text. A comprehensive coverage of the subject is found in reference (1), while other recent volumes have provided more concise accounts (2—4). For up-to-date reports on the mass spectrometric literature, with relevant sections on natural products, two biennial reviews are recommended (5, 6).

The remainder of this review is divided into three sections. In Sections II and III, various mass spectrometric techniques are reviewed, with heavy emphasis on those which are providing new information in natural product chemistry. Section IV deals with compound classes individually and those natural products which have hitherto proved to be difficult to handle by MS, because of their involatility or thermal instability, receive special attention.

It is emphasized at the outset of this review that MS is not an indispensable technique for solving problems in natural product chemistry. It certainly provides a unique combination of selectivity and sensitivity, but it can be even more powerful when used in conjunction with complementary spectroscopic (*e.g.* NMR, UV, IR) and chromatographic (*e.g.* HPLC, GC, TLC) methods.

II. Methods of Ionization

II.1. Electron Impact

Ionization of the sample by impact with $50 - 70$ eV electrons in a low-pressure ion-chamber is still the commonest means of generation of positively-charged molecular ions. Provided that thermal degradation of the sample does not occur prior to ionization, the abundance of the molecular ion peak is usually high for compounds containing aromatic moieties (*e.g.* porphyrins, aromatic steroids, indole alkaloids), particularly those bearing electron-donating groups and/or incorporating polyaromatic functions. Fragmentation of the molecular ion occurs, usually to yield structurally significant ions, and this is of course the basis of the normal EI mass spectrum.

If the molecular sample can be induced to form a molecular anion by electron capture, then the negative ion spectrum so obtained can be recorded with high sensitivity simply by reversing various electrical potentials in the mass spectrometer. To achieve this sensitivity the sample must possess, or acquire via a suitable derivative, a positive electron affinity. Halogenated compounds, quinones and nitro compounds commonly possess positive electron affinities but many natural products require derivatization if a negative ion electron impact spectrum is to be obtained. Perfluorinated derivatives are commonly employed for OH, NH_2 and CO_2H functions [see refs. (5 – 7)] and further reference to their use will be found below for specific natural product classes. An important condition which elicits efficient production of molecular anions is the high pressure of a neutral gas in the ion source, favouring a high concentration of low energy electrons. Electron capture spectra show little fragmentation, but are useful where the generation of an abundant molecular ion is required (*e.g.* in quantitative measurements).

Since the advent of "soft-ionization" techniques, the practical usefulness of negative ion mass spectrometry, as applied to natural products, has increased. Negative ions can be produced efficiently both by chemical ionization and fast atom bombardment (see later sections of this review).

The same classes of compound that generate abundant singly-charged molecular cations on electron impact also tend to produce a significant proportion of multiply-charged ions [*e.g.* porphyrins and other polycyclic aromatics, see reference (8)]. Since these ions are found at fractions (*i.e.* 1/2, 1/3 etc.) of the m/z values for the singly-charged molecular ions, it is possible to characterize compounds of high molecular weight by their observation. It appears that a low ion source temperature favours the production of double vs. single ionization and this effect can be attributed to the increase in vibrationally-induced autoionization (favouring M^+ ion production) as the temperature is increased (8).

II.2. Chemical Ionization

A molecular sample can be ionized in the mass spectrometer via a gas phase ion-molecule reaction. This process is known as chemical ionization (CI) and evolved as an alternative to the formation of molecular ions *via* direct EI. Because the ionization process involves a chemical reaction, there are potentially many variations of the technique, employing different reagent ions adapted to the properties of the molecular sample. Consequently, a variety of CI reagent gases have been developed for use with different natural product classes (*4–6*).

Ionization of the sample molecules occurs in the ion chamber by collision with an excess of reagent ions at relatively high pressure ($\sim$ 1 torr). The reagent ions are usually, but not always, positively charged and are generated either from a pure sample gas, or mixture of gases. A list of some reagent gases and their corresponding positive reagent ions is shown in Table 1.

Table 1. *Common CI Reagent Gases and Ions*

Reagent gas(es)	H_2	CH_4	H_2O	$CHMe_3$	NH_3	N_2/H_2	CO_2/H_2	CO/H_2	N_2O	NO
Main reagent ion	H_3^+	CH_5^+	H_3O^+	CMe_3^+	NH_4^+	N_2H^+	CO_2H^+	HCO^+	NO^+	NO^+

In positive ion CI, the two commonest types of ionizing reaction are (a) proton transfer and (b) reagent ion capture

(a) $M + [B + H]^+ \rightarrow [M + H]^+ + B$

(b) $M + X^+ \quad\quad \rightarrow MX^+$

Many natural products, containing electron-donating groups, are amenable to protonation *via* reaction (a). The internal energy of the protonated molecular ion $[M + H]^+$, and hence the extent of its fragmentation, is influenced strongly by the proton affinity of the conjugate base B. For example, hydrogen CI spectra exhibit more abundant fragment ions than isobutane spectra, whereas methane spectra represent an intermediate situation. The respective proton affinities of H_2, CH_4 and $Me_2C=CH_2$ are 420, 530 and $815 \, kJ \, mol^{-1}$. However, it appears that whatever the CI reagent gas, there is a high relative abundance of the protonated molecular ion. In recent years, mixtures of reagent gases have been employed to generate reagent ions which yield the optimum degree of fragmentation of the sample molecule (*i.e.* a considerable amount of "fine tuning" is possible).

In addition to its protonating function, the NH_4^+ ion can solvate the sample molecules [reaction (b) above], producing an abundant $[M + 18]^+$ peak for some natural product classes (particularly carbohydrates and their derivatives). Ammonia is now well established in such cases as an alternative to methane and isobutane as a positive CI reagent gas (*9*). Nitrous oxide has also been used with success to generate abundant $[M + NO]^+$ ions from a variety of compounds, including natural terpenes (*10*). The fragmentations were characteristic.

Negative ion CI has found only limited applicability in natural-product chemistry, because of the requirement for molecular or fragment ions to stabilize a negative charge. More success has been found in environmental chemistry where polyhalogenated compounds (possessing high electron affinities) are commonplace (*11*). Nevertheless, negative ion CI has proved to be a sensitive and selective technique for certain natural product classes. Basic reagent ions (*e.g.* OH^- generated from a CH_4/N_2O mixture) can efficiently generate $[M - H]^-$ ions from natural products containing OH, CO_2H or amide functional groups. In certain compound classes [for example, sterols (*12*) and terpene alcohols (*13*)], the decomposition of the $[M - H]^-$ ions has proved sensitive to the stereochemistry of the parent molecule.

The main advantages of chemical ionization are summarized as follows:

(1) Many polar natural products (*e.g.* carbohydrates and their derivatives) which do not give EI mass spectra of value are amenable to CI.

(2) Ions which determine the molecular weight (*e.g.* $[M + H]^+$, $[M + NH_4]^+$, $[M - H]^-$) are usually more abundant than the molecular ions obtained in EI spectra.

(3) These abundant "pseudomolecular" ions can be collisionally activated to generate structurally informative fragment ions which are identified *via* a suitable MS/MS technique (see III.2).

(4) CI spectra are appropriate for quantitative studies since the abundances of protonated or solvated molecular ions are usually reproducible and not subject to rapid fluctuation.

(5) Since CI produces stable, low energy ions, it is sometimes capable of distinguishing between isomers (including stereoisomers) where EI mass spectrometry fails.

Despite the above positive features of CI, it remains a complementary technique to EI since the latter technique is usually easier to perform and frequently yields more detailed structural information.

II.3. Desorption Chemical Ionization

Compounds which will not vaporize without decomposition provide a challenge for mass spectrometry and many natural products fall into this category. For compounds on the verge of decomposing at their vaporization temperatures this problem can sometimes be overcome by shortening the path lengths between sample and the ion beam by the use of extended probes and the methodology has been reviewed (*14*). Used in conjunction with chemical ionization the technique has been termed "desorption chemical ionization" (DCI). However the same principle applies in the EI mode and successful examples are given later. It seems likely that the main ion-producing mechanisms are those of conventional EI and CI and that the enhanced parent ion intensities result from reduced thermal decomposition rather than from a different ionization process.

Where the sample will not vaporize, even through a short path length, without decomposing one must use one of the methods which produce ions from the condensed phase. The collective term desorption ionization techniques has been used (*15*) for these methods and biological applications up to mid-1980 are reviewed in ref. (*4*).

II.4. Field Desorption

When placed in a strong electrostatic field $(10^7 - 10^8 \text{ V cm}^{-1})$ a molecule can lose an electron and form a positive charged ion. In the earlier technique of field ionization mass spectrometry the sample was first vaporized prior to exposure to the field. Ionization occurs in the region between an emitter wire, the field anode, and a counterelectrode maintained at a potential of 10 kV relative to the emitter. In field desorption mass spectrometry (FDMS) the sample is coated on to a specially prepared emitter and is desorbed by the high field without prior vaporization. This modification of field ionization, introduced by BECKEY in 1969 (16) made possible the study of polar molecules of biological interest and in this first study showed that the extensive, sequential thermal losses of water from the $[M + H]^+$ ion of D-glucose which occurred in the conventional field ionization spectrum obtained after vaporization in the ion source were negligible when D-glucose was field desorbed from the emitter. The development of FD as a useful technique in biochemical studies has been reviewed (17).

Field desorption spectra are usually simple since the field imparts little excess internal energy to the desorbed ions. The mass spectra often contain only the molecular ion $(M^{+\cdot})$ or protonated molecular ion $([M + H]^+)$. Which of these species is formed depends on the nature of the sample. Lipophilic or aprotic molecules tend to form $M^{+\cdot}$ ions whereas $[M + H]^+$ ions form from polar molecules such as peptides and carbohydrates by intermolecular proton transfer. The intensity of $[M + H]^+$ ions can often be enhanced in FD as well as in other desorption ionization techniques by treatment with acid. Because of the ambiguity as to whether $M^{+\cdot}$ or $[M + H]^+$ is formed, precise molecular weight determination of an unknown sample by FD can be a problem. By mixing the sample with a metal salt, for example lithium chloride, $[M + cation]^+$ ions may be produced, defining the molecular weight unambiguously (18).

II.5. Secondary Ion Mass Spectrometry and Fast Atom Bombardment Mass Spectrometry

These techniques are discussed together because the principles behind them are similar. In secondary ion mass spectrometry (SIMS) (19), the ions are produced by bombarding the sample molecules with a beam of fast ions. When the ion beam is replaced by a beam of fast atoms the technique is called fast atom bombardment mass spectrometry (FABMS) (20). Introduced in its modern form by BARBER and co-workers (21), FABMS has proved a versatile method for examining natural products. The production of ions in FABMS is illustrated in Fig. 1.

In earlier applications of SIMS and FABMS the sample was deposited as a thin film on the probe resulting often in cationization by the surface metal to form $[M + cation]^+$. Latterly, solutions in a liquid matrix coated on the probe surface have been customarily used, a method which allows continuous exposure to the ion or atom beam of fresh solute molecules.

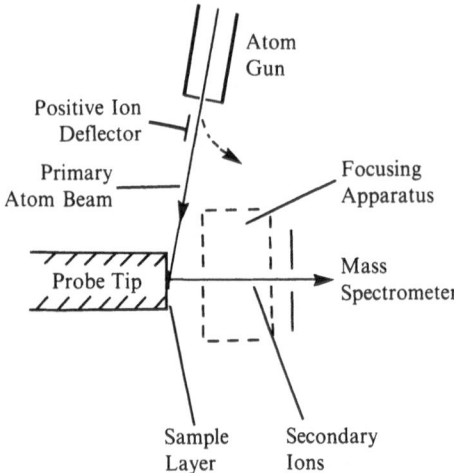

Fig. 1. Schematic representation of a fast atom bombardment source

Glycerol is most commonly used as it combines low volatility with good solvent properties for polar substances. The glycerol matrix affords low abundance ions at every value of m/z as well as useful oligomer ions in both positive and negative ion modes which assist the assignment of correct m/z values to the sample peaks. In SIMS and FABMS, an abundant molecular ion, usually a protonated species $[M + H]^+$, is often accompanied by a structure defining fragmentation, in contrast with FDMS. Consequently FABMS and FDMS have sometimes been used as complementary techniques, the latter giving unambiguous assignments of the molecular weight and the former affording other structural information (22).

II.6. Plasma Desorption Mass Spectrometry

This technique, also called fission fragment-induced desorption, was first reported by TORGERSON and co-workers in 1974 (23). Samples coated on to a thin nickel foil are aligned with a ^{252}Cf fission fragment source. Fission fragments from the californium pass through the foil generating localized temperatures of $2-3 \times 10^4 \, °C$ for ca. 10^{-11} seconds. The high temperatures vaporize ion impurities in the nickel (e.g. hydrogen and sodium ions) and these secondary ions react with heated sample molecules to produce $[M + H]^+$ or $[M + Na]^+$ ions. These are analyzed in a time-of-flight tube. The time taken by an ion to traverse this tube uniquely determines the mass-to-charge ratio and is measured by determining the interval between a "start" pulse generated by the fission fragment and a "stop" pulse produced when an ion is detected at the collector.

In the time-of-flight methodology the detection of ions of high mass is not limited as it is with magnet sector instruments by the present capabilities of high field magnets. Because of its use in conjunction with plasma desorption mass spectrometry (PDMS), this is still supreme among the desorption ionization techniques in its capacity for handling compounds of high molecular weight ($> 10^4$ daltons) (24). A drawback of the methodology is the intrinsically low resolving power, which produces, at *ca.* 10^4 daltons an uncertainty of about ± 10 a.m.u. in the mass assignment. Even so, this degree of precision is often sufficient for a decision to be made between alternative structures determined by other physical methods, and examples are given in section IV.

II.7. Laser Desorption Mass Spectrometry

The energy imparted by powerful sub-microsecond laser pulses ($10^6 - 10^8$ W cm^{-2}) is responsible for the formation of the ions in laser desorption mass spectrometry (LDMS) (25). The technique has aspects in common with each of the foregoing desorption ionization methods. Thus cationization by surface metal to produce, *e.g.,* $[M + Ag]^+$ ions and by added salts to produce, *e.g.,* $[M + Li]^+$ ion has parallels with the previously cited results obtained using SIMS and FDMS respectively. The method of producing the ions by using a pulsed energy source is reminiscent of the mechanism whereby ions are produced in PDMS and likewise time-of-flight instrumentation, and more recently Fourier transform methodology (26), have been used to advantage. Very recently LDMS, used in conjunction with thermospray sample deposition has shown promise in the analysis of biomolecules by coupled liquid chromatography-mass spectrometry (27) (see III.3).

II.8. General Comments and Future Prospects

Positive and negative ion mass spectra are equally accessible in all the desorption ionization modes with the exception of FDMS. Here the problem has been to achieve desorption of the negative ions before field induced emission of electrons occurs (28). The fact that many natural products contain ionized acidic functions undoubtedly assists the formation of negative ions from them and the negative ion spectra are frequently more useful than their positive ion counterparts.

New ionization techniques continue to be developed. One of the most promising is the thermospray method introduced by VESTAL and co-workers (29) in which the effluent from a liquid chromatograph is partially flash vaporized and the resulting aerosol directed on to a heated metal plate. The method is discussed further in Section III.3.

Another technique which is in principle highly compatible with liquid chromatography is atmospheric pressure ionization mass spectrometry. Ion-molecule reactions initiated by electrons from [63]Ni ionize the sample at atmospheric pressure and the ions pass through a pinhole into the high vacuum region of the mass spectrometer (30). The problem of ionizing non-volatile compounds in this source was overcome by a nebulizing method which produces minute droplets from solutions of the sample (31). The method was demonstrated for nucleosides, amino acids and oligosac-charides. Finally, in a very recent new method, also operating at atmospheric pressure and termed liquid ionization mass spectrometry (32), metastable argon atoms ionize samples on a needle tip to which a high voltage is applied. Nucleosides, sucrose and phospholipids were examined.

III. Other Techniques in Mass Spectrometry

III.1. High-Mass Spectra

Some of the ionization techniques described above (particularly FAB) are capable of generating ions having masses in the range 3000 to 10000 daltons and even higher. This breakthrough, achieved in the last few years, has extended the applications of MS into an area encompassing a large number of important biological macromolecules. Accompanying the developments in ionization techniques have been the necessary advances in the technology of mass analysis so that ions of high mass can be effectively separated and sensitively detected.

In order to offset the decrease in sensitivity with increasing mass, it is necessary to operate at high accelerating voltage, and values of $8-10$ keV are now commonly employed (33, 34). Furthermore, the technology of magnet design has improved so that large radius magnets are available that can achieve high mass ranges at high accelerating voltages. Mass resolution is often sacrificed when ions of high mass are being detected and in such cases a choice is made between sensitivity and resolution. Advances in technology have also led to extensions of the mass range for measurements by quadrupole and time-of-flight mass analyzers, but the resolution remains poor. Satisfactory mass measurement can be difficult to achieve at high mass but the use of fomblin (35) and caesium iodide (36) clusters as mass-markers has helped to alleviate the problem.

III.2. Tandem Mass Spectrometry

A collection of different tandem instrumental arrangements are now available in which the ionic products [usually formed by collisionally-activated dissociation (CAD)] from a mass-selected ion are separated and identified by one or more further analyzers. Such techniques are often known collectively as mass spectrometry/mass spectrometry (MS/MS). The whole combination can involve two or more analyzers from the list: magnetic sector, electric sector, and quadrupole mass filter. There is now a large variety of such instruments and their relative merits have been discussed on several occasions (5, 6, 37).

Undoubtedly these techniques provide a number of positive features for structural identification:

(1) They often provide a unique fingerprint for an ion structure.
(2) Both molecular and fragment ions can be characterized.
(3) Mixtures of compounds can be analyzed without prior chromatographic separation.
(4) Sometimes, the measurement of translational energy released in an ionic decomposition can be made. This is a structurally related parameter.

However, such techniques are not as powerful as the above list indicates, with reference to the structural identification of unknown natural products. In this section a brief assessment of the current status of the techniques will be attempted, particularly when used in conjunction with modern ionization methods.

The EI spectra of steroids provided an early example of the utility of mass spectrometry in structural elucidation. The CAD spectra (obtained by tandem MS/MS techniques) of various steroids have now been investigated and yield additional structural information (38–41). Items 1, 2 and 4 in the positive features of structural identification listed above provided information which could be used to distinguish between epimeric structures and to identify constant structural features in different steroids.

Since the recent development of FABMS permits the production of long-lasting, stable and abundant ion currents in the molecular ion region, the combination of this technique with those of tandem mass spectrometry suggests itself as a potential tool for structural studies on involatile natural products. There have been several reports of successful FAB/MS/MS measurements. For example, nanomole amounts of the neuropeptide, substance P, were sufficient to generate abundant $[M + H]^+$ ions (m/z 1347) (42). This selected ion yielded sequence-ion information following CAD and analysis *via* the B/E linked scan technique (which generates well-resolved daughter ions). CAD spectra have been obtained from even higher-mass ions produced by FAB ionization of the peptide gastrin (MW 2096) and of vitamin B$_{12}$ (43). Again the normal FAB mass spectrum was almost devoid of fragment ions, but those fragment ions produced following CAD were structurally significant. One drawback in measurements on high-mass precursor ions can be the lack of precision in mass measurement of the CAD daughter ions.

In conclusion, the FAB/tandem MS combination appears to offer some promise for structural studies on involatile natural products. Negative ionization and chemical ionization have also been employed to provide precursor ions for analysis and it appears that structural information complementary to that obtained by normal mass analysis is forthcoming for selected compounds.

III.3. Coupled Chromatography

The direct combination of packed column gas chromatography and mass spectrometry (GC/MS), usually with computer attachment, has been used routinely for over a decade to identify and quantify mixture components. Capillary GC/MS, with its high chromatographic resolution has now also become an established technique and refinements in all aspects of the combination are continually being reviewed (5, 6).

In natural product investigations, the most crucial decisions usually involve the choice of (i) derivative and (ii) ionization technique. The factors governing the choice of ionization method are the requirements for an abundant molecular ion (particularly in quantitative studies) and for diagnostic fragment ions. The derivative is chosen to fit in with the ionization method (e.g. aromatic substituents to confer molecular stability in positive-ionization if necessary and perfluorinated derivatives to confer electron capture properties in negative ionization) and also, of course, to give suitable GC properties. Further discussion can be found in the sections on individual natural product classes below.

High performance liquid chromatography is well established as a tool for the separation of mixtures of labile biological substances, but its combination with mass spectrometry (LC/MS) remains fraught with difficulties as a valid analytical procedure. Several new types of direct liquid introduction interfaces (as opposed to transport interfaces) have been developed recently and appear to offer promise for the future. The thermospray interface (44) has been used successfully for amino acids, small peptides, nucleosides, antibiotics and glucuronides. The spectra are the relatively simple ones commonly obtained from soft ionization methods and resemble those obtained by FABMS. A gas nebulizer interface has enabled representative spectra (resembling those generated from DCI) to be obtained on catecholamines (45). This method seems to be limited at present in its applicability to thermally labile compounds.

Thin-layer chromatography (TLC) is frequently employed to separate non-volatile or thermally labile materials prior to characterization by mass spectrometry. A new procedure has been described which avoids the elution step in the usually adopted sequence (46). The technique does not incorporate an interface but involves a rapid and simple procedure for transfering the TLC adsorbent (located as usual by UV fluorescence) to an FAB probe-tip, coated with double-edged masking tape. Satisfactory spectra were obtained for several coccidiostats, including septamycin.

III.4. Computers

The utilization of computers in MS, both to control the acquisition of spectra and to handle the data, is now almost universal. However, it is not intended here to dwell on the many refinements that are continually being made in this field. Rather, the current status of computers in (a) analysis by GC/MS and (b) high-resolution mass measurements will be briefly presented, as aids to structural elucidation in natural product chemistry. Comprehensive coverage of the subject can be found in other texts (5, 6).

One of the most elegant and useful applications of computers in MS has been the capability to resolve overlapping GC peaks and subtract background spectra so that almost "pure" spectra of components are extracted. Simple algorithms are available that perform the operation in real time, but the extraction of good quality spectra requires sophisticated programming and hence takes longer (47).

In those cases where a laboratory is routinely handling natural products that have been seen before in the same laboratory, a self-constructed library of mass spectra is a valuable aid to identification. Such a system is particularly useful in the GC/MS analysis of body fluids (in biomedical research and clinical diagnosis) and in the perfume and flavor industries (e.g. terpene identification). Metabolic profiling by GC/MS (48) of the components of urine or serum provides an instance where computer techniques are a necessity and have become quite sophisticated.

For the identification of unknown natural products, a precise mass measurement of the molecular ion, on a high resolution magnetic instrument, is an essential parameter in order to suggest an elemental composition. Mass markers are available (e.g. C_2I_4 and PFK) that provide "coarse" and "fine" adjustment of curve-fitting equations and enable dynamic high-resolution measurements to be made during the scan. Accuracies better than 10 ppm are possible in commercial packages employing such methods, which in GC/MS analyses are frequently the only ones that serve the purpose. Manual peak-matching techniques still provide the highest accuracy (better than 1 ppm) in those instances where the sample can remain in the mass spectrometer for longer periods. However, mass measurements to accuracies of 5 ppm have been described, utilizing a computer-controlled peak-matching system and operating under GC/MS conditions (49).

III.5. Quantitative Measurements

There are many instances in which quantitative measurements are required in natural product chemistry, particularly in the biomedical field. MS is a well-established tool in this area and offers a unique combination of sensitivity, specificity and dynamic range. In the past GC/MS has usually been the method of choice and procedures and problems involved in such analyses are well documented (1, 2, 50). Frequently, the most difficult item in the whole procedure is the choice, and possible synthesis of an appropriate internal standard (homologue, isomer, closely related analogue, or isotopically labelled analogue). Furthermore, the selection of a derivative with suitable GC and MS properties for both sample and standard is required.

The main question posed recently in quantitative mass spectrometry is whether the newer techniques of sample introduction, ionization and ion analysis can be reliably utilized. Since FABMS provides a means for analysis of underivatized polar and high-molecular weight materials (many of which are important to the clinical chemist and drug analyst), attempts have been made to obtain good quantitative measurements for such compounds. Some early encouraging results have been obtained for low molecular weight compounds (51). However, for compounds of high molecular weight the cluster of ions observed in the molecular ion region does not provide the optimum conditions for quantitative measurements. Nevertheless, by use of ^{18}O-labelled internal standard, it has been reported that endogenous amounts of individual peptides in biological tissue can be quantified by FAB/MS at the part per billion level (52).

IV. Applications of the Newer Techniques in Mass Spectrometry to Specific Classes of Natural Product

Because of the vast scope of this topic, attention is concentrated, where possible, on applications in which mass spectrometry has contributed vitally to solving the structure of a natural product or to an understanding of its biological function. This is not, however, to belittle the part which studies on model compounds have played in developing the newer techniques, and these are also exemplified, particularly where they concern newer methods not yet widely applied to compounds of unknown structure, or types of natural product which are particularly difficult to analyze by mass spectrometry.

IV.1. Peptides and Proteins

Electron impact (EI) mass spectrometry was first employed to sequence peptides more than twenty years ago. Due to the involatility of all but the smallest peptides derivatization was required and various such techniques evolved. One traditional method has been the use of N-acetyl permethyl derivatives, which yield characteristic "sequence" ions in their EI mass spectra. The technique has some advantages over conventional "wet" chemical sequencing methods for N-terminally blocked peptides and their mixtures (2). The sequencing procedure can be carried out on only 10 – 30 nmol of peptide and a molecular length of between 4 and 9 amino acid residues can be typically handled.

The advent of "soft" ionization techniques, such as field- or plasma-desorption, provided useful advances, since molecular weight information could be readily obtained from the spectra of underivatized peptides. However, more dramatic advances have been forthcoming in the last few years, due to the developments of FABMS and high-mass magnets. The most significant advantages of FABMS for peptide sequencing are (1) the ability to study underivatized peptides, (2) the acquisition of molecular weight *and* sequence information, (3) a reduction in sample size and (4) the facility to study larger peptides (up to 20 amino acid residues routinely and higher under suitable conditions). However, it should be emphasized at the outset that it is often advantageous to employ complementary biochemical techniques, such as Edman degradation and carboxypeptidase digestion (53).

The discussion in this section concentrates on the structural elucidation of peptides and proteins involving (a) modern ionization techniques (particularly FAB), (b) current GC/MS methods and (c) any new chemical or biochemical processes used to augment the overall mass spectrometric analytical procedure.

The structural determination of a toxic octapeptide from the sawfly larva neatly illustrates the role of FABMS, together with chemical and other instrumental procedures, in the identification of the sequence and chirality of the constituent amino acids (54).

Prior to the advent of FABMS, conventional amino acid analyzers had estimated the molar ratios of the constituent amino acids, and EIMS had

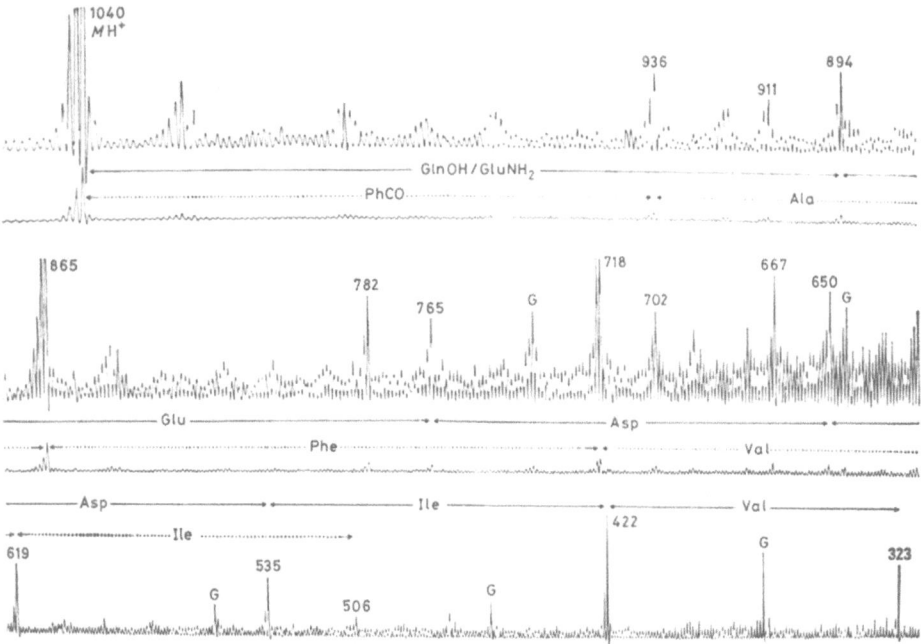

Fig. 2. The positive ion FAB mass spectrum of the octapeptide 6. The peaks marked G
represent protonated polymers of glycerol.
[Reproduced with permission from ref. (54)]

established benzoyl as an N-terminal blocking group and determined the
partial sequence PhCO-Ala-Phe-Val-Ile-X.

The partial positive ion FAB mass spectrum (obtained by Xe bombardment) is shown in Fig. 2 and incorporates an abundant protonated
molecular ion at m/z 1040. Three clearly defined series of diagnostic
fragment ions, corresponding to the general structures (1) (N-terminal), (2)
(N-terminal) and (3) (C-terminal) identify the sequence as PhCO-Ala-Phe-
Val-Ile-Asp-Asp-Glu-Gln (or *iso*-Gln). The negative ion FAB spectrum
showed an abundant $[M - H]^-$ ion at m/z 1038 and the series of ions
corresponding to structure (4) was able to confirm the sequence of the six
C-terminal amino acids.

RCO$^+$ (RCONH$_2$)H$^+$ (H$_2$NR)H$^+$ RCONH$^-$

(1)   (2)   (3)   (4)

$$R - CH - COOCH(CH_3)_2$$
$$NH - COCF_3$$

(5)

The ambiguity of the C-terminal amino acid structure was resolved by a Hofmann degradation, followed by hydrolysis and amino acid analysis. The absolute configuration of the amino acids were determined from the GC/MS (EI and CI) properties of their N-trifluoroacetyl isopropyl esters (5) on a chiral column. This procedure revealed that the two aspartate residues had different configurations and the ambiguity of position of these two residues in the peptide was resolved by analyzing products of partial acid hydrolysis. The structure of the toxin was determined as (6).

PhCO- D-Ala- D-Phe- L-Val- L-Ile- D-Asp- L-Asp- D-Glu- L-Gln

(6)

The following conclusions emerge from this model study:

(1) EIMS was unable to elucidate the sequence of amino acids in this particular octapeptide.
(2) Conventional amino acid analysis assisted the subsequent mass spectrometric investigation.
(3) The positive ion FAB spectrum determined the molecular weight and contained several series of fragment ions that identified the sequence of amino acids.
(4) The negative ion FAB spectrum confirmed the molecular weight and partially the sequence of amino acids.
(5) Wet chemical methods, together with GC/MS on a chiral column, were necessary to determine the absolute configuration of the amino acids and to resolve an isomeric ambiguity.

Structures of cyclic peptides have also been evaluated with the aid of FABMS (55). The positive and negative ion FAB spectra of various lipophilic cyclic peptides derived from marine organisms gave abundant parent ions in the molecular ion region and informative series of fragment ions formed by segmental amino acid loss. FAB mass spectra of hydrolysis products (and their CD_3 esters to identify C-terminal-containing sequences) aided the identification and NMR assignments were particularly valuable for the unusual amino acids. The structure of one of the peptides was revised on evidence of the FAB spectrum, compared with a previous EI spectrum.

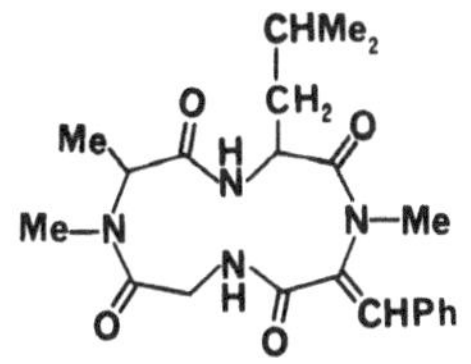

(7)

In contrast, it has been argued that the cyclic tetrapeptide tentoxin (**7**) yields an EI mass spectrum (containing abundant molecular and sequence ions) that rivals the FAB spectrum (*56*). However, this situation is unlikely to obtain where the cyclic peptide mass is substantially higher.

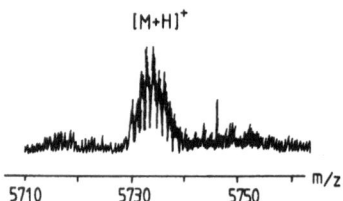

(**8**)

FABMS has been employed to identify and quantify dansyl amino acids (**8**) formed in a procedure for the N-terminal analysis of proteins (*57*). This sensitive method involves the reaction of the protein with dansyl chloride, followed by acid hydrolysis. Previously the characteristic involatile N-terminal amino acid derivative (**8**) was analyzed either by TLC or high-voltage electrophoresis but FAB ionization now permits rapid identification of the protonated molecular ion. The responses of this and other characteristic ions were monitored employing dansyl amino butyric acid as an internal standard, enabling the dansyl derivatives (**8**) to be determined quantitatively at a level of 0.1 nmol.

$[M+H]^+$

			m/z
5710	5730	5750	

Fig. 3. FAB mass spectrum of the molecular ion region of bovine insulin obtained using a 50 sec scan over 50 mass units
[Reproduced with permission from ref. (*59*)]

It is now commonplace for underivatized peptides having molecular weights up to 2000 daltons to be sequenced by FABMS, and the development of high-field magnets has enabled spectra to be obtained with high sensitivity at full accelerating voltage (e.g. 8 kV) up to about 3500 daltons. It has been shown that proteins in the mass range of 5000 – 6000 daltons are amenable to analysis using high-field magnets and operating at 4 kV accelerating voltage (*58 – 60*). For example, Fig. 3 shows the FAB spectrum obtained over a 50 a.m.u. range of bovine insulin (MW 5729 daltons) (*59*). In this particular scan, the resolving power of the mass spectrometer was increased (sacrificing sensitivity) sufficiently to obtain unit mass separation. The peptide gave a clearly identifiable protonated molecular ion cluster. For molecules of this size, ^{13}C-signals predominate over the ^{12}C-signal and, employing caesium iodide clusters as mass markers, it is possible to identify the molecular weight to within 2 a.m.u. It has been pointed out (60) that the bovine insulin molecular ion envelope is probably constituted from several overlapping clusters. Three reduced forms of insulin may be present, incorporating different numbers of intact cystines. It is evident that FAB ionization employed with high field magnets provides an effective combination for screening and establishing the primary structure of proteins. A new procedure, known as FAB-mapping (61), has been described which seeks to achieve this aim for recombinant DNA protein products.

Because FAB ionization generates a stable protonated molecular ion from underivatized high-mass peptides, it became opportune to investigate the structural information achievable by collisionally-activated dissociation (CAD) of these ions, employing an appropriate tandem mass spectrometer. Early results (43) indicate that some sequence information is obtainable which was not extractable from the normal FAB spectrum. Ions of mass greater than 2000 daltons were selected for collisional activation and potentially the technique can elucidate detailed fragmentation patterns of peptides and their mixtures. Future developments depend on improvements in sensitivity and resolution.

The outstanding success of FABMS has tended to obscure the earlier advances in other soft ionization techniques in obtaining molecular weight information on underivatized peptides. Field desorption (FD) and plasma desorption (PD) have been particularly successful in this respect.

The advantages and drawbacks of FDMS have been well documented over the last 10 years and a recent review has covered the important biochemical applications (17). It is generally accepted that protonated molecular ions can be readily identified from underivatized peptides, but that sequence information is usually limited. Optimum ionization conditions can be difficult to attain, but it has been shown that structural details can be obtained from peptides by careful control of the emitter heating current, together with photographic or fast electrical recording of spectra (62). For example, the FD spectrum of the nonapeptide bradykinin showed $[M + H]^+$ and $[M + Na]^+$ ions at m/z 1060 and 1082 respectively. The spectrum was complex but all fragments from cleavage of the peptide bonds were recorded with the exception of those of both terminal amino acids.

Plasma desorption mass spectrometry has been employed in only a few laboratories for peptide identification. Nevertheless, the techniques can yield useful structural information [see ref. (1), p. 1209]. For example, in the positive ^{252}Cf spectrum of the 31-residue peptide β-endorphin, an $[M + Na]^+$ ion at m/z 3487 was readily identifiable. In a more recent example, when a 90 MeV beam of $^{127}I(+20)$ ions was used to desorb bovine insulin, a molecular peak at m/z 5730 was observed in the positive-ion spectrum and fragment ions were observed in the correct mass ranges for the α- and β-chains (63). It has been pointed out that the characteristic tailing observed in the peaks of plasma desorption spectra is reproducible and is due to the superposition of signals from ions which retain their identity during flight on those which undergo metastable decompositions. It was shown that the ^{252}Cf spectra obtained for the 19-residue peptide alamethicin I were strikingly similar for natural and synthetic peptide (64).

GC/MS analysis of volatile peptide derivatives lends itself well to the rapid survey of sequence segments in a protein, to define an N-terminus, to check a DNA reading-frame, or to generate enough sequence information to construct a complementary DNA-probe. A strategy has been developed, employing computer-aided GC/EIMS analysis, for the rapid and accurate determination of the amino acid sequence of large proteins (65). This strategy involves combining DNA sequencing of the gene for the protein of interest with GC/MS identification of tetra- and pentapeptides in partial hydrolysates of the entire protein or very large fragments thereof. These peptides are matched to blocks of codons at locations scattered throughout the entire structural gene.

The approach has been employed to determine the amino acid sequence in *E. coli* alanine t-RNA synthetase. After partial hydrolysis, the peptide mixtures were converted to their methyl esters, N-trifluoroacetylated, reduced with B_2D_6 and converted to their O-TMS derivatives (9). These compounds possess good GC properties and yield simple, yet informative fragmentations. A variety of derivatives have been explored for sensitive sequence analysis of small peptides and a new permethylation procedure has been described for trifluoroacetylated peptides at the $2-10$ nmol level. Mixtures of these derivatives, ideally of di- to hexa-peptides are readily analyzed by GC/MS (66).

Many individual procedures have been devised which make use of isotopic labels to assist in particular peptide sequencing studies by mass spectrometry. Two further recent examples, in addition to those reported above, are mentioned here.

$$CF_3CD_2NH\text{-----}\overset{\overset{\displaystyle R}{|}}{C}HCD_2NH\overset{\overset{\displaystyle R'}{|}}{C}H\text{-----}CD_2OSiMe_3$$

(9)

$$CH_3\overset{\overset{\displaystyle NH_2}{|}}{C}DCH_2\overset{}{C}HCOOH$$
$$\overset{\overset{}{|}}{C}H_2D$$

(10)

The differentiation between isomeric leucine and isoleucine has long been a problem in mass spectrometric peptide sequencing. However, a simple method has been developed, employing leucine containing a non-labile isotopic label, to distinguish between these two amino acids (67). The technique involves growing a bacterial or cell culture in a medium containing d_2-leucine (10). The labelled residue is incorporated into the cell's proteins and the resulting mass spectra of leucine-containing peptides (following hydrolysis) cleanly exhibit sequence ions 2 mass units higher than the corresponding isoleucine peptides. The utility of the technique was demonstrated by work on the sequence of alanine t-RNA synthetase, a protein 875 amino acids in length.

A novel mass spectrometric method has been described that permits the identification of the C-terminal peptide of a protein (68). The technique involves the incorporation of ^{18}O into all α-CO_2H groups released during enzyme-catalysed partial hydrolysis (e.g. with trypsin) of the protein in ^{18}O-enriched water ($^{18}O:^{16}O$; 50:50). Individual peptides were analyzed by GC/EIMS as their trifluoroacetyl permethyl derivatives and the C-terminal peptide was characterized as the one that did *not* incorporate ^{18}O. (The non-C-terminal peptides were recognized by the contrived 1:1 intensity ratios 2 mass units apart in their mass spectra.) The technique was used to identify the C-terminal peptide in a bacterially-produced interferon.

IV.2. Carbohydrates

After peptides, probably the most studied natural products using the newer ionization techniques have been carbohydrates, particularly oligo- and polysaccharides. Although EIMS and CIMS have proved useful in determining molecular weight and sequence of oligosaccharides up to ca. 2000 daltons [see ref. (69) and refs. cited therein] a disadvantage of these methodologies applied to such compounds results from the widespread use of permethylation to make them sufficiently volatile for analysis. Many oligosaccharides naturally contain methylated monosaccharide subunits and permethylation can interfere with the ability to distinguish these from unmethylated counterparts (70). Thus FDMS has been used (70) to determine molecular weight and sequence in methylmannose polysac-charides isolated from *Mycobacterium smegmatis* and including homo-logous series comprised of $\alpha 1 \rightarrow 4$-linked mannose and 3-O-methylmannose units. Besides the ions $[M + H]^+$ and $[M + Na]^+$, systematic cleavages of hexose units starting from the non-reducing end of the chain in smaller homologues and from both ends of the chain in larger homologues, for example $Man_1 MeMan_9$-OCH_3 (11), defined the sequence. A substance

(11)

formerly thought to have the structure $Man_1MeMan_7\text{-}OCH_3$ was shown in this study to comprise 2 components, one having this structure, the other, $Man_2MeMan_6\text{-}OCH_3$, differing from it by a single methyl group.

The foregoing study clearly showed that FDMS can afford sequence information though its authors had no explanation for the mechanism of the processes which caused fragmentation. However this subject has been explored (71–73) by Komori, Schulten et al. who observed analogies between field-induced fragmentations and the acid-catalysed cleavage of glycoside linkages in solution. For example (71) the fungistatic agent tomatine (Fig. 4), a saponin isolated from the leaves of *Lycopersicon pimpinellifolium*, fragmented under FD in a manner predictable from its solution chemistry. Thus, in solution β-D-xylosides are hydrolyzed about 4.8 times faster than β-D-glucosides. Analogously (Fig. 4) loss with hydrogen transfer of the xylose residue (133 daltons) from the parent ions of tomatine, namely $[M + Na]^+$ *(m/z 1056)* and $[M + H]^+$ *(m/z 1034)*, gave respectively the ions *m/z* 924 and 902 and was favoured over the loss of the glucose residue to give *m/z* 894 and 872.

The advantages and drawbacks of FDMS in the analysis of natural oligosaccharides and their derivatives are discussed in ref. (73). Whilst the simple cleavages of the glycosidic linkages are a help in sequence

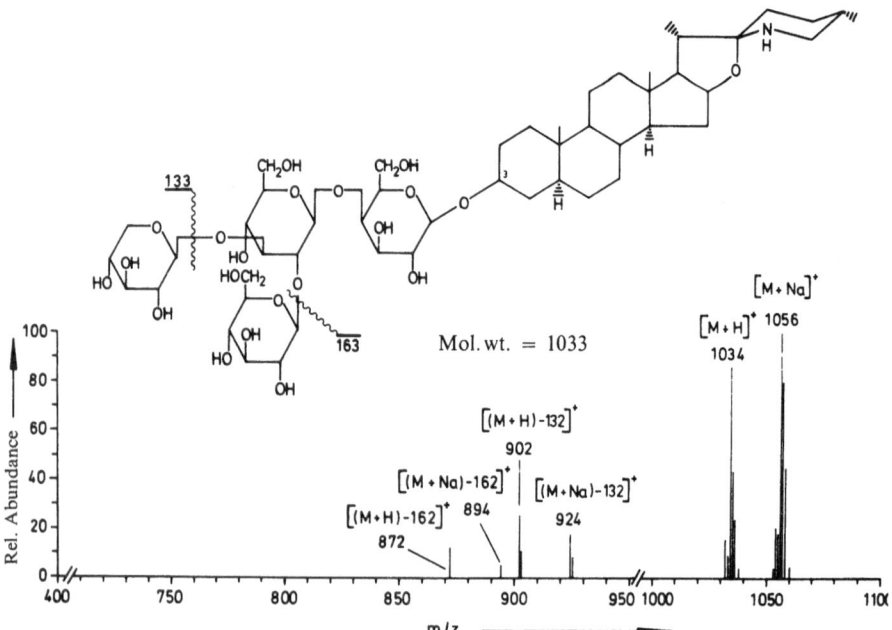

Fig. 4. FD mass spectrum of tomatine. The ratio of the loss of glucose:xylose is about 1:3.6. [Reproduced with permission from ref. (71)]

determination, there are no cleavages within the sugar moieties, which would help to define the branching points in the oligosaccharide chain or the point of attachment of other substituents. Also stereoisomers cannot be distinguished.

FDMS and FABMS were both used (74) to analyze a 6-O-methylglucose polysaccharide from *Mycobacterium smegmatis* and led, in conjunction with other physical and chemical methods, to a revision of its structure. This polysaccharide (12) was not desorbed intact in FDMS and only its permethylated derivative gave a FD spectrum. In contrast FABMS readily gave cationized molecular ions ($[M + Na]^+$ and $[M + K]^+$) and sequence information. A detailed discussion (75) of this FAB spectrum gave insight into the comparative merits of the two ionization techniques in the analysis of polysaccharides. Both types of spectrum contained sequence-defining fragments due to cleavages between the monosaccharide residues. However only FABMS afforded cleavages within sugar moieties to give ions such as (13) and (14), giving valuable information about the position of methyl substituents. Such cleavages were also seen in the negative ion FAB spectrum which also contained a doubly-charged molecular anion $[M - 2H]^{2-}$ at m/z 1425, confirming the molecular weight of (12).

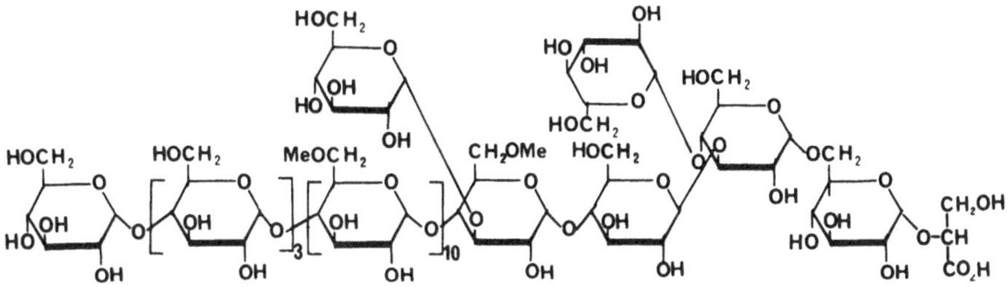

(12)

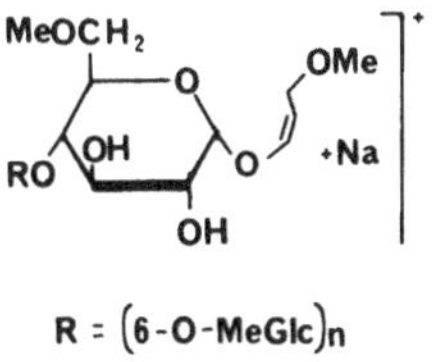

R = $(6\text{-}O\text{-MeGlc})_n$

(13)

R^1 = Glc; R^2 = $(6\text{-}O\text{-MeGlc})_{10}$

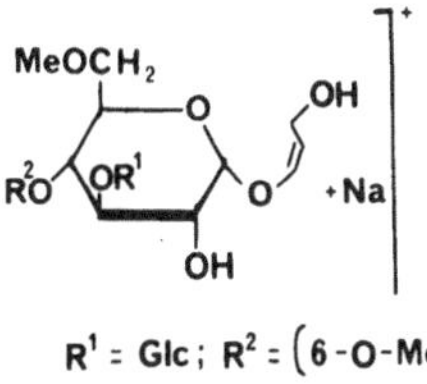

(14)

SIMS was used (76) in identifying a pentasaccharide [α-L-Fuc(1→3)-β-D-GalNAc(1→3)-β-D-Gal(1→4)-β-D-Gal(1→3)-D-2-acetamido-2-deoxygalactitol] (15) containing a novel disaccharide unit, α-L-Fuc→D-GalNAc, and obtained from salmon egg polysialoglycoproteins by alkali-

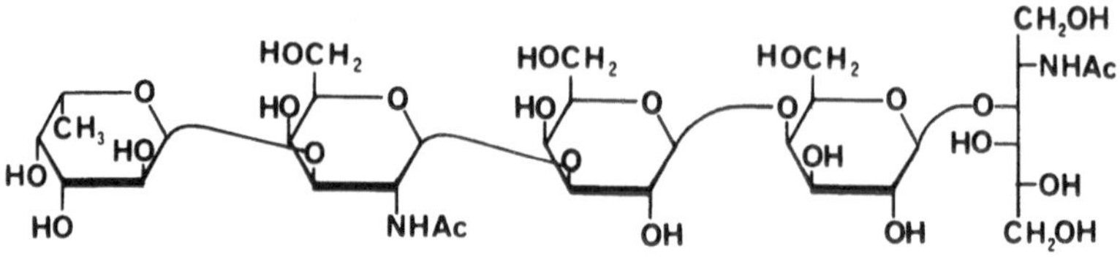

(15)

borohydride treatment. Enhancement of the mass spectrum using 0.1 M
HCl revealed the molecular weight through $[M + H]^+$, $[M + Na]^+$ and
$[M + K]^+$ ions, loss of ketene from $[M + H]^+$ (also characteristic of N-
acetyl derivatives on electron impact) and sequence ions including two
corresponding to ionized Fuc-O-GalNAc-O-Gal and Gal-O-Gal-O-
GalNAcol. A study (77) with the emphasis on confirming molecular weight
by exploiting matrix effects in SIMS concerns the viridopentaose (16), a
constituent of the antibiotic sporaviridine. Coated as a film on a solid
(silver) support it afforded characteristic $[M + Ag]^+$ ions as a doublet with
components of almost equal intensity due to the Ag-isotopes (107 and 109).
In glycerol, it gave mainly $[M + H]^+$, and in diethanolamine,
$[M + diethanolamine]^+$. The molecular weight of such compounds was
difficult to obtain using FDMS.

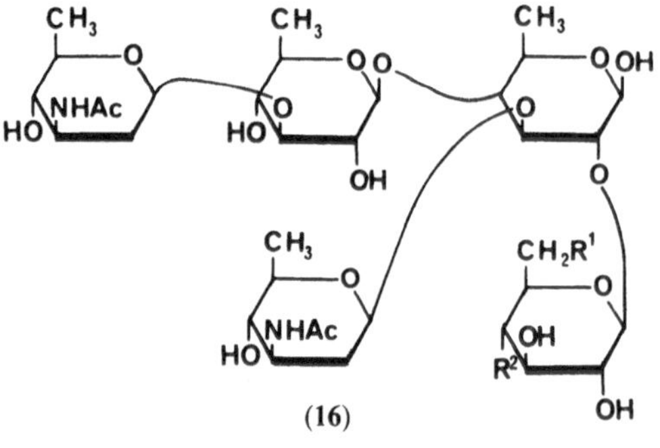

(16)

IV.3. Nucleotides

In contrast to the real achievements with peptides and oligosaccharides,
the newer techniques in mass spectrometry have so far proved of limited use
for determining molecular weight and sequences of oligonucleotides of
natural origin, but have been applied with success to synthetic compounds
of known sequence. Partly, this is due to the even greater polarity and higher
molecular weight of the constituent monomers (nucleotides) and partly to

the existence of alternative methods of greater efficiency. Protection of polar functional groups and analysis by EIMS or CIMS has been exemplified for small oligonucleotides but a major problem is the very large increase in molecular weight on introducing the necessary protecting groups such that even a protected diribonucleoside phosphate has a molecular weight exceeding 1000 daltons (78). The main thrust of work using the conventional ionization modes has been in pyrolysis studies aimed at identifying and quantifying minor bases in DNA, a topic which will be referred to later.

Field desorption spectra of unprotected diribonucleoside phosphates afforded (79) both the 5'- and 3'-mononucleotide as fragment ions. Further cyclizations gave 3',5'-cyclic phosphates from the 5'-nucleotides and both 3',5'- and 2',3'-cyclic phosphates from the 3'-nucleotides (Scheme 1).

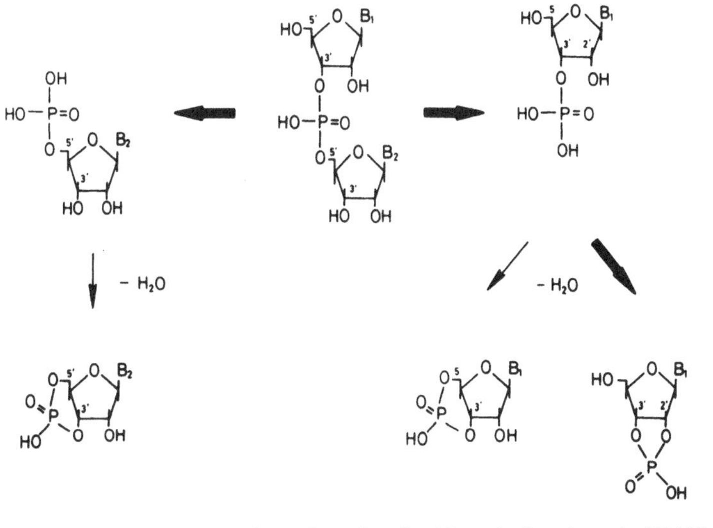

Scheme 1. Sequence-specific formation of nucleoside cyclophosphates in FDMS [Reproduced with permission from ref. (79)]

However the 5'-nucleotides cyclized preferentially and gave the more abundant cyclic phosphate ion in all the possible pairs of combinations studied (i.e. ApC, CpA; ApG, GpA etc.) defining the sequence in these diribonucleoside phosphates. The positive ion FAB spectrum of a simple dideoxyribonucleoside phosphate, TpT (80) was complex, containing peaks due to degradation of sugar and nucleoside residues. The negative ion spectrum was much simpler and the MIKE spectrum of the parent ion $[TpT - H]^-$ contained only 3 daughter ions $[(TpT - H) - B]^-$, $[TMP - H]^-$ and $[B - H]^-$. This simple fragmentation is reflected in the

negative ion FAB spectra of more complex oligomers. Thus the tetra-
deoxyribonucleotides d(G-G-T-A) and d(T-A-C-C) gave (81), by stepwise
degradation at the phosphodiester linkages, prominent sequence defining
fragments. Indeed, negative ion FABMS could be used to sequence
unprotected oligodeoxyribonucleotides of up to 10 nucleotide units in
length (82). Two types of sequence ion were discriminated. Bonds
connecting oxygen to C-3' on ribose were more labile than bonds to C-5'
and produced more intense sequence ions. This is illustrated (Fig. 5) for the
decadeoxyribonucleotide d(G-A-A-G-A-T-C-T-T-C) where C-3'-O cleavage ions (m/z 2775, 2462, 2149 etc.) were invariably more intense than their
C-5'-O cleavage counterparts (m/z 2815, 2511, 2207 etc.). Although the
foregoing study concerned the analysis of synthetic oligodeoxyribonucleotides the conclusions clearly apply equally to natural unprotected
oligomers.

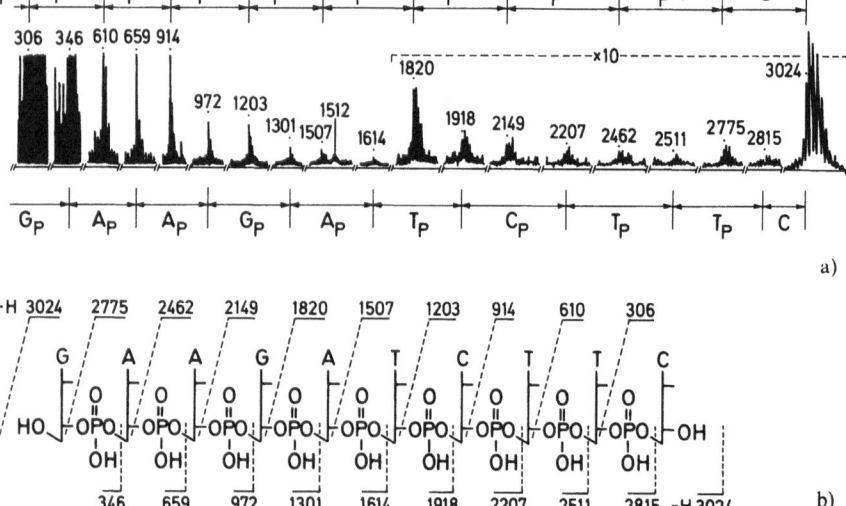

Fig. 5. Negative ion FAB mass spectrum (a) of d(G-A-A-G-A-T-C-T-T-C) and the corresponding shorthand structure (b) with those bonds marked which on breakage give rise to
the main fragment ions.
[Reproduced with permission from ref. (82)]

Unfortunately the extensive studies on oligonucleotides using PDMS
refer to *protected* synthetic oligomers and a detailed discussion of them is
therefore outside the scope of this Review. However an exciting pointer to
future developments is the observation (24) of the molecular ion for a dimer
(m/z 12637) of a protected dodecanucleotide, which points to the possible
use of PDMS to detect "duplex" formation in oligonucleotides from native
DNA and RNA.

Mononucleotides are less of a problem in mass spectrometric analyses and are often accessible by conventional EIMS or CIMS of volatile derivatives. Even with derivatization, the "in beam" technique of DCI was needed to detect (*83*) the cytokinin 9-β-D-ribofuranosylzeatin-5′-phosphate (**17**) in a partially purified extract from *Datura innoxia* crown gall tissue and to quantify it using a deuterium-labelled internal standard (**18**).

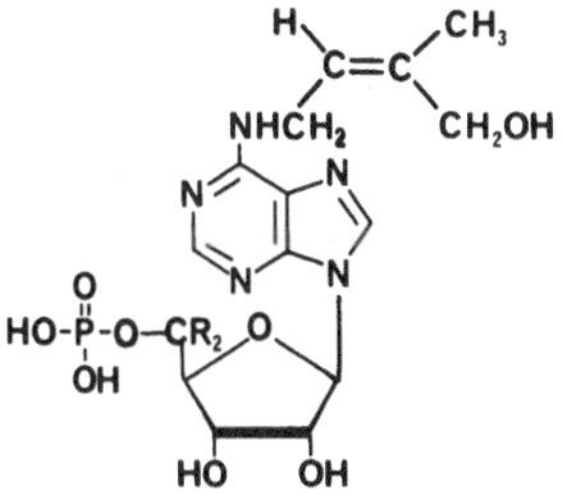

(17); R = ¹H
(18); R = ²H

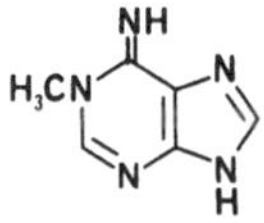

Finally, a study (*84*) using SIMS of methylated ribonucleosides and deoxyribonucleosides such as might be obtained by enzymatic degradation of RNA and DNA is relevant inasmuch as methylated bases occur naturally and not only as products of exposure to chemical methylating agents. The approach was compared with pyrolysis MS/MS which afforded evidence (*85*) that 1-methyladenine (**19**) was a minor base in salmon sperm DNA, and the cautionary point was made that pyrolysis can afford artefacts through migration of methyl substituents.

IV.4. Tetrapyrroles

Among natural products in this category, vitamin B_{12} and chlorophylls in particular have been extensively examined using the newer ionization techniques. Again the emphasis has been on model studies on known structures though some of these have relevance to aspects of their biosynthesis or the mechanism of their biological activities.

The analysis of intact cobalamines by mass spectrometry requires desorption ionization methods and the first spectrum of vitamin B_{12} (**20**) containing molecular ion species was obtained in 1978 using FDMS (*86*). By a careful analysis of the relative abundances of the ^{12}C- and ^{13}C-containing molecular ions in the FD spectrum of a salt of aquo-cyano-cobyrinic acid heptamethyl ester derived from natural vitamin B_{12} these authors demonstrated (*87*) a depletion of ^{13}C of 0.0428% with respect to the predicted

^{13}C abundance of 1.112%. This reflected the $^{13}C/^{12}C$ ratio for the carbon source (beet sugar molasses) for the microbiological synthesis of the original vitamin B_{12}.

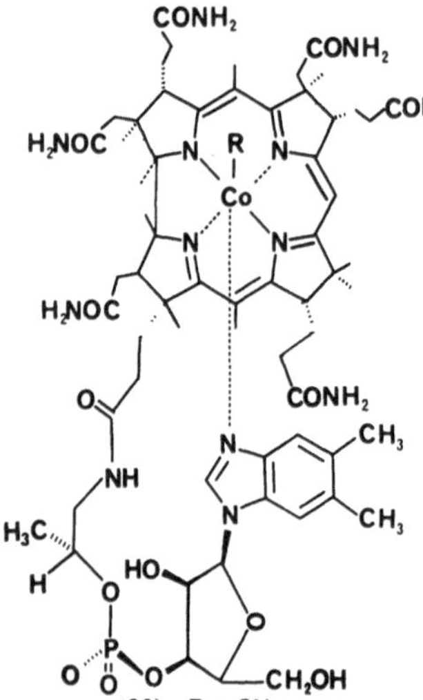

(20); R = CN

(21); R = 5'-adenosyl

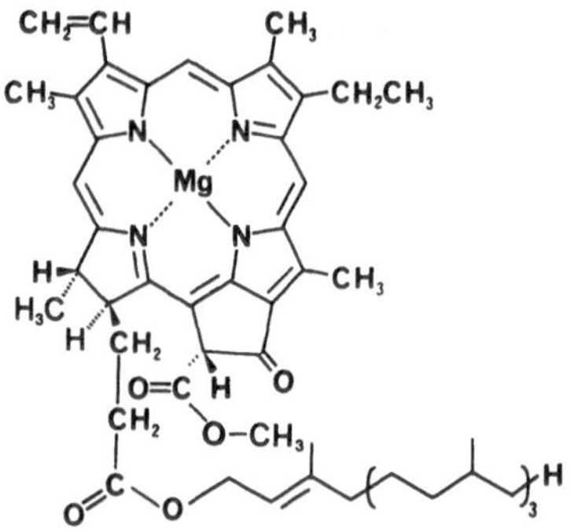

(22)

Vitamin B_{12} itself has proved an easier subject for FABMS than for FDMS and FABMS has even given (88) a molecular protonated ion $[M + H]^+$ at m/z 1579 for coenzyme B_{12} (21). The fragmentation pathway of the cobalamins under positive and negative ion FABMS has been studied using collisional activation of $[M + H]^+$ ions in a tandem mass spectrometer (43) and by high resolution measurements (89). A point of ambiguity resolved by both techniques was the origin of the fragment ion corresponding to $[M + H]^+ - CN - 59$, proved to be formed by the loss of acetamide and not the central cobalt atom.

Two aspects of the chemistry of chlorophyll which are important to an understanding of its role in photosynthesis are its ability to form hydrates and the self-association resulting in part from $\pi - \pi$ interactions. In the FD mass spectrum (90) of anhydrous chlorophyll a (22), M^+ at m/z 892 was the base peak but hydration altered the spectrum which then contained ions of higher m/z value, including the molecular ion of the hydrate (m/z 910). Aggregate ions in PD mass spectra were mentioned earlier in connection with oligonucleotides (24) and chlorophyll a oligomers up to the heptamer were recorded (91). Surprisingly, success in obtaining mass spectra of intact chlorophyll was not confined to the desorption ionization methods. When chlorophyll a was coated on a gold wire and introduced into an electron impact source near to the ion beam a spectrum containing the intact M^+ ion as the sole species was fleetingly observed (92).

IV.5. Fatty Acids, Alcohols, and Esters

The determination of double-bond position in unsaturated fatty acids has traditionally seemed to be an intractable problem for MS alone. Recourse to GC/MS and comparison of the column retention times of unknowns with those of standards has usually solved the problem. However, several techniques have recently managed to overcome the difficulties without using GC.

CIMS and FABMS have been the methods of choice for revealing minor structural differences because of the low energy ionization processes involved. If the molecular ion (or closely related ion) so generated is passed through a collision cell and undergoes CAD, then differences between the spectra obtained can be quite striking. For example, FAB negative ion tandem mass spectrometry has been employed to locate the double bond position in mono-unsaturated fatty acids (93). Negative ion FAB of such compounds generates almost exclusively the $[M - H]^-$ ion, but the CAD spectra of isomeric compounds are quite distinct. For example, ions A and B are prominent in the spectrum of 9-octadecenoic acid (23).

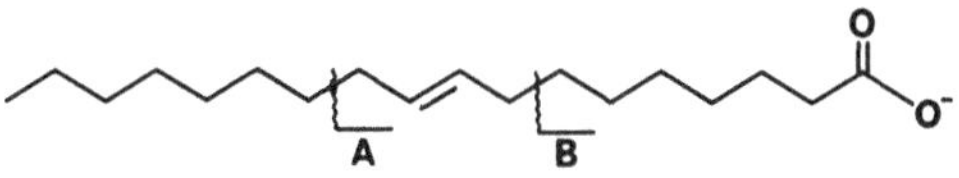

(23)

A procedure has been devised (94) to locate the double-bond position in fatty acids (including mixtures), which involves derivatization, ammonia CI and CAD. The fatty acids contained between 16 and 22 carbon atoms and were esterified and derivatized to yield a mixture of two products (Scheme 2; 24→25).

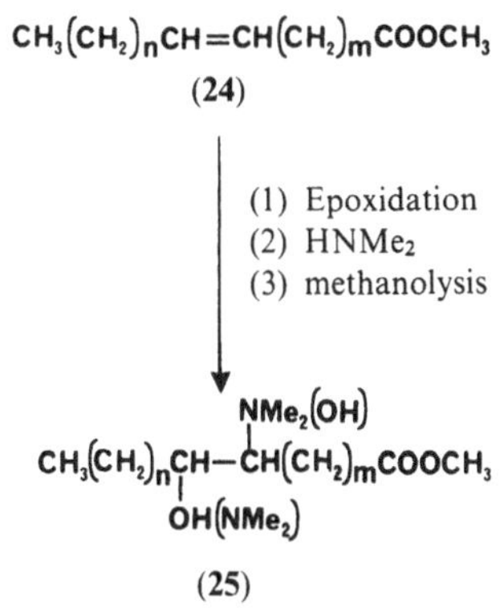

$$CH_3(CH_2)_nCH=CH(CH_2)_mCOOCH_3$$

(24)

(1) Epoxidation
(2) $HNMe_2$
(3) methanolysis

$$\underset{OH(NMe_2)}{\overset{NMe_2(OH)}{CH_3(CH_2)_nCH-CH(CH_2)_mCOOCH_3}}$$

(25)

Scheme 2. Location of the double bond in an unsaturated fatty acid ester: formation of an appropriate derivative (94)

The ammonia-CI spectrum exhibits an abundant $[M + H]^+$ and two prominent fragment ions which correspond to the ions (26, 27) retaining the NMe_2 group. The double bond position is therefore readily identifiable and if CAD/tandem MS analysis is further used, each component of a fatty acid mixture can be thus identified. The approach was employed to analyze a mixture of monosubstituted fatty acids from *Mycobacterium phlei*.

$$Me_2\overset{+}{N}=CH\,(CH_2)_n\,CH_3$$

(26)

$$Me_2\overset{+}{N}=CH\,(CH_2)_m COOCH_3$$

(27)

It is usually impossible to determine by EIMS the double bond position(s) in unsaturated fatty acids, esters or alcohols merely by comparison of spectra with literature values. However, empirical methods can be employed to achieve this aim under carefully controlled mass spectrometric conditions, particularly where the mass spectrometer is dedicated to a particular compound class or problem.

This principle has been neatly applied to the spectra of straight chain $C_{10} - C_{18}$ acetates and alcohols, some of which are sex attractant phero-mones in Lepidoptera (95). EI mass spectra of these compounds are repeatedly measured under reproducible GC/MS conditions. The ratio of intensities, $[m/z\,55]:[m/z\,54]$ increases linearly as the double bond is located more distally from the acetate or alcohol functional group. As a check, the intensity of $m/z\,61$ is also correlated with the double bond position. The reproducibility of measurements was such that this position was predictable within ± 1 carbon atom.

FABMS has been shown to distinguish readily between some simple unsaturated dicarboxylic acids (96). The $[M + H]^+$ ions from the Z-acids (28) are much less prone to fragment than those from the corresponding E-acids (29). The Z-acids probably form the stable proton-bridged complex (30). In contrast, protonation of the E-acids on the carbon-carbon double bond becomes an energetically feasible process relative to carbonyl protonation. The ion formed can readily undergo the observed elimination of H_2.

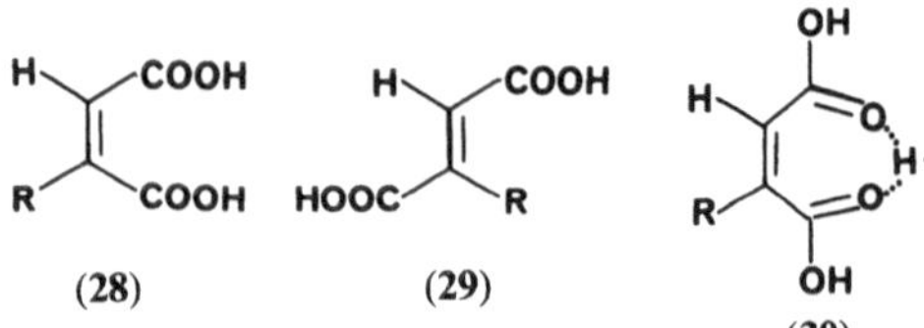

(28)   (29)   (30)

IV.6. Complex Lipids

Many compounds of this class combine lipophilic residues such as fatty acids with hydrophilic polar residues such as carbohydrates, amino acids and phosphate and hence are suitable subjects for study by the newer ionization methods.

Dipalmitoylphosphatidylcholine (31) has been quantified in amniotic fluid by FABMS using a d_9-labelled standard (97). Such measurements were an alternative to the determination of lecithin-sphingomyelin ratios as a predictive test for lung viability in mid-term and late foetal development. This study is also noteworthy inasmuch as quantification using FABMS has not been widely exemplified. An advantage of the mass spectrometric method is its specificity for the dipalmitoyl derivative (MW 735 daltons). The FAB spectrum also afforded evidence that the mixed palmitoyl oleoyl (MW 761) and tetradecanoyl (MW 707) esters were also present. Homologues were also detected (98) by FABMS of zwitterionic ornithine containing lipids (32) from *Thiobacillus thiooxidans* and were further discriminated by their CAD spectra.

$CH_3(CH_2)_n CO_2CH_2$
    $HCOCO(CH_2)_n CH_3$
$(CH_3)_3\overset{+}{N}CH_2CH_2O\overset{O}{\overset{\|}{P}}OCH_2$
               O^-

(31) ; n = 14
(37) ; n = 20

$COCH(CH_2)_6CH—CH(CH_2)_5CH_3$
  CO_2^- O OH CH_2
$H_3\overset{+}{N}(CH_2)_3CHNHCOCH_2CH(CH_2)_{12}CH_3$

(32)

fatty acyl-D-Phe-D-*allo*Thr-D-Ala-L-alaninol-O-(6-deoxysugar)
    |
O-(6-deoxysugar)

(33)

$CH_3(CH_2)_nCO_2CH_2$
    $CHOCO(CH_2)_nCH_3$
    $RCH_2O\overset{O}{\overset{\|}{P}}OCH_2$
          O^-

(34) ; R = $(CH_3)_3\overset{+}{N}CH_2$
(35) , R = NH_2CH_2
(36) ; R = $HOCH_2CH(OH)$

The identity of polysaccharides from *Mycobacterium smegmatis* was mentioned earlier (70). C-Mycosides from these bacteria contain glycosidic, peptidic and fatty acid moieties combined in one structure [of type (33)] and these have been investigated (99) using FDMS and EIMS giving molecular weight and substructure information respectively. To obtain good FD spectra cationization using sodium iodide [*cf.* ref. (18)] was necessary to obtain molecular ions ([M + Na]$^+$) and again afforded strong evidence that homologues were present. The EI spectra contained no molecular ions but fragments confirming the nature of the amino acid and sugar residues. FDMS has been used to achieve fractionation of various types of lipid from crude extracts of blood plasma by controlled heating of the emitter enabling detection of the different classes of lipid present, and quantification of cholesterol and of triolein using ^{13}C-labelled internal standards (*100*).

Evidence that FABMS may not be wholly suitable for the analysis of highly lipophilic compounds comes from a study (*101*) of a series of phospholipids of increasing lipophilicity. Thus the quality of these spectra decreased in the increasingly lipophilic series phosphatidylcholine (**34**), phosphatidylethanolamine (**35**), phosphatidylglycerol (**36**). Also in varying the chain length of the fatty acid residues dibehenoylphosphatidylcholine (**37**), having the longest chain length, gave the weakest FAB spectrum. A change in the matrix from glycerol can be advantageous in improving FAB spectra. Thus gangliosides from human brain were successfully analyzed (*102*) using triethanolamine containing 1,1,3,3-tetramethylurea, but not using the conventional glycerol matrix.

IV.7. Eicosanoids

Prostaglandins and related compounds, particularly those containing hydroxyl groups, tend to be thermally labile and mass spectrometric procedures for their analysis have been developed accordingly. GC/MS methods have been widely employed (*103*) and a range of derivatives have been synthesized for the purpose. Positive ion EIMS is the commonest technique, often employing trimethylsilyl (TMS) and related derivatives, but it is notable that negative ion MS has recently found favour, particularly where sensitive detection is required. Electron capture properties have been conferred by derivatization with fluorinated groups. For example, ketone functions have been converted to pentafluorobenzyl (PFB) oximes and terminal carboxyl groups to PFB esters (*103*).

The involatility and thermal lability of eicosanoids render them appropriate candidates for FABMS, and this technique has been used to compare directly underivatized leukotriene C$_4$ (**38**) a slow-reacting substance of anaphylaxis, from biological and synthetic origin (*104*). In the negative ion FAB spectrum, prominent [M − H]$^-$ and [M − 2H + Na]$^-$ ions were observed at m/z 624 and 646 respectively and diagnostic fragment ions at m/z 306 and 319 as indicated.

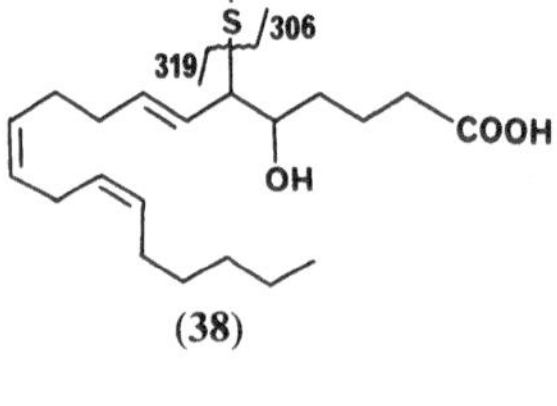

(**38**)

IV.8. Steroids

The correlation between the electron impact mass spectra of steroids and their structure has been a well-researched topic over the last twenty years. The success of mass spectrometry in this area owes much to the detailed analyses by DJERASSI and co-workers. For unconjugated steroids, EIMS has been able to answer many structural questions.

Modern ionization techniques have had their biggest impact in the steroid field in the analysis of steroid conjugates. Whereas glucuronides can be converted into derivatives suitable for analysis by EIMS and CIMS (*105*) there have been almost no reports of derivatives of sulphate conjugates which retain the sulphate moiety and these usually have to be identified by hydrolysis with specific enzymes coupled with analysis of the liberated steroid. The introduction of soft ionization methods has offered an alternative to hydrolysis (with possible further derivatization) of these conjugates prior to characterization by mass spectrometry. In particular, secondary ion mass spectrometry (SIMS), using a Cs^+ beam for ionization, has been found to give excellent results for steroid sulphates and glucuronides (*106*). FAB spectra, using a Xe beam, gave similar results. In both techniques, samples were analyzed in a liquid glycerol matrix. The most useful results were obtained from the negative ion spectra, which were dominated by $[M - H]^-$ ions, with little fragmentation. The positive ion spectra incorporated cationized (Na^+ or K^+) steroid conjugates.

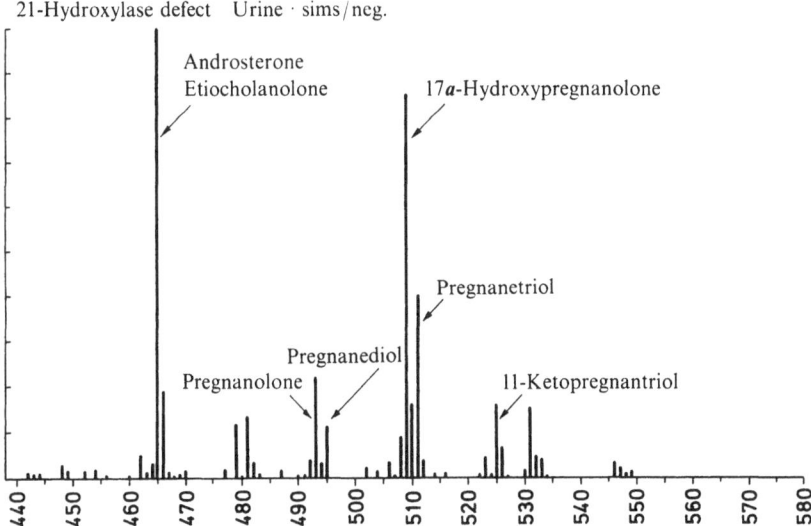

Fig. 6. Negative SIMS profile of steroid glucuronides from a 21-hydroxylase deficiency urine. [Reproduced with permission from ref. (*106*)]

These ionization techniques show considerable promise for the profiling of steroid conjugates (*106, 107*), without resort to hydrolysis or chromatographic separation of individual steroids. Clinically, the observation of abnormal profiles can lead to rapid diagnoses of metabolic disorders. One potential drawback compared with GC/MS analysis is the inability to separate isomeric steroids, but the main advantage is the speed and ease of the measurements. The total steroid profile of steroid glucuronides from a 21-hydroxylase deficiency urine is shown as an example in Fig. 6 (*106*). The profile is notable for the lack of cortisol (**39**) metabolites (m/z 539, 541) and the presence of prominent ions (m/z 509, 511) formed by the glucuronides of 17-hydroxypregnanolone and pregnanetriol. Another significant feature was the close resemblance of Fig. 6 to the corresponding GC profile, obtained after hydrolysis and derivatization. The sulphate conjugate dehydroepiandrosterone sulphate (**40**), is an abundant metabolite formed in the steroidogenic pathways leading from cholesterol to the androgens and the oestrogens in humans. It has been quantified in plasma by FABMS (*108*) using the 7,7-dideuterated analogue (**41**) as internal standard, and conversion into a pentafluorobenzyloxime derivative (**42**) to increase the molecular weight beyond the region where glycerol cluster ions interfere.

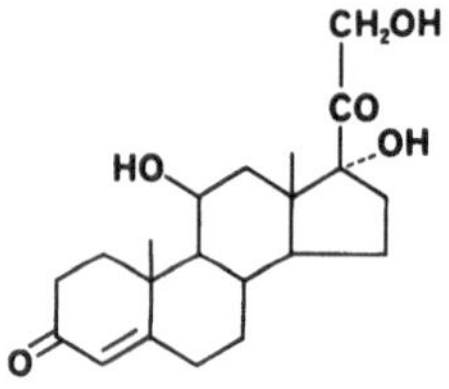

(**39**)

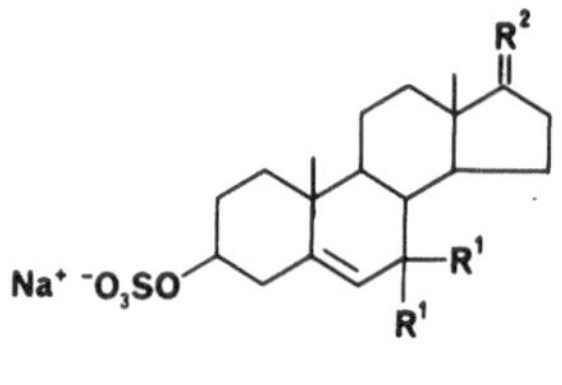

(**40**); $R^1 = H$, $R^2 = O$

(**41**); $R^1 = {}^2H$, $R^2 = O$

(**42**); $R^1 = {}^1H$ or ${}^2H$; $R^2 = NOCH_2C_6F_5$

The formation of trimethylsilyl molecular adduct ions under conditions suitable for desorption chemical ionization (DCI) appears to be a promising technique for the identification of certain non-volatile compounds, including polar steroids (*109*). The preferred gas phase reagent is tetramethylsilane with admixed nitrogen at a source pressure of $0.5-0.7$ torr. Under DCI conditions, the only apparent requirement for Me_3Si^+ molecular adduct formation is the presence of a functional group with an available electron pair. The number of additional functional groups, the molecular size and the degree of structural complexity are not highly important factors in formation of these $[M + 73]^+$ ions. If the Me_4Si reagent is diluted (1:1) with $d_{12} - Me_4Si$, another ion is formed at $[M + 82]^+$ from addition of $d_9 - Me_3Si^+$. This ion-pair formation 9 mass units apart enables molecular weights to be readily determined.

This technique has been successfully applied to various polar steroids, some of which do not yield satisfactory EI spectra. The most notable was the spectrum obtained from ecdysterone (43), an insect moulting hormone, which contains 6 hydroxyls and a keto group. The $SiMe_3^+$ adduct ion was the most abundant under DCI conditions. Note also that the positive ion FAB spectrum of (43) gave a strong $[M + H]^+$ ion at m/z 481, which eliminated successively 3 H_2O molecules. FABMS also helped to characterize (110) two conjugates related to (43) and isolated from the eggs of the desert locust *Schistocerca gregaria*. The negative ion spectra contained $[M - H]^-$ ions of ecdysone-22-phosphate (44) and 2-deoxyecdysone-22-phosphate (45).

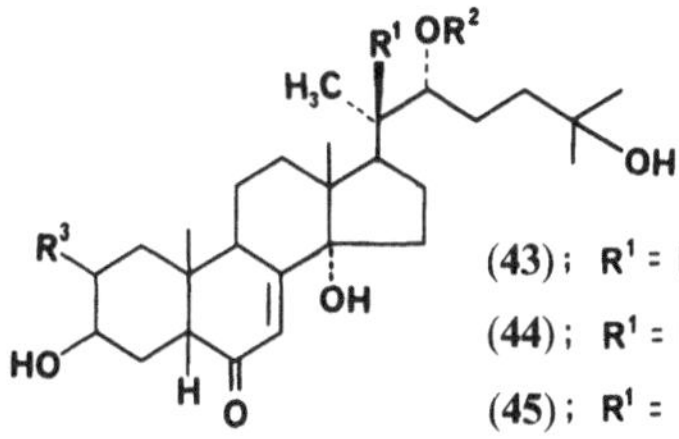

(43); $R^1 = R^3 = OH$, $R^2 = H$
(44); $R^1 = H$, $R^2 = PO_3H_2$, $R^3 = OH$
(45); $R^1 = R^3 = H$, $R^2 = PO_3H_2$

The scope of conventional EI mass spectrometry in distinguishing between epimeric steroids is limited. However, it has been shown that more stereochemical information is forthcoming from low energy unimolecular metastable reactions, analyzed in a reversed geometry instrument (40). This is because (i) the spectra are simpler than those formed by ion source reactions, being only the reaction products of one ion (*e.g.* the molecular ion) and (ii) the occurrence of unimolecular reactions of metastable ions is very sensitive to changes in the critical energies of such reactions. The study showed that pairs of hydrocarbon and ketone steroid isomers, differing only in the stereochemistry at the A/B or C/D ring junction, possessed substantially different unimolecular metastable spectra which could be rationalized in terms of differences in strain energy. Fig. 7 shows these spectra for the molecular ions of 5α- and 5β-androstan-3,11,17-trione (46).

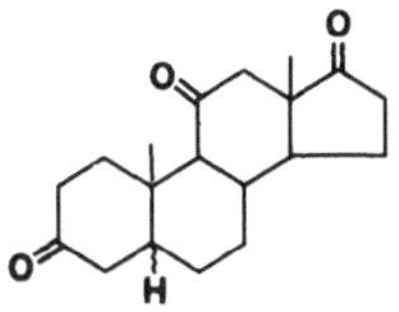

(46)

The enhanced loss of CH_3 from the 5β isomer is explained in terms of the lower critical energy for this reaction (*cf.* the 5α-isomer) due to the increased steric strain at the A/B ring junction. It has further been shown (*38*) that steroid epimers show differences in their collisionally activated dissociation spectra. In this study, steroid fragment ions were employed as targets for collision.

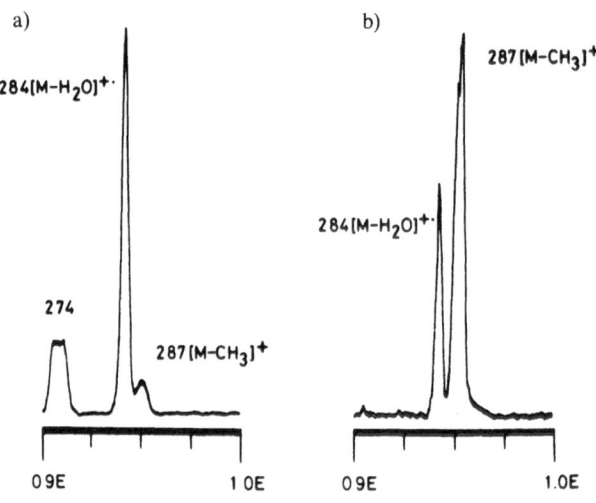

Fig. 7. The partial MIKE spectra of (a) 5α-androstan-3,11,17-trione and (b) 5β-androstan-3,11,17-trione showing the principal unimolecular reactions of M^+ (*m/z* 302).
[Reproduced with permission from ref. (*40*)]

IV.9. Antibiotics

The structures of natural products in this category are diverse and can include structural elements of classes of natural products already discussed, particularly peptides and carbohydrates, often of unusual type. The main role of mass spectrometry in structural elucidation of the more complex examples has been in verifying their molecular weights, so that structures proposed on the basis of other physical and chemical techniques can be validated or refuted. An example of the use of PDMS in this context was given (*64*) in Section IV.1. As a further example, the proposed tetracyclic structure (**47**) for the aglycone portion of ristocetin A, a member of the vancomycin group of glycopeptide antibiotics, corresponded to a molecular weight of 2066 daltons and was confirmed to within a few a.m.u. by observing the $[M + Na]^+$ ion (*m/z* 2086 ± 5 daltons) in its ^{252}Cf PD mass spectrum (*111*).

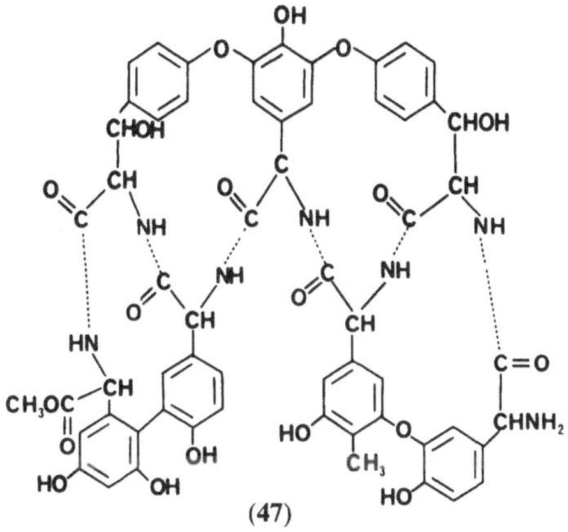

(47)

FABMS has proved useful for the analysis of the bleomycins, a family of glycopeptide-derived antibiotics which have found use in the treatment of cancer. Following the successful use (35) of FDMS to obtain the molecular weight of bleomycin B$_2$ (48), this and the more difficult bleomycin A$_2$ (49), containing a positively charged dimethylsulphonium cation instead of the guanidine residue were examined (22) both by FDMS and FABMS. The methods were complementary in that FAB spectra were easier to obtain, but FD spectra afforded fewer fragment ions and gave more reliable information on molecular weight.

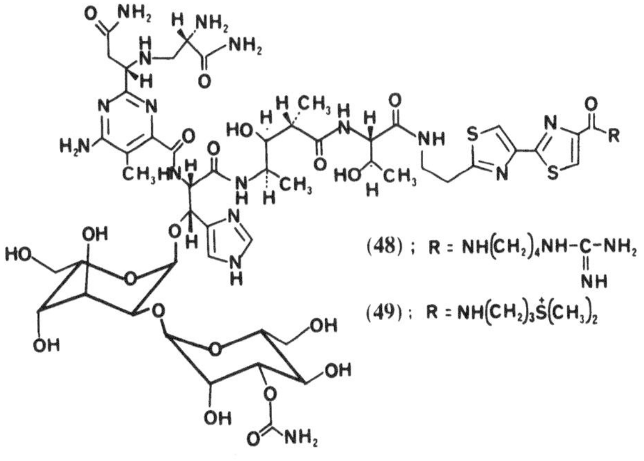

(48) ; R = NH(CH$_2$)$_4$NH-C-NH$_2$
                              ‖
                              NH

(49) ; R = NH(CH$_2$)$_3$Ṡ(CH$_3$)$_2$

A number of antibiotics were examined using both FDMS and SIMS (112). The techniques were about equally effective for the easier examples, but for the more polar molecules SIMS was more effective. Thus phosphoramidone (50) containing three types of polar moiety (amino acid, carbohydrate and ionized phosphate) gave no molecular ion in FDMS but a small [M + 2Na − H]$^+$ ion in SIMS. In using SIMS for the structural determination (*113*) of antrimycin A (**51**), an antibiotic isolated from a culture filtrate of *Streptomyces xanthocidicus,* an intense [M + H]$^+$ ion was observed for more than 10 minutes, whereas using FDMS it was seen for only *ca.* 20 seconds. Moreover only the SIMS spectrum gave sequence-defining fragment ions [see (**51**)].

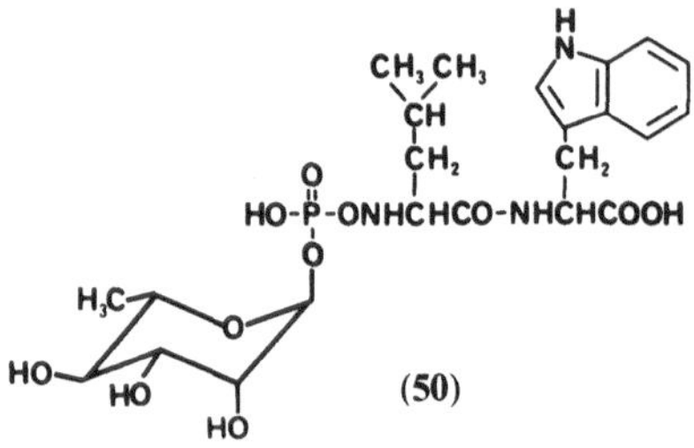

(50)

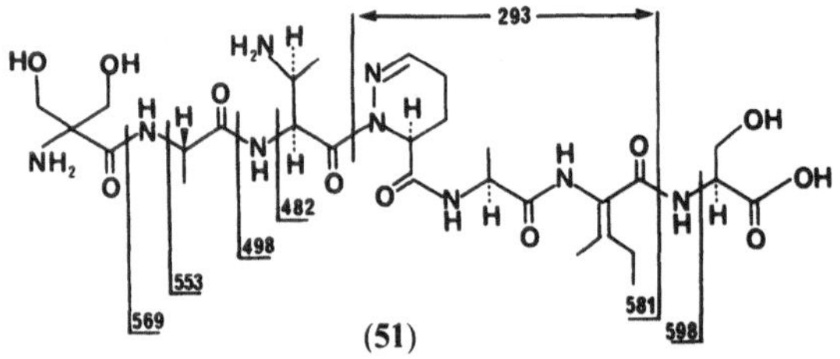

(51)

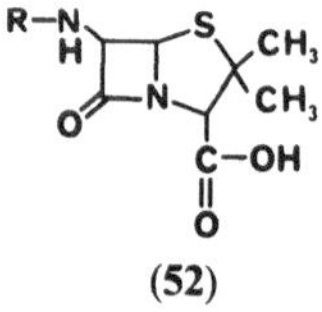

(52)

The penicillins have relatively simple structures (**52**) and the free acids have given [M + H]$^+$ ions using in-beam EIMS (*114*). Their alkali metal salts have been studied using FABMS and the positive ion spectra contained [2M + cation]$^+$ and the negative ion spectra, [2M − cation]$^-$ species as well as [M + X]$^+$ and [M − X]$^-$ molecular ions (*115*).

IV.10. Miscellaneous Natural Products

The mycotoxins occur as natural contaminants in foodstuffs and include the carcinogenic aflatoxins. Although most of these compounds give molecular ions on EI, extensive fragmentation occurs and a method was required which would detect a number of these substances in a mixture.

FDMS is highly suitable for this purpose since molecular ions are often the only components of the mass spectrum. For example (*116*) *ca.* 15 ppb aflatoxin B_1 (**53**) and 2.2 ppb aflatoxin B_2 (**54**) were clearly detectable in a crude extract of corn.

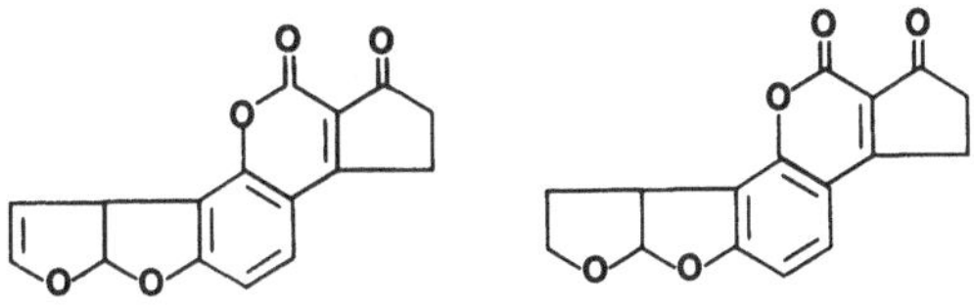

(53)   (54)

Choline (**55**) and acetylcholine (**56**) are physiologically important examples of naturally occurring quaternary ammonium salts, a class of compound intrinsically intractable to analysis other than by the desorption ionization methods. They have been quantified in rat brain tissue using FDMS (*117*). Using laser desorption (LD) combined with MS/MS the quaternary alkaloid candicine chloride (**57**) was identified in crude extracts from cacti and two new quaternary alkaloids, O-methylcandicine (**58**) and the N,N,N-trimethyl-β-methoxyphenylethylammonium cation (**59**) were discovered (*118*).

$ROCH_2CH_2\overset{+}{N}(CH_3)_3$

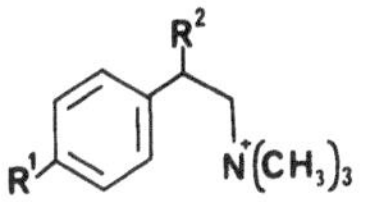

(55); R = H

(56); R = CH₃CO

(57); R¹ = OH, R² = H

(58); R¹ = OCH₃, R² = H

(59); R¹ = H, R² = OCH₃

The structure (**60**) was proposed on the basis of ^{1}H NMR spectroscopy for the complex flower pigment platyconin, an anthocyanin isolated from *Platycodon grandiflorum,* as a revision of an earlier structure, lacking the two glucose moieties, and derived from spectroscopic and chemical analysis. FABMS coupled with accurate mass determination confirmed (*119*) that the molecular weight and fragmentation pattern were consistent with this revised structure. FDMS gave (*120*) unambiguous information on the molecular weight of a polymeric polyhydroxyphenol (**61**) from brown algae. The peracetyl derivative showed sequential losses, starting from $[M + Na]^+$ of all 19 acetyl groups, confirming the number of hydroxy groups in (**61**).

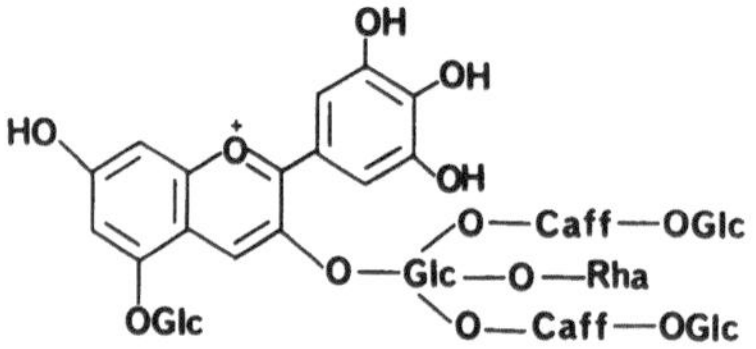

(60)

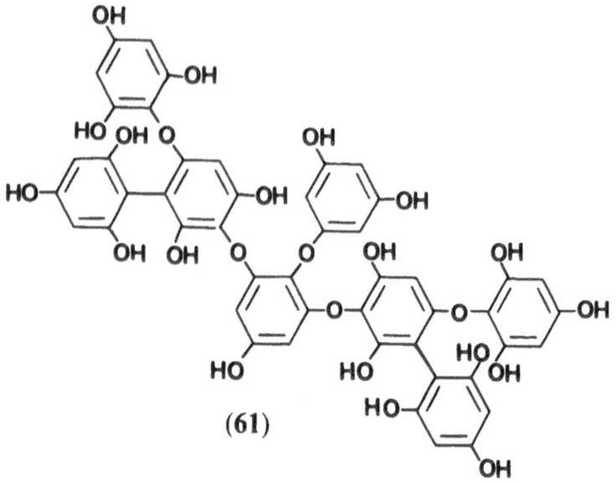

(61)

References

1. WALLER, G. R., and O. C. DERMER, eds.: Biochemical Applications of Mass Spectrometry, First Supplementary Volume. New York: Wiley-Interscience. 1980.
2. HOWE, I., D. H. WILLIAMS, and R. D. BOWEN: Mass Spectrometry: Principles and Applications. McGraw-Hill. 1981.
3. ROSE, M. E., and R. A. W. JOHNSTONE: Mass Spectrometry for Chemists and Biochemists. Cambridge University Press. 1982.
4. MORRIS, H. R., ed.: Soft Ionization Biological Mass Spectrometry. London: Heyden. 1981.
5. JOHNSTONE, R. A. W., ed.: Specialist Periodical Reports, Mass Spectrometry, vol. 7. The Royal Society of Chemistry. London: 1984.
6. BURLINGAME, A. L., J. O. WHITNEY, and D. H. RUSSELL: Mass Spectrometry. Analyt. Chemistry **56,** 417R (1984).
7. HUNT, D. F., and F. W. CROW: Electron-Capture Negative Ion Chemical Ionization Mass Spectrometry. Analyt. Chemistry **50,** 1781 (1978).
8. SCHILDCROUT, S. M.: Temperature-Dependent Single vs. Double Ionization in the Mass Spectra of Phthalocyanine and its Metal (II) Complexes. J. Amer. Chem. Soc. **105,** 3852 (1983).
9. KEOUGH, T., and A. J. DESTEFANO: Factors Affecting Reactivity in Ammonia Chemical Ionization Mass Spectrometry. Org. Mass Spectrom. **16,** 527 (1981).
10. POLLEY, C. W., and B. MUNSON: Nitrous Oxide as Reagent Gas for Positive Ion Chemical Ionization Mass Spectrometry. Analyt. Chemistry **55,** 755 (1983).
11. DOUGHERTY, R. C.: Negative Chemical Ionization Mass Spectrometry. Analyt. Chemistry **53,** 625A (1981).
12. BELOEIL, J. C., M. BERTRANNE, D. STAHL, and J. C. TABET: Stereochemistry of Gaseous Anions: Hydroxide Ion Negative Chemical Ionization of Androstanediols. J. Amer. Chem. Soc. **105,** 1355 (1983).
13. BRUINS, A. P.: Negative Ion Chemical Ionization Mass Spectrometry in the Determination of Components in Essential Oils. Analyt. Chemistry **51,** 967 (1979).
14. COTTER, R. J.: Mass Spectrometry of Nonvolatile Compounds. Desorption from Extended Probes. Analyt. Chemistry **52,** 1589A (1980).
15. BUSCH, K. L., S. E. UNGER, A. VINCZE, R. G. COOKS, and T. KEOUGH: Desorption Ionization Mass Spectrometry: Sample Preparation for Secondary Ion Mass

Spectrometry, Laser Desorption, and Field Desorption. J. Amer. Chem. Soc. **104**, 1507 (1982).

16. BECKEY, H. D.: Field Desorption Mass Spectrometry: A Technique for the Study of Thermally Unstable Substances of Low Volatility. Int. J. Mass Spectrom. Ion Phys. **2**, 500 (1969).

17. SCHULTEN, H.-R., U. BAHR, and P. B. MONKHOUSE: Biochemical Application of Field Desorption Mass Spectrometry. J. Biochem. Biophys. Methods **8**, 239 (1983).

18. RÖLLGEN, F. W., and H.-R. SCHULTEN: A New Method for Surface Ionization of Organic Molecules by Attachment of Alkali Ions in Moderate Electric Fields. Z. Naturforsch. **30a**, 1685 (1975).

19. LIU, L. K., K. L. BUSCH, and R. G. COOKS: Matrix-Assisted Secondary Ion Mass Spectra of Biological Compounds. Analyt. Chemistry **53**, 109 (1981).

20. RINEHART, K. L.: Fast Atom Bombardment Mass Spectrometry. Science **218**, 254 (1982).

21. BARBER, M., R. S. BORDOLI, R. D. SEDGWICK, and A. N. TYLER: Fast Atom Bombardment of Solids (F. A. B.): a New Ion Source for Mass Spectrometry. J. Chem. Soc. (London), Chem. Commun. **1981**, 325.

22. DELL, A., H. R. MORRIS, M. D. LEVIN, and S. M. HECHT: Field Desorption and Fast Atom Bombardment Mass Spectrometry of Bleomycins and their Derivatives. Biochem. Biophys. Res. Comm. **102**, 730 (1981).

23. TORGERSON, D. F., R. P. SKOWRONSKI, and R. D. MACFARLANE: New Approach to the Mass Spectrometry of Non-Volatile Compounds. Biochem. Biophys. Res. Comm. **60**, 616 (1974).

24. MCNEAL, C. J., and R. D. MACFARLANE: Observation of a Fully Protected Oligonucleotide Dimer at m/z 12 637 by ^{252}Cf-Plasma Desorption Mass Spectrometry. J. Amer. Chem. Soc. **103**, 1609 (1981).

25. POSTHUMUS, M. A., P. G. KISTEMAKER, H. L. C. MEUZELAR, and M. C. TEN NOEVER DE BRAUW: Laser Desorption-Mass Spectrometry of Polar Nonvolatile Bio-Organic Molecules. Analyt. Chemistry **50**, 985 (1978).

26. MCCRERY, D. A., E. B. LEDFORD, Jr., and M. L. GROSS: Laser Desorption Fourier Transform Mass Spectrometry. Analyt. Chemistry **54**, 1435 (1982).

27. HARDIN, E. D., T. P. FAN, C. R. BLAKLEY, and M. L. VESTAL: Laser Desorption Mass Spectrometry with Thermospray Sample Deposition for Determination of Nonvolatile Biomolecules. Analyt. Chemistry **56**, 2 (1984).

28. OTT, K. H., F. W. ROELLGEN, J. J. ZWINSELMAN, R. H. FOKKENS, and N. M. M. NIBBERING: Field Desorption Mass Spectrometry of Negative Ions (NFD-MS) of Salts. Angew. Chem. **93**, 96 (1981).

29. BLAKLEY, C. R., J. J. CARMODY, and M. L. VESTAL: A New Soft Ionization Technique for Mass Spectrometry of Complex Molecules. J. Amer. Chem. Soc. **102**, 5931 (1980).

30. HORNING, E. C., M. G. HORNING, D. I. CARROLL, I. DZIDIC, and R. N. STILLWELL: New Picogram Detection System Based on a Mass Spectrometer with an External Ionization Source at Atmospheric Pressure. Analyt. Chemistry **45**, 936 (1973).

31. KAMBARA, H.: Sample Introduction System for Atmospheric Pressure Ionization Mass Spectrometry of Nonvolatile Compounds. Analyt. Chemistry **54**, 143 (1982).

32. TSUCHIYA, M., and H. KUWABARA: Liquid Ionization Mass Spectrometry of Nonvolatile Organic Compounds. Analyt. Chemistry **56**, 14 (1984).

33. MCLAFFERTY, F. W., P. J. TODD, D. C. MCGILVERY, and M. A. BALDWIN: High-Resolution Tandem Mass Spectrometer (MS/MS) of Increased Sensitivity and Mass Range. J. Amer. Chem. Soc. **102**, 3360 (1980).

34. MORRIS, H. R.: Biomolecular Mass Spectrometry. Int. J. Mass Spectrom. Ion Physics **45**, 331 (1982).

35. MORRIS, H. R., A. DELL, and R. A. MCDOWELL: Extended Performance Using a High Field Magnet Mass Spectrometer. Biomed. Mass Spectrom. **8**, 463 (1981).

36. Buko, A. M., L. R. Phillips, and B. A. Fraser: Peptide Studies Using a Fast Atom Bombardment High Field Mass Spectrometer and Data System: 1-Sample Introduction, Data Acquisition and Mass Calibration. Biomed. Mass Spectrom. 10, 324 (1983).
37. McLafferty, F. W., ed.: Tandem Mass Spectrometry. New York: Wiley. 1983.
38. Cheng, M. T., M. P. Barbalas, R. F. Pegues, and F. W. McLafferty: Tandem Mass Spectrometry: Structural and Stereochemical Information from Steroids. J. Amer. Chem. Soc. 105, 1510 (1983).
39. Larka, E. A., I. Howe, J. H. Beynon, and Z. V. I. Zaretskii: Translational Energy Released in Decompositions of Metastable Ions: A Simple Criterion for Distinction Between Steroid Epimers by Mass Spectrometry. Org. Mass Spectrom. 16, 465 (1981).
40. — — — — The Determination of Ring Junction Stereochemistry in Steroids using Mass-Analyzed Ion Kinetic Energy Spectrometry. Tetrahedron 37, 2625 (1981).
41. Gaskell, S. J., A. W. Pike, and D. S. Millington: The Fragmentation of Stereoisomeric Androstanediol t-Butyldimethylsilyl Ethers. A study of Linked-Field Scanning. Biomed. Mass Spectrom. 6, 78 (1979).
42. Desiderio, D. M., and I. Katakuse: Fast Atom Bombardment-Collision Activated Dissociation-Linked Field Scanning Mass Spectrometry of the Neuropeptide Substance P. Analyt. Biochem. 129, 425 (1983).
43. Amster, I. J., M. A. Baldwin, M. T. Cheng, C. J. Proctor, and F. W. McLafferty: Tandem Mass Spectrometry of Higher Molecular Weight Compounds. J. Amer. Chem. Soc. 105, 1654 (1983).
44. Liberato, D. J., C. C. Fenselau, M. L. Vestal, and A. L. Yergey: Characterization of Glucuronides with a Liquid Chromatography/Mass Spectrometry Interface. Analyt. Chemistry 55, 1741 (1983).
45. Apffel, J. A., U. A. T. Brinkman, R. W. Frei, and E. A. I. M. Evers: Gas-Nebulized Direct Liquid Introduction Interface for Liquid Chromatography/Mass Spectrometry. Analyt. Chemistry 55, 2280 (1983).
46. Chang, T. T., J. O. Lay, and J. F. Rudolph: Direct Analysis of Thin-Layer Chromatography Spots by Fast Atom Bombardment Mass Spectrometry. Analyt. Chemistry 56, 109 (1984).
47. Hargrove, W. F., D. Rosenthal, and P. C. Cooley: Improvement of Algorithm for Peak Detection in Automatic Gas Chromatography-Mass Spectrometer Data Processing. Analyt. Chemistry 53, 538 (1981).
48. Hunt, D. F., A. B. Giordani, G. Rhodes, and D. A. Herold: Metabolic Profiling of Urinary Carboxylic Acids. Clin. Chem. (Winston-Salem, N. C.) 28, 2387 (1982).
49. Tondeur, Y., J. R. Hass, D. J. Harvan, and P. W. Albro: Computer-Assisted Determination of Masses in High-Resolution Mass Spectrometry with Selected Ion Monitoring. Analyt. Chemistry 56, 373 (1984).
50. Falkner, F. C.: Comments on Some Common Aspects of Quantitative Mass Spectrometry. Biomed. Mass Spectrom. 8, 43 (1981).
51. Beckner, C. F., and R. M. Caprioli: Quantitative Aspects of Fast Atom Bombardment Mass Spectrometry. Biomed. Mass Spectrom. 11, 60 (1984).
52. Desiderio, D. M., and M. Kai: Preparation of Stable Isotope-Incorporated Peptide Internal Standards for Field Desorption Mass Spectrometry Quantification of Peptides in Biologic Tissue. Biomed. Mass Spectrom. 10, 471 (1983).
53. Bradley, C. V., D. H. Williams, and M. R. Hanley: Peptide Sequencing Using the Combination of Edman Degradation, Carboxypeptidase Digestion and Fast Atom Bombardment Mass Spectrometry. Biochem. Biophys. Res. Comm. 104, 1223 (1982).
54. Williams, D. H., S. Santikarn, P. B. Oelrichs, F. de Angelis, J. K. Macleod, and R. J. Smith: The Structure of a Toxic Octapeptide, Containing 4 D-amino acids, from the Larvae of a Sawfly. Lophyrotoma Interrupta. J. Chem. Soc. (London), Chem. Commun. 1982, 1394.

55. WASYLYK, J. M., J. E. BISKUPIAK, C. E. COSTELLO, and C. M. IRELAND: Cyclic peptide structures from the Tunicate *Lissoclinum Patella* by FAB mass spectrometry. J. Organ. Chem. (USA) **48**, 4445 (1983).

56. MEYER, W. L.: On the Mass Spectrometric Structure Determination of the Cyclic Tetrapeptide Tentoxin. Tetrahedron Letters **24**, 2163 (1983).

57. BECKNER, C. F., and R. M. CAPRIOLI: Protein N-Terminal Analysis Using Fast Atom Bombardment Mass Spectrometry. Analyt. Biochemistry **130**, 328 (1983).

58. BARBER, M., R. S. BORDOLI, G. J. ELLIOTT, R. D. SEDGWICK, A. N. TYLER, and B. N. GREEN: Fast Atom Bombardment Mass Spectrometry of Bovine Insulin and Other Large Peptides. J. Chem. Soc. (London), Chem. Commun. **1982**, 936.

59. DELL, A., and H. R. MORRIS: Fast Atom Bombardment-High Field Magnet Mass Spectrometry of 6000 Dalton Polypeptides. Biochem. Biophys. Res. Comm. **106**, 1456 (1982).

60. BUKO, A. M., L. R. PHILLIPS, and B. A. FRASER: Peptide Studies Using a Fast Atom Bombardment-High Field Mass Spectrometer and Data System: 2-Characteristics of Positive Ionization Spectra of Peptides, m/z 858 to m/z 5729. Biomed. Mass Spectrom. **10**, 408 (1983).

61. MORRIS, H. R., M. PANICO, and G. W. TAYLOR: FAB-Mapping of Recombinant-DNA Protein Products. Biochem. Biophys. Res. Comm. **117**, 299 (1983).

62. PROME, J. C., J. ROUSSEL, B. CALDAS, J. MERY, J. PARELLO, and D. PATOURAUX: In: Recent Developments in Mass Spectrometry in Biochemistry and Medicine 6 (A. FRIGERIO and M. McCAMISH, eds.), p. 35. Amsterdam: Elsevier. 1980.

63. HAAKANSSON, P., I. KAMENSKY, B. SUNDQVIST, J. FOHLMAN, P. PETERSON, C. J. McNEAL, and R. D. MACFARLANE: Iodine-127-plasma Desorption Mass Spectrometry of Insulin. J. Amer. Chem. Soc. **104**, 2948 (1982).

64. CHAIT, B. T., B. F. GISIN, and F. H. FIELD: Fission Fragment Ionization Mass Spectrometry of Alamethicin I. J. Amer. Chem. Soc. **104**, 5157 (1982).

65. HERLIHY, W. C., N. J. ROYAL, K. BIEMANN, S. D. PUTNEY, and P. R. SCHIMMEL: Mass Spectra of Partial Protein Hydrolysates as a Multiple Phase Check for Long Polypeptides Deduced from DNA Sequences: NH_2-Terminal Segment of Alanine t-RNA Synthetase. Proc. Nat. Acad. Sci. (USA) **77**, 6531 (1980).

66. ROSE, K., M. G. SIMONA, and R. E. OFFORD: Amino Acid Determination by GLC-Mass Spectrometry of Permethylated Peptides. Optimisation of the Formation of Chemical Derivatives at the 2 – 10 nmol Level. Biochem. J. **215**, 261 (1983).

67. HERLIHY, W. C., D. KIDWELL, B. MEEUSEN, and K. BIEMANN: Mass Spectrometric Differentiation of Leucine and Isoleucine in Proteins Derived from Bacteria or Cell Culture. Biochem. Biophys. Res. Comm. **102**, 335 (1981).

68. ROSE, K., M. G. SIMONA, R. E. OFFORD, C. P. PRIOR, B. OTTO, and D. R. THATCHER: A New Mass Spectrometric C-terminal Sequencing Technique Finds a Similarity Between γ-Interferon and α_2-Interferon and Identifies a Proteolytically Clipped γ-Interferon That Retains Full Antiviral Activity. Biochem. J. **215**, 273 (1983).

69. BREIMER, M. E., G. C. HANSSON, K.-A. KARLSSON, G. LARSON, H. LEFFLER, W. PIMLOTT, B. E. SAMUELSSON, N. STRÖMBERG, S. TENEBERG, and J. THURIN: Sequencing of Large Oligosaccharides by Direct Inlet Mass Spectrometry. Application to Cell Surface Glycolipids. Int. J. Mass Spectrom. Ion Physics. **48**, 113 (1983).

70. LINSCHEID, M., J. D'ANGONA, A. L. BURLINGAME, A. DELL, and C. E. BALLOU: Field Desorption Mass Spectrometry of Oligosaccharides. Proc. Nat. Acad. Sci. (USA) **78**, 1471 (1981).

71. KOMORI, T., M. KAWAMURA, K. MIYAHARA, T. KAWASAKI, O. TANAKA, S. YAHARA, and H.-R. SCHULTEN: Field Desorption Mass Spectrometry of Physiologically Active Steroid- and Dammarane-Saponins. Z. Naturforsch. **34c**, 1094 (1979).

72. KOMORI, T., I. MAETANI, N. OKAMURA, T. KAWASAKI, T. NOHARA, and H.-R. SCHULTEN:

Zur Analogie der Zuckerabspaltung aus oligoglykosidischen Naturstoffen bei der Säurehydrolyse und der Felddesorptions-Massenspektrometrie. Liebigs Ann. Chem. **1981**, 683.

73. Schulten, H.-R., T. Komori, T. Kawasaki, T. Okuyama, and S. Shibata: Confirmation of New, High-mass Saponins from *Gleditsia Japonica* by Field Desorption Mass Spectrometry. Planta Medica **46**, 67 (1982).

74. Forsberg, L. S., A. Dell, D. J. Walton, and C. E. Ballou: Revised Structure for the 6-O-Methylglucose Polysaccharide of *Mycobacterium Smegmatis*. J. Biol. Chem. **257**, 3555 (1982).

75. Dell, A., and C. E. Ballou: Fast Atom Bombardment Mass Spectrometry of a 6-O-Methylglucose Polysaccharide. Biomed. Mass Spectrom. **10**, 50 (1983).

76. Shimamura, M., T. Endo, Y. Inoue, and S. Inoue: A Novel Neutral Oligosaccharide Chain Found in Polysialoglycoproteins Isolated from Pacific Salmon Eggs. Structural Studies by Secondary Ion Mass Spectrometry, Proton Nuclear Magnetic Resonance Spectroscopy, and Chemical Methods. Biochemistry **22**, 959 (1983).

77. Harada, K.-I., M. Suzuki, and H. Kambara: Structural Characterization of Viridopentaoses and Their Related Saccharides by Matrix-Assisted Molecular SIMS. Tetrahedron Letters **23**, 2481 (1982).

78. Dolhun, J. J., and J. L. Wiebers: Mass Spectrometry of Trimethylsilyl Derivatives of Nucleoside and Dinucleotide Phenylboronates. Application to Oligonucleotide Sequence Analysis. J. Amer. Chem. Soc. **91**, 7755 (1969).

79. Schulten, H.-R., and H. M. Schiebel: Sequence Specific Fragments in the Field Desorption Mass Spectra of Dinucleoside Phosphates. Nucleic Acids Res. **3**, 2027 (1976).

80. Sindona, G., N. Uccella, and K. Weclawek: Structure Determination of Isomeric Oligodeoxynucleotide Salts by Fast-Atom Bombardment Mass Spectrometry. J. Chem. Res. (S) **1982**, 184.

81. Panico, M., G. Sindona, and N. Uccella: Bioorganic Applications of Mass Spectrometry. 3. Fast-Atom-Bombardment-Induced Zwitterionic Oligonucleotide Quasimolecular Ions Sequenced by MS/MS. J. Amer. Chem. Soc. **105**, 5607 (1983).

82. Grotjahn, L., R. Frank, and H. Blöcker: Ultrafast Sequencing of Oligo-deoxyribonucleotides by FAB-Mass Spectrometry. Nucleic Acids Res. **10**, 4671 (1982).

83. Summons, R. E., L. M. Palni, and D. S. Letham: Determination of Intact Zeatin Nucleotide by Direct Chemical Ionization Mass Spectrometry. FEBS Lett. **151**, 122 (1983).

84. Unger, S. E., A. E. Schoen, R. G. Cooks, D. J. Ashworth, J. D. Gomes, and C.-J. Chang: Identification of Modified Nucleosides by Secondary-Ion Mass Spectrometry. J. Organ. Chem. (USA) **46**, 4765 (1981).

85. Schoen, A. E., R. G. Cooks, and J. L. Wiebers: Modified Bases Characterized in Intact DNA by Mass-Analyzed Ion Kinetic Energy Spectrometry. Science **203**, 1249 (1979).

86. Schulten, H.-R., and H. M. Schiebel: Principle and Technique of Field-Desorption Mass Spectrometry. Analysis of Corrins and Vitamin B_{12}. Naturwiss. **65**, 223 (1978).

87. Schiebel, H. M., and H.-R. Schulten: Depletion of ^{13}Carbon in the Biosynthesis of Vitamin B_{12}. Naturwiss. **67**, 256 (1980).

88. Barber, M., R. S. Bordoli, D. Sedgwick, and A. N. Tyler: Fast Atom Bombardment Mass Spectrometry of Cobalamines. Biomed. Mass Spectrom. **8**, 492 (1981).

89. Grotjahn, L., V. B. Koppenhagen, and L. Ernst: Fast Atom Bombardment Mass Spectrometry of the Vitamin B_{12} Analogues Hydrogenocobalamin and Cupribalamin. Z. Naturforsch. **39 b**, 248 (1984).

90. Dougherty, R. C., P. A. Dreifuss, J. Sphon, and J. J. Katz: Hydration Behaviour of Chlorophyll *a*: A Field Desorption Mass Spectral Study. J. Amer. Chem. Soc. **102**, 417 (1980).

91. HUNT, J. E., R. D. MACFARLANE, J. J. KATZ, and R. C. DOUGHERTY: Self Assembled Chlorophyll *a* Systems as Studied by Californium-252 Plasma Desorption Mass Spectroscopy. Proc. Nat. Acad. Sci. (USA) **77**, 1745 (1980).

92. CONSTANTIN, E., Y. NAKATANI, G. TELLER, R. HUEBER, and G. OURISSON: Electron-Impact and Chemical Ionization Mass Spectrometry of Chlorophylls, Phaeophytins and Phaeophorbides by Fast Desorption on a Gold Support. Bull. soc. chim. France II **1981**, 303.

93. TOMER, K. B., F. W. CROW, and M. L. GROSS: Location of Double Bond Position in Unsaturated Fatty Acids by Negative Ion MS/MS. J. Amer. Chem. Soc. **105**, 5487 (1983).

94. CERVILLA, M., and G. PUZO: Determination of Double Bond Position in Monosaturated Fatty Acids by Mass Analyzed Ion Kinetic Energy Spectrometry/Collision Induced Dissociation After Chemical Ionization of Their Amino Alcohol Derivatives. Analyt. Chemistry **55**, 2100 (1983).

95. LEONHARDT, B. A., E. F. DEVILBISS, and J. A. KLUN: Gas Chromatographic Mass Spectrometric Indication of Double Bond Position in Monosaturated Primary Acetates and Alcohols without Derivatization. Org. Mass Spectrom. **18**, 9 (1983).

96. DALLINGA, J. W., N. M. M. NIBBERING, J. VAN DER GREEF, and M. C. TEN NOEVER DE BRAUW: A Fast Atom Bombardment and Field Ionization/Field Desorption Study of Some Isomeric Unsaturated Dicarboxylic Acids. Org. Mass Spectrom. **19**, 10 (1984).

97. HO, B. C., C. FENSELAU, G. HANSEN, J. LARSEN, and A. DANIEL: Dipalmitoyl-phosphatidylcholine in Amniotic Fluid Quantified by Fast-Atom-Bombardment Mass Spectrometry. Clin. Chem. (Winston-Salem, N. C.), **29**, 1349 (1983).

98. TOMER, K. B., F. W. CROW, H. W. KNOCHE, and M. L. GROSS: Fast Atom Bombardment and Mass Spectrometry/Mass Spectrometry for Analysis of Ornithine-Containing Lipids from *Thiobacillus Thiooxidans*. Analyt. Chemistry **55**, 1033 (1983).

99. DAFFE, M., M. A. LANEELLE, and G. PUZO: Structural Elucidation by Field Desorption and Electron-Impact Mass Spectrometry of the C-Mycosides Isolated from *Mycobacterium Smegmatis*. Biochim. Biophys. Acta **751**, 439 (1983).

100. LEHMANN, W. D., and M. KESSLER: Characterization and Quantification of Human Plasma Lipids from Crude Lipid Extracts by Field Desorption Mass Spectrometry. Biomed. Mass Spectrom. **10**, 220 (1983).

101. FENWICK, G. R., J. EAGLES, and R. SELF: Fast Atom Bombardment Mass Spectrometry of Intact Phospholipids and Related Compounds. Biomed. Mass Spectrom. **10**, 382 (1983).

102. ARITA, M., M. IWAMORI, T. HIGUCHI, and Y. NAGAI: 1,1,3,3-Tetramethylurea and Triethanolamine as a New Useful Matrix for Fast Atom Bombardment Mass Spectrometry of Gangliosides and Neutral Glycosphingolipids. J. Biochemistry (Tokyo) **93**, 319 (1983).

103. BLAIR, I. A.: Measurement of Eicosanoids by Gas Chromatography and Mass Spectrometry. Brit. Med. Bulletin **39**, 223 (1983).

104. MURPHY, R. C., W. R. MATHEWS, J. ROKACH, and C. FENSELAU: Comparison of Biological-Derived and Synthetic Leukotriene C_4 by Fast Atom Bombardment Mass Spectrometry. Prostaglandins **23**, 201 (1982).

105. FENSELAU, C., and L. P. JOHNSON: Analysis of Intact Glucuronides by Mass Spectrometry and Gas Chromatography-Mass Spectrometry. Drug Metab. Dispos. **8**, 274 (1980).

106. SHACKLETON, C. H. L., and K. M. STRAUB: Direct Analysis of Steroid Conjugates: the Use of Secondary Ion Mass Spectrometry. Steroids **40**, 35 (1982).

107. SHACKLETON, C. H. L.: Inborn Errors of Steroid Biosynthesis: Detection by a New Mass Spectrometric Method. Clin. Chem. (Winston-Salem, N. C.) **29**, 246 (1983).

108. Gaskell, S. J., B. G. Brownsey, P. W. Brooks, and B. N. Green: Fast Atom Bombardment Mass Spectrometry of Steroid Sulphates: Qualitative and Quantitative Analyses. Biomed. Mass Spectrometry 10, 215 (1983).
109. Stillwell, R. N., D. I. Carroll, J. G. Nowlin, and E. C. Horning: Formation of Trimethylsilyl Molecular Adduct Ions in Desorption Chemical Ionization Mass Spectrometry of Non-volatile Organic Compounds. Analyt. Chemistry 55, 1313 (1983).
110. Isaac, R. E., M. E. Rose, H. H. Rees, and T. W. Goodwin: Identification of Ecdysone-22-phosphate and 2-Deoxyecdysone-22-phosphate in Eggs of the Desert Locust, Schistocerca Gregaria, by Fast Atom Bombardment Mass Spectrometry and N.M.R. Spectroscopy. J. Chem. Soc. (London), Chem. Commun. 1982, 249.
111. Williams, D. H., V. Rajananda, and J. R. Kalman: On the Structure and Mode of Action of the Antibiotic Ristocetin A. J. Chem. Soc. (London), Perkin Trans. 1, 1979, 787.
112. Kambara, H., S. Hishida, and H. Naganawa: Comparative Study of Field Desorption and Secondary Ion Mass Spectra for Antibiotics. J. Antibiot. 35, 67 (1982).
113. Morimoto, K., N. Shimada, H. Naganawa, T. Takita, H. Umezawa, and H. Kambara: Minor Congeners of Antrimycin: Application of Secondary Ion Mass Spectrometry (SIMS) to Structure Determination. J. Antibiot. 35, 378 (1982).
114. Ohashi, M., R. P. Barron, and W. R. Benson: In-Beam Electron Ionization Mass Spectra of Penicillins. J. Pharm. Sci. 72, 508 (1983).
115. Barber, M., R. S. Bordoli, R. D. Sedgwick, A. N. Tyler, B. N. Green, V. C. Parr, and J. L. Gower: Fast Atom Bombardment Mass Spectrometry of Some Penicillins. Biomed. Mass Spectrom. 9, 11 (1982).
116. Sphon, J. A., P. A. Dreifuss, and H.-R. Schulten: Field Desorption Mass Spectrometry of Mycotoxins and Mycotoxin Mixtures, and its Application as a Screening Technique for Foodstuffs. J. Assoc. Off. Anal. Chem. 60, 73 (1977).
117. Lehmann, W. D., H.-R. Schulten, and N. Schröder: Determination of Choline and Acetyl Choline in Distinct Rat Brain Regions by Stable Isotope Dilution and Field Desorption Mass Spectrometry. Biomed. Mass Spectrom. 5, 591 (1978).
118. Davis, D. V., R. G. Cooks, B. N. Meyer, and J. L. McLaughlin: Identification of Naturally Occurring Quaternary Compounds by Combined Laser Desorption and Tandem Mass Spectrometry. Analyt. Chemistry 55, 1302 (1983).
119. Saito, N., C. F. Timberlake, O. G. Tucknott, and I. A. S. Lewis: Fast Atom Bombardment Mass Spectrometry of the Anthocyanins Violanin and Platyconin. Phytochemistry 22, 1007 (1983).
120. Grosse-Damhues, J., K.-W. Glombitza, and H.-R. Schulten: An Eight-Ring Phlorotannin from the Brown Alga Himanthalia Elongata. Phytochemistry 22, 2043 (1983).

(Received June 6, 1984)

Chemical Synthesis of the Trichothecenes

By P. G. McDougal, Department of Chemistry, Georgia Institute of Technology, Atlanta, Georgia, U.S.A., and N. R. Schmuff*, Laboratory of Chemistry, National Heart, Lung, and Blood Institute, National Institutes of Health, Bethesda, Maryland, U.S.A.

With 1 Figure

Contents

* Current Address: Questel, Inc., 1625 Eye Street, N.W., Washington, DC 20006, U.S.A.

I. General Introduction

That research in the trichothecene field is an area of intense interest cannot be disputed. In the years 1982 and 1983, there were more than 500 publications dealing with this subject, including more than 30 reviews. In addition, a comprehensive monograph has recently appeared (*151*). Virtually every aspect of the problem has been reviewed, including isolation and structure determination (*8, 9, 70, 75, 81, 136*), biological and biochemical studies (*11, 39, 136, 155*), toxicology (*110, 154, 156*), quantitative methods for determination (*7, 108, 126, 153*), and biosynthesis (*35, 137*). Efforts directed at the chemical synthesis of trichothecenes have received less attention (*47, 112, 138, 141*), primarily due to the relative scarcity of published material on the topic prior to 1980. However, since that time, more than ten total syntheses of naturally occuring trichothecenes have been reported, including some of the structurally more complex members of the family. The data base Federal Research in Progress reveals that for fiscal year 1982, there were nine research groups with U.S. government funding pursuing the chemical synthesis of trichothecenes. In light of this proliferation of synthetic work, the present review will concentrate on that area. One deviation from this plan will be an exhaustive compilation of known naturally occurring compounds with trichothecenoid structures. This listing will be an update of an earlier tabulation by Tamm (*137*) and will generally follow Tamm's previous format.

II. Origin and Biological Activity

The trichothecenes are a group of closely related sesquiterpenoids produced by various species of imperfect fungi. The first compounds of this class were discovered at Imperial Chemical Industries in 1946 during an extensive search for new antibiotics. Historically, glutinosin (**25**) was the first trichothecene isolated, but this was later shown to be a mixture of verucarrins A (**37**) and B (**39**) (see Table IV) (*63, 65*). Freeman and Morrison (*50*) are credited with the first isolation and purification of a trichothecene, trichothecin (**26**) from the fungus *Trichothecium roseum* in 1949. Since that date, over 80 trichothecenes have been identified from nine genera of fungi: *Fusarium, Myrothecium, Trichothecium, Trichoderma, Cephalosporium, Cyclindrocarpen, Stachybotrys, Verticimonosporium* and *Calonectria*.

A wide array of biological activities has been ascribed to this family of mycotoxins (*11, 39*). These include antifungal, antibacterial and antiviral activity as well as insecticidal and phytotoxic behavior. An extensively studied and potentially important property of the trichothecenes is their cytostatic activity (*39*). In 1962, Härri and co-workers found that verrucarin A (**37**) caused 50% inhibition of mouse tumor cell *P-815* growth at a concentration of 0.6 ng/ml, making it one of the most active cytostatic agents known (*65*). This activity is not limited to the macrocycles, as anguidine (**9**) has exhibited cytopathogenic effects against baby hamster kidney cells at a concentration of 1.5 ng/ml (*88*). In fact, the majority of the trichothecenes have been shown to possess *in vitro* cytotoxic activity. Anguidine has recently completed Phase II in a clinical trial against cancer of the colon and breast conducted by the National Cancer Institute (*4*). Yet, despite this multiplicity of activity, no trichothecene or related synthetic analog has so far reached the market place.

Consonant with this cytostatic behavior, studies have shown that both protein and DNA synthesis are inhibited by the trichothecenes (*159*). An extensive investigation into their mechanism of action has divided the trichothecenes into three classes based on their mode of protein synthesis inhibition; elongation inhibitors (E-types), initiation inhibitors (I-types) and termination inhibitors (T-types) (*36*). Although the molecular basis of their action is still unclear, Ueno (*155*) has presented evidence that the trichothecenes react with thiol residues in the enzyme peptidyl transferase. The 12,13-epoxide in the trichothecanoid framework (see Fig. 1 for numbering of the trichothecene skeleton) is thought to be the electrophilic site responsible for this reactivity, as removal of this epoxide nullifies the cytostatic behavior (*39*). Interestingly, the few trichothecenes tested have failed to demonstrate significant mutagenicity (*155*).

As with most cytostatic agents, the trichothecenes are for the most part acutely toxic in whole animal studies, with verrucarin A (**37**) recording one of the lowest LD-50 values (0.5 mg/kg, ip, mouse) of any non-nitrogen containing natural product (*154*). Other toxic manifestations, short of death, include gastrointestinal disturbance, hypotension, anemia and lymphoid necrosis (*29, 155*). These mycotoxins are also severe skin irritants, causing inflammation and scabbing. The implication of trichothecenes as the causative agent in a number of animal and human toxicoses, has provided further impetus for the study of this intriguing class of compounds (*31*).

III. Structure

The initial work on the structure determination of the trichothecenes was reported from the laboratories of Freeman (*49*), Tamm (*65*), and Fishman (*45*). Based on extensive chemical studies these groups arrived at a similar but incorrect carbon skeleton, now known as the apotrichothecene

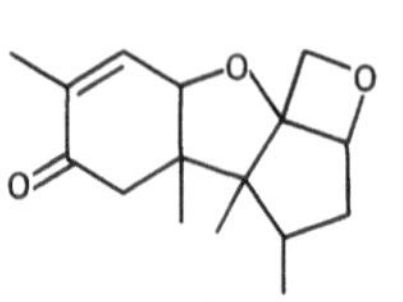

Proposed Structure
of Trichothecolone

80

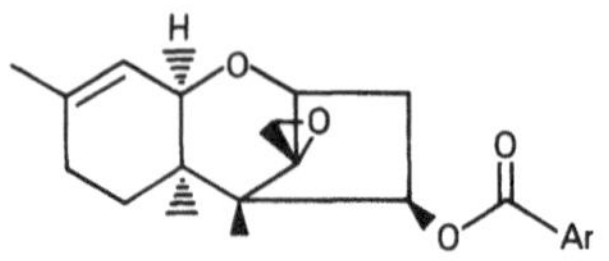

Ar=p-BrC₆H₅

81

skeleton, exemplified by FREEMAN's proposed structure (**80**) for tri-chothecolone. However, in 1964, single-crystal X-ray analysis of the p-bromobenzoate derivative of trichodermol (**81**) showed the structure to be the intriguing tetracyclic epoxide characteristic of all the trichothecenes (*1*). Other workers in the field quickly revised their previous structural assignments and all then known trichothecenes were correlated with trichodermol (**2**) (*38, 128*).

It was soon realized that a growing number of fungal metabolites possessed the same framework as trichodermol (**2**). In 1967, GODTFREDSEN, GROVE and TAMM (*54*) proposed the name trichothecane for this new sesquiterpene family. At the same time, they introduced the nomenclature apotrichothecane to describe the skeleton obtained from the rearrangement of the trichothecanes. As most members of the trichothecane family contain an unsaturation between C-9 and C-10 these compounds are most often referred to as the trichothecenes.

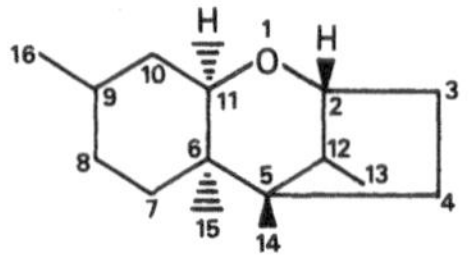

Trichothecane Skeleton

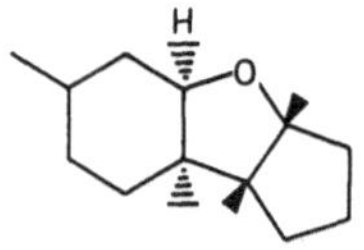

Apotrichothecane Skeleton

Fig. 1

To date, over 80 naturally occurring trichothecenes have been identified and can be classified into three distinct structural groups: Simple tri-chothecenes, macrocyclic trichothecenes and the recently discovered tri-choverroids. The simple trichothecenes, with 38 members, contain the basic mono- or polyhydroxylated sesquiterpene skeleton, with none, one, or more of the hydroxy groups esterified by acetic, crotonic, isovaleric, lactic or β-hydroxyisovaleric acid. This grouping can be further subdivided based on the oxidation level of C-8. Table I contains those simple sesquiterpenes in which C-8 is fully reduced, Table II is comprised of those members which possess an 8α-hydroxy group, while Table III lists all the compounds in which C-8 is at the oxidation level of a ketone. This subdivision has its foundation in the biogenesis of the trichothecenes, the C-8 saturated compounds serving as biosynthetic precursors for the 8-oxo-trichothecenes *via* the 8-hydroxylated species (*137*). Although, historically, chemical degradation and correlation played a major role in the structure determination, today the site and stereochemistry of hydroxylation is more easily deduced from the ^{1}H NMR spectrum. Due to the rigid nature of the tetracycle, a number of diagnostic proton-proton couplings exist (*8*).

Table I

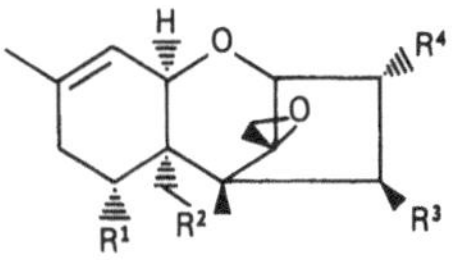

	Name	R¹	R²	R³	R⁴	Ref.
1	12, 13-Epoxytrichothec-9-ene	H	H	H	H	97
2	Trichodermol (Roridin C)	H	H	OH	H	1,2,65
3	Trichodermin	H	H	OAc	H	55,56
4	15-Deacetylcalonectrin	H	OH	H	OAc	52
5	Calonectrin	H	OAc	H	OAc	43,52
6	Di-O-acetylverrucarol	H	OAc	OAc	H	111
7	Scirpentriol	H	OH	OH	OH	113,158
8	15-Acetoxyscirpenol	H	OAc	OH	OH	37,113
9	Anguidine (Diacetoxyscirpenol)	H	OAc	OAc	OH	38,46,128
10	7-Hydroxyanguidine	OH	OAc	OAc	OH	71,160
11	4-Acetoxyscirpendiol	H	OH	OAc	OH	69,132
12	Triacetoxyscirpenol	H	OAc	OAc	OAc	132
13	7-Hydroxyscirpentriol	OH	OH	OH	OH	14

In the second structural group, the macrocyclic trichothecenes, the hydroxy groups at C-4 and C-15 of the simple trichothecene skeleton are bridged by a di- or trilactide ribbon. The verrucarins (triesters, Table IV), the roridins (diesters, Table V) and the baccharins (diesters, Table VI) are the major structural subunits of this group. Vertisporin and the satratoxins complete this class, and can be found at the bottom of Table V. Verrucarol (**82**) is most often the sesquiterpene onto which the macrocycle is attached. Verrucarin K (**43**) is a notable exception and was the first trichothecene known which lacks the 12,13-epoxide unit. A number of macrocyclic

Table II

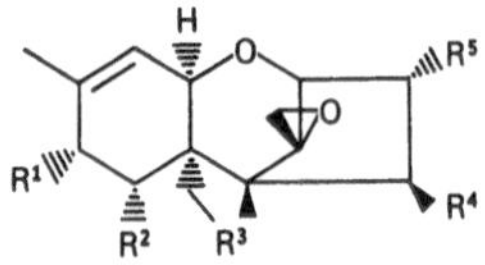

	Name	R¹	R²	R³	R⁴	R⁵	Ref.
14	4β, 8a-Dihydroxy-12,13-epoxytrichothec-9-ene	OH	H	H	OH	H	97
15	T-2-Tetraol	OH	H	OH	OH	OH	105
16	HT-2-Toxin	OX^a	H	OAc	OH	OH	10,163
17	T-2-Toxin (fusariotoxin T2)	OX^a	H	OAc	OAc	OH	10,169
18	Acetyl-T-2-toxin	OX^a	H	OAc	OAc	OAc	90
19	Solaniol (neosolaniol)	OH	H	OAc	OAc	OH	72,157
20	NT-1-Toxin	OAc	H	OH	OAc	OH	68
21	7,8-Dihydroxyanguidine	OH	OH	OAc	OAc	OH	71,160
22	8-Acetoxy-7-hydroxyanguidine	OAc	OH	OAc	OAc	OH	61
23	Neosolaniol monoacetate	OAc	H	OAc	OAc	OH	73,96
24	NT-2-Toxin	OH	H	OH	OAc	OH	73
		OH	OAc	OAc	OH	OAc	127
		OY^b	H	OAc	OH	OH	32
		OH	H	OAc	OH	OH	32
		OY^b	H	OH	OH	OH	32

a. $X = -COCH_2CH(CH_3)_2$
b. $Y = -COCH_2C(OH)(CH_3)_2$

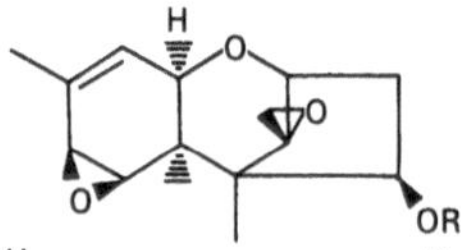

R = -H   Crotocol   Ref. 127
R = $-COCH = CHCH_3$(Z-isomer) Crotocin

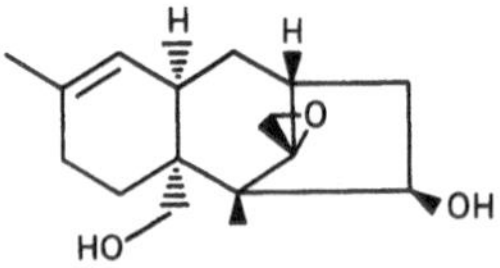

Verrucarol

82

Table III

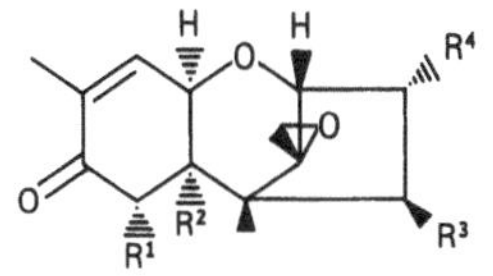

	Name	R^1	R^2	R^3	R^4	Ref
25	Trichothecolone	H	H	OH	H	3
26	Trichothecin	H	H	OCr^a	H	16,49,55
27	Vomitoxin (deoxynivalenol)	OH	OH	H	OH	161,162,171
28	Deoxynivalenol monoacetate	OH	OH	H	OAc	17
29	Nivalenol	OH	OH	OH	OH	139,62
30	Fusarenone	OH	OH	OAc	OH	139,152,62
31	Nivalenol diacetate	OH	OAc	OAc	OH	37,62
32	4-O-Acetyltrichothecolone	H	H	OAc	H	53
33	CBD$_2$	H	H	$OLac^b$	H	167
34	3,15-Dihydroxy-12,13-epoxytrichothec-9-en-8-one	H	OH	H	OH	15
35	3,15-Diacetyldeoxynivalenol	OH	OAc	H	OAc	170
36	4-O-Cinnamoyltrichothecolone	H	H	OX^c	H	53

a. Cr = $-COCH=CHCH_3$ (Z-isomer)
b. Lac = $-COCH(OH)CH_3$ (R-isomer)
c. X = $-COCH=CHPh$ (E-isomer)

trichothecenes have a more highly oxygenated sesquiterpenoid backbone, most notably the baccharins. These compounds, originally isolated from the Brazilian shrub *Baccharis megapotamica* (*93, 94*), contain a β-orientated 8-hydroxy group in their sesquiterpene portion, in contrast to the α-orientation of the C-8 hydroxy groups found in both the simple and other macrocyclic trichothecenes. Additional stereochemical anomalies, including the unusual *R* configuration at C-2' in baccharinols B3 (**64**) and B7 (**65**), led JARVIS and co-workers (*81*) to reexamine the source of these supposed plant metabolites. Their studies have shown that the baccharins are in fact fungal metabolites absorbed from the soil and subsequently biotransformed by the plant, a phenomenon which accounts for the unusual oxidation pattern. Interestingly, the baccharins possess greater antileukemic activity than either the verrucarins or the roridins. (*79*).

Due to the lack of diagnostic proton couplings in the NMR spectra of the macrocyclic trichothecenes, assignment of stereochemistry in the macrocyclic chain has been troublesome in this class of compounds. As a result, X-ray crystallography has played a major role in structure de-

Table IV. *Verrucarins*

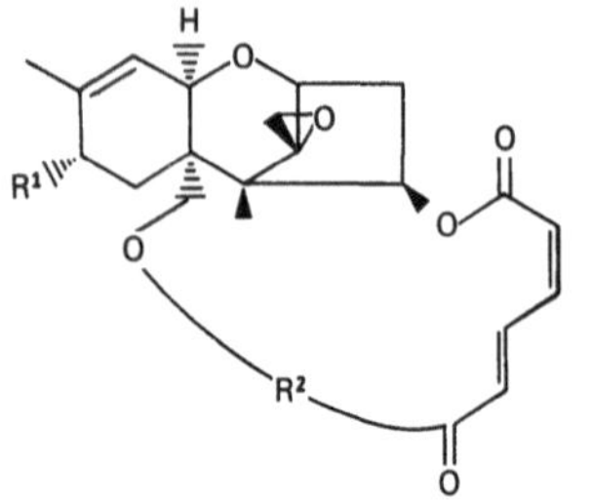

	R¹	R²	Reference
(37) Verrucarin A	H	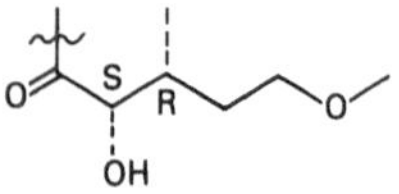	59,103
(38) 2'-Dehydro Verrucarin A	H	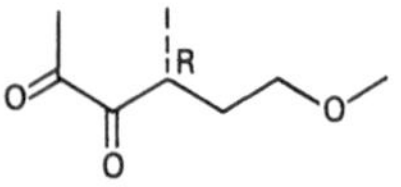	172
(39) Verrucarin B	H	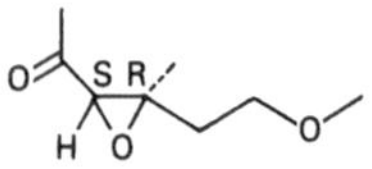	23,60
(40) Verrucarin J	H	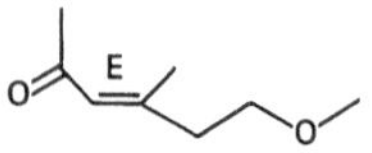	44
(41) Verrucarin L	OH	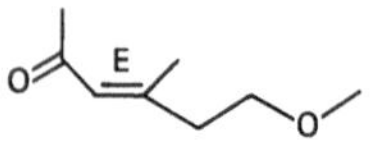	76,83
(42) Verrucarin L Acetate	OAc	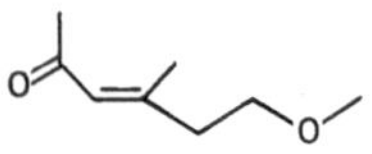	76,83

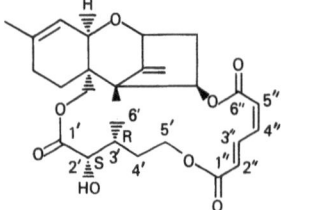

Ref. 24
Verrucarin K
(43)

Table V. *Roridins*

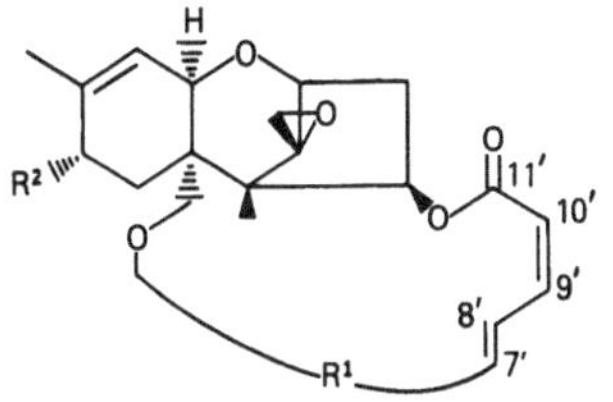

	R²	R¹	Ref.
(44) Roridin A	H	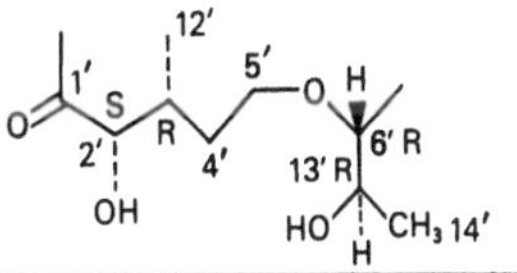	19,21,82,83
(45) Isororidin A	H	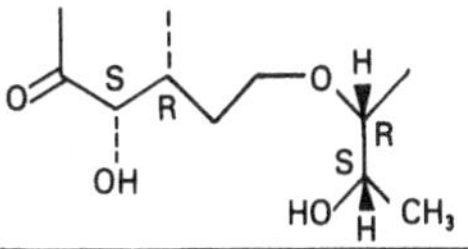	82
(46) Roridin D	H	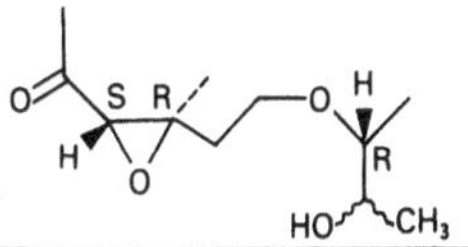	20,21,83
(47) Roridin E Satratoxin D	H	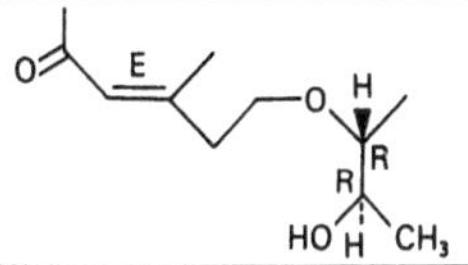	64,131,142
(48) Isororidin E	H	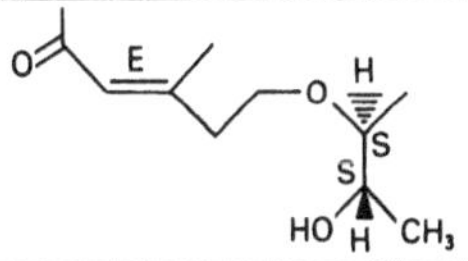	83,101,102
(49) Roridin H	H	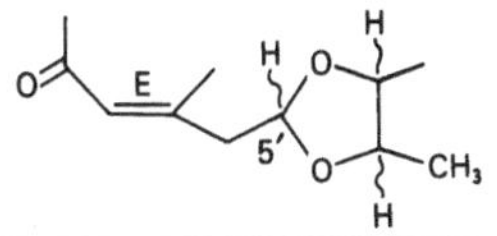	102,142
(50) Roridin J	H	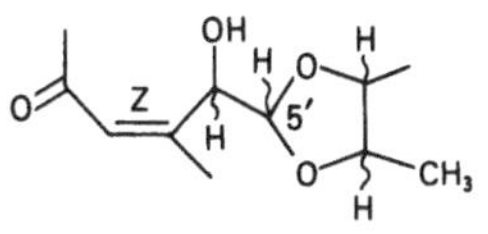	80
(51) Roridin K acetate	α-OAc	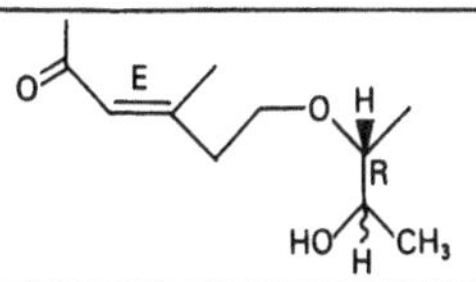	83

Table V *(continued)*

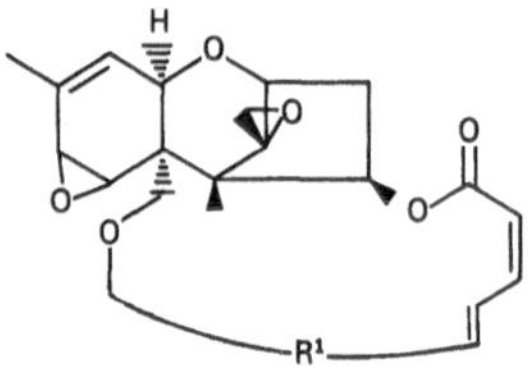

		R¹	Ref.
(52)	7β,8β-Epoxyroridin H	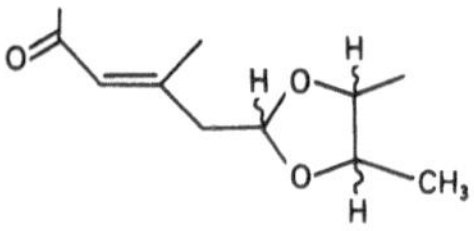	100,101,102
(53)	7β,8β-2′,3′-Diepoxy-roridin H	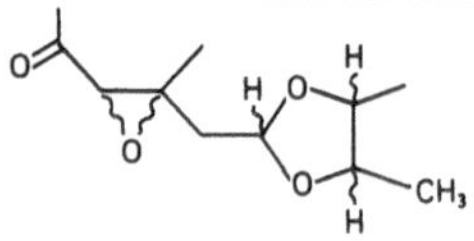	100,101
(54)	7β,8β-Epoxyiso-roridin E	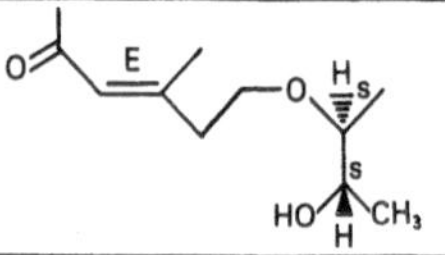	100,101

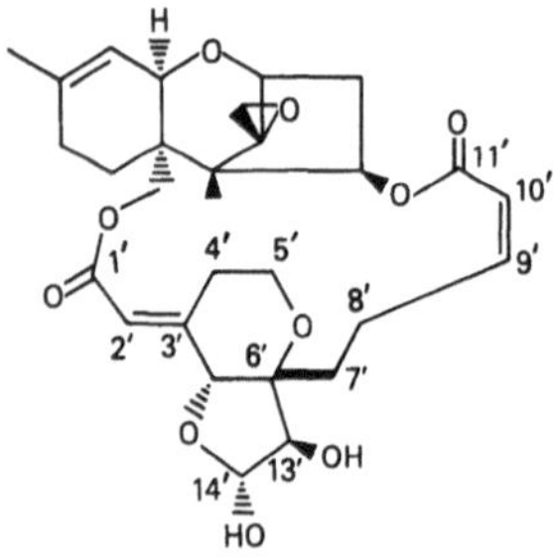

(55) Vertisporin Ref. 66,104

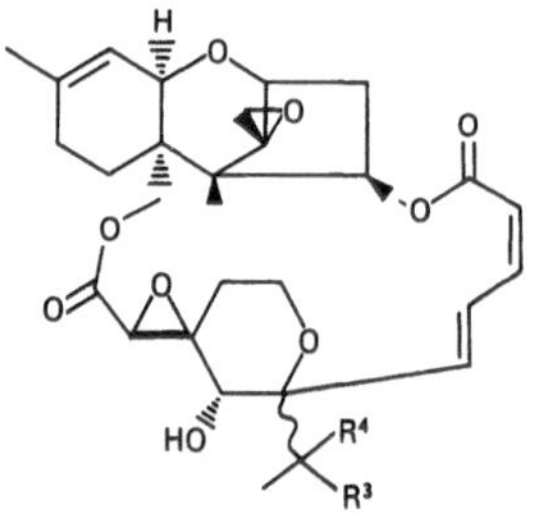

(56) Satratoxin H Ref.

(57a) R³= R⁴=O Satratoxin F
(57b) R³=H,R⁴=OH Satratoxin G
Ref. 40

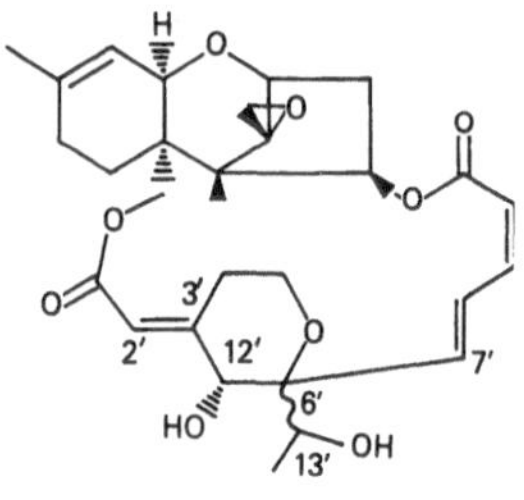

termination of the macrocycles. Even within the less complex verrucarins, which lack the two centers of asymmetry common to the rest of the macrocyclic group, problems have arisen. Although the structure of

References, pp. 211—219

Table VI. *Baccharins*

Baccharins

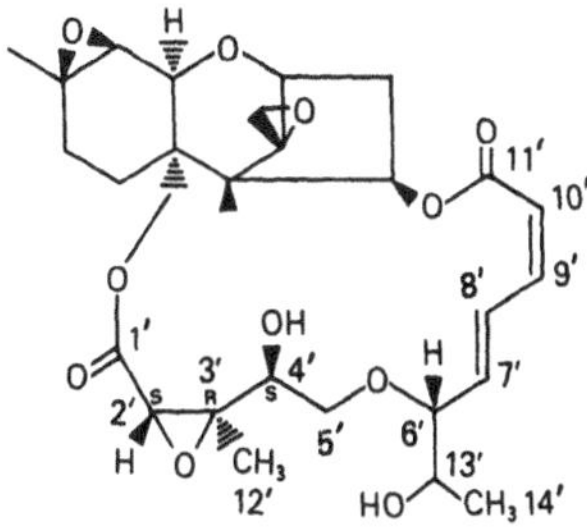

(59) Baccharin B5(13'-R)
(60) Baccharin B8(13'-S)   Ref. 93,94

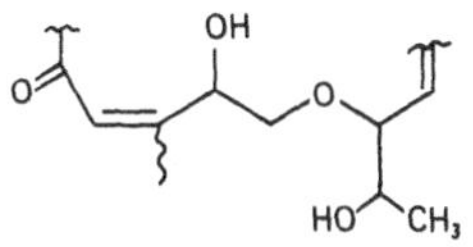

(61) Baccharene Ref. 95

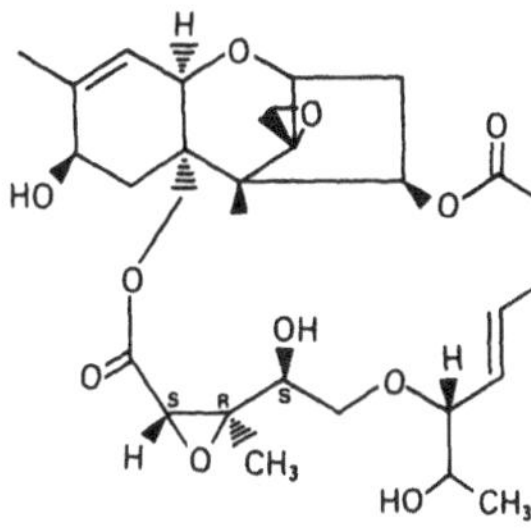

(62) Baccharinol B4 (13'-R)
(63) Baccharinol B6 (13'-S)   Ref. 94

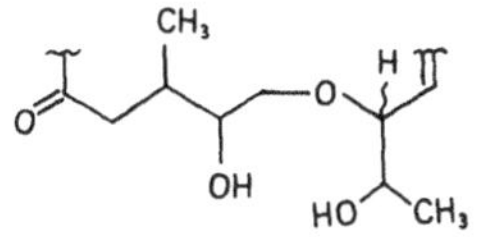

(58 a,b) Baccharinol B1, B2

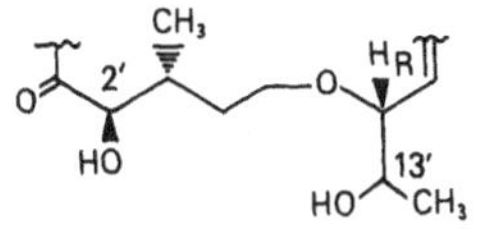

(64) Baccharinol B3 (13'-R)
(65) Baccharinol B7 (13'-S)   Ref. 74, 95

verrucarin A was correctly deduced as early as 1965, a deduction later confirmed by X-ray analysis (*103*), the configuration of the 2′,3′-epoxide in verrucarin B (**39**) was established only recently (*23*). Furthermore, the original assignment of geometry to the 2′,3′-double bond in verrucarin J (**40**) has had to be reversed as a result of several later studies (*42, 102*). Today, the complete structures of all the verrucarins are known. Unfortunately, this statement does not hold true for all known macrocyclic trichothecenes. In particular, the stereochemistry at C-6′ and C-13′, the two asymmetric centers missing in the verrucarin family, are difficult to ascertain by chemical or spectroscopic methods (*75*). To further complicate matters, it is not uncommon for a pair of trichothecanoid metabolites to be isolated which differ only in their configuration at C-6′ and/or C-13′; e. g., roridin A (**44**) and isororidin A (**45**) (*82*). This stereochemical pattern was first uncovered within the baccharins, where four such diastereomeric pairs exist (see Table VI).

Baccharin (**60**) was assigned the absolute configuration 6′-*R*, 13′-*S* by X-ray crystallography (*93*). Subsequent chemical studies related the structures of isobaccharin (**59**), baccharinol (**63**), and isobaccharinol (**62**) back to baccharin (**60**) (all four compounds are now named as shown in Table VI), thereby establishing their absolute configuration (*94*). The pair of isomers baccharinol B3 (**64**) and B7 (**65**) was isolated more recently; again an X-ray study on baccharinol B3 established the configurations (*74*). The stereochemical features of other baccharins such as baccharene (**61**) and baccharinol B1 and B2 (**58a, b**) have not yet been fully delineated (*74, 95*). From the initial structural studies of KUPCHAN, it seemed possible that the absolute configuration at C-13′ could be assigned using ^{13}C NMR shifts of C-12′ and C-13′. Inspection of Table VII shows that for the baccharin group where C-6′ has the *R* configuration (see Table VI), both C-13′ and C-12′ are shifted upfield in the 13′-*S* epimer relative to the 13′-*R* epimer. Unfortunately, the correlation does not hold for all macrocyclic trichothecenes, as evidenced by the isomer pair roridin A (**44**) and isororidin A (**45**).

Table VII. *C-13 NMR Data*

	C-6′	C-13′	C-12′
(**59**) Baccharin B5	(*R*)	71.0 (*R*)	17.7
(**62**) Baccharinol B4	(*R*)	71.0 (*R*)	17.8
(**60**) Baccharin B8	(*R*)	68.8 (*S*)	15.6
(**63**) Baccharinol B6	(*R*)	69.0 (*S*)	15.8
(**44**) Roridin A	(*R*)	70.4 (*R*)	18.0
(**45**) Isororidin A	(*R*)	70.0 (*S*)	17.9

Of the eleven roridins isolated so far six have had their structures, with stereochemistry, determined. Although the basic skeleton of the roridins was established by TAMM (*19*) in 1966, the stereochemistry at C-6′, C-13′

Table VIII. *Trichoverroids*

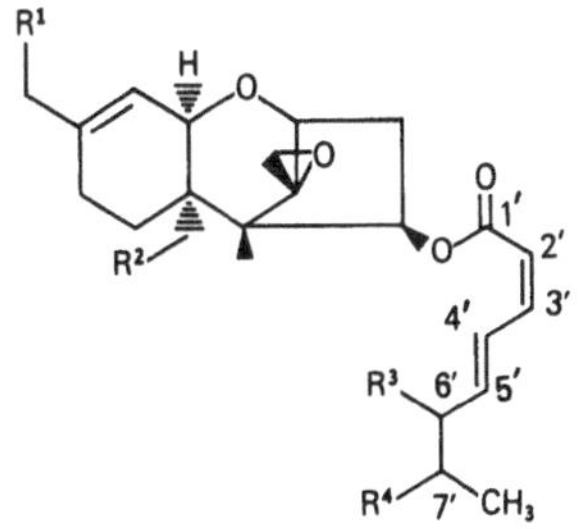

		R_1	R_2	R_3	R_4	Ref.
66	Trichodermadienediol A	H	H	OH(S)	OH(S)	78,83
67	Trichodermadienediol B	H	H	OH(S)	OH(R)	78,83
68	Trichoverrol A	H	OH	OH(S)	OH(S)	78,83
69	Trichoverrol B	H	OH	OH(S)	OH(R)	78,83
70	Trichoverrin A	H	X[a]	OH(S)	OH(S)	78,83
71	Trichoverrin B	H	X[a]	OH(S)	OH(R)	78,83
72	Trichodermadiene	H	H	O (R,R)		77,83
73	Roridin L2	H	OH	Y[b](R*)	OH(R*)	18
74	16-OH-Trichodermadienediol A	OH	H	OH(S)	OH(S)	84
75	16-OH-Trichodermadienediol B	OH	H	OH(S)	OH(R)	84
76	16-OH-Roridin L2	OH	OH	Y[b](R*)	OH(R*)	86
77	Trichoverritone	H	X[a]	Y[b](R*)	OH(R*)	86

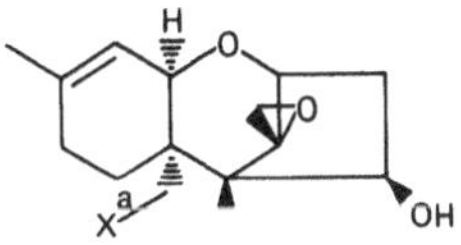

Verrol Ref. 85

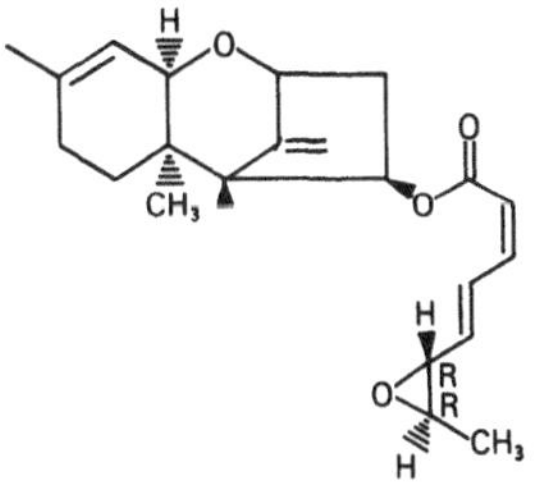

Deoxytrichodermadiene Ref. 85

a. X = -OCOCH≡C(CH₃)CH₂CH₂OH

b. Y = O

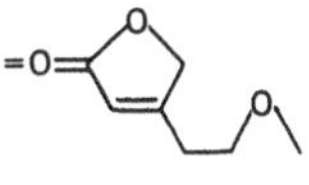

could be assigned with certainty to any member of the roridin family only after an X-ray analysis had established the structure of roridin A (**44**) in 1982 (*82*). Based solely on ^{1}H NMR chemical shifts and coupling constants, the 6',13'-configurations of roridin E (**47**), isororidin E (**48**) and epoxyisororidin E (**54**) were supposedly established by correlation with the baccharins (*102*), but subsequent X-ray studies showed the assignments for epoxyisororidin E and isororidin E to be in error (*82*). The configuration assigned to roridin E has recently been confirmed by total synthesis (*135*). With the configuration of isororidin E confirmed as 6′-*S*, 13′-*S*, this substance does show some interesting variations in the ^{1}H and ^{13}C shifts of C-7′ and C-8′ relative to the other macrocycles, including an abnormally large 6 Hz coupling between H-6′ and H-7′. If these changes are in fact diagnostic of all 6′-*S* isomers, this would be a great aid in determining stereochemistry at C-6′.

Among the remaining roridins, roridin J (**50**), roridin H (**49**), 7β,8β-epoxyroridin H (**52**) and 7β,8β,2′,3′-diepoxyroridin H (**53**) are noteworthy due to the presence of an additional ring in the macrocycle. In these compounds, the macrocycle is connected by way of an acetal linkage which creates an additional asymmetric center at C-5′. To date no study has addressed itself to determining the stereochemistry of these roridins. It has been established by n.O.e. experiments that roridin J possesses *Z* geometry at the 2′,3′-double bond (*80*). Although this geometry is unique among the verrucarins and roridins, it does match the configuration found in the more complex compounds vertisporin and satratoxins.

As with these roridins, vertisporin and the satratoxins have additional rings within the macrocycle. In these cases, the ring results from a new bond formed between C-6′ and C-12′. Vertisporin (**55**) has an additional ring, a cyclic hemiacetal linking C-12′ and C-14′.

The last group of trichothecenes are the trichoverroids which appear in Table VIII. These *seco* macrocyclic trichothecenes have either partial or complete carbon chains at C-4 and C-15 characteristic of the macrocyclic compounds, but lack the requisite ring-forming bond. The first of these, trichodermadiene (**72**), was reported by Jarvis and co-workers in 1980 (*77*). In this compound a diene fragment is affixed to C-4 in a manner reminiscent of the macrocycles, but the C-15 hydroxyl group remains unesterified. Subsequently, Jarvis (*78, 83, 86*) discovered other such *seco* compounds which possess the complete carbon chains of the macrocyclic trichothecenes. Evidence has been presented suggesting that the trichoverroids are intermediates in the biosynthesis of the macrocycles (*78*).

Very recently (*105a*) sambucinol and sambucoin, two compounds possessing a modified trichothecene structure have been described.

Sambucinol   Sambucoin

IV. Synthesis of the Simple Sesquiterpenoid Trichothecenes

1. Introduction

In 1971, COLVIN, RAPHAEL and ROBERTS (*33*) reported the first synthesis of a trichothecene, the monohydroxylated trichodermin (**3**). The remainder of the decade saw a number of successful approaches to the trichothecene skeleton which culminated in an alternative synthesis of trichodermin (**3**) by STILL and TSAI (*134*) in 1980. It was not until 1982, nearly 20 years after their structure was firmly established, that synthetic chemists successfully assembled the polyhydroxylated trichothecenes. This milestone was achieved simultaneously by SCHLESSINGER and NUGENT (*125*) with their synthesis of verrucarol (**82**) and KRAUS and co-workers (*92*) with their synthesis of calonectrin (**5**). Very soon thereafter, TROST and McDOUGAL (*145*) as well as ROUSH and D'AMBRA (*117*) published alternative syntheses of verrucarol (**82**). More recently, BROOKS and co-workers (*27*) have completed an enantioselective synthesis of anguidine (**9**), the first tri-oxygenated trichothecene to be prepared by total synthesis. With this recent explosion of work, a number of efficient synthetic sequences leading to the trichothecenes now exist, which should prove valuable in exploring the biological profile of this intriguing class of compounds.

Approaches to the simple sesquiterpenoid trichothecenes can be categorized by the type of reaction used to form the ring system. Primarily, four bond associations have been used to construct the trichothecene skeleton as depicted in Scheme 1. Groups 1 and 2 (X = O, Y or Z = OH)

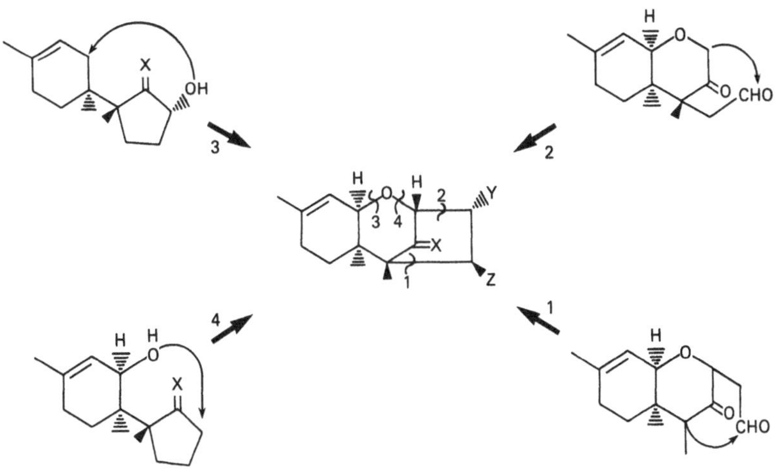

Group 3,4 Biomimetic Approach   Group 1,2 Aldol Approach

Scheme 1

form a subset which will be referred to as the aldol approach. Groups 3 and 4 (any number of variations at X, Y and Z) form a subset which will be termed the biomimetic approach. It should be noted that while in the past only the group 3 cyclization has been referred to as biomimetic, the exact events responsible for the biosynthesis of the trichothecenes are still uncertain. For the purpose of this review, inclusion of both group 3 and group 4 in one subset emphasizes the synthetic strategies shared by these two approaches.

Although the group 1 aldol approach formed the basis of Raphael's landmark synthesis of trichodermin (3) (33), subsequent studies have shown the other three groups to be more flexible and higher yielding strategies. The group 3 biomimetic approach has enjoyed the most prominent role in trichothecene synthesis with trichodermin (3) (134), 12,13-epoxytrichothec-9-ene (1) (98), verrucarol (82) (117, 125) and anguidine (9) (27) all being assembled by this methodology. The alternative biomimetic approach, group 4, has been utilized in a synthesis of verrucarol (82) and has also yielded the first 11-epitrichothecene having a trans-fused AB ring junction (145). Successes for the group 2 aldol approach, an excellent strategy for synthesizing the 3-oxytrichothecenes (Scheme I, Y = OH), include syntheses of 12,13-epoxytrichothec-9-ene (1) (51) and calonectrin (5) (92). All three of these latter approaches have been the subject of numerous model studies. A number of model studies fall outside the scope of the four methodologies depicted in Scheme 1.

In reviewing the synthetic efforts to date, work related to the total synthesis of naturally occurring trichothecenes, as opposed to simpler model systems, will be stressed. Although the schemes will present the synthetic sequence in detail, the text will be used mainly to highlight key reactions within the scheme and to interrelate the strategies employed by the various groups working within the field. Carbon atoms will be referred to in the text by the number which would represent their position within the trichothecene skeleton (see Fig. 1 for the trichothecene numbering system). This convention will allow early synthetic intermediates to be easily related to the final trichothecene structure, thereby revealing developing synthetic strategy.

2. The Aldol Approach

As mentioned in the Introduction, it was the group 1 aldol approach which provided the first entry into the trichothecene skeleton with Raphael's (33) classic synthesis of trichodermin (3). The synthesis, predicated on inducing the keto aldehyde (87) to cyclize to the 13-nortrichothecene (89), is shown in Scheme 2. Construction of the key

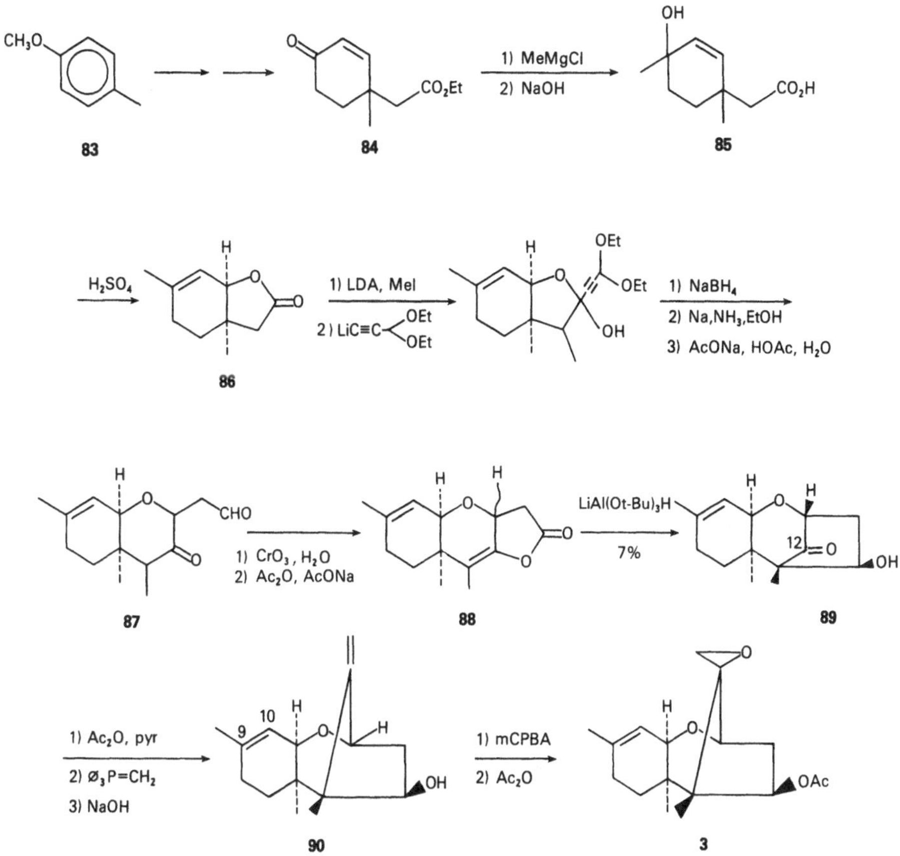

Scheme 2

intermediate (**87**) was smoothly accomplished from p-methylanisole (**83**). Importantly, acid catalyzed dehydration of hydroxy acid (**85**) forms only the desired *cis*-lactone (**86**), a result productively utilized in subsequent trichothecene syntheses. Unfortunately, all efforts to effect direct cyclization of aldehyde (**87**) to the tricycle (**89**) failed. However, a clever modification in which the two reacting centers were tied together prior to the aldol reaction gave better results; i.e., treatment of the enol ester (**88**) with one equivalent of lithium tri-*t*-butoxyaluminium hydride gave the desired aldol product (**90**), albeit in only a 7% yield. Completion of the synthesis relied on the hydroxy-directed epoxidation of diene (**90**) to establish the natural 12,13-epoxide stereochemistry. Reaction of the 12-keto group with the methylene transfer reagent trimethylsulphonium methylide yielded the unnatural 12,13-epoxide, a result substantiated by other researchers (*51*). As will be seen, epoxidation of dienes related to (**90**) is often complicated by concomitant oxidation of the 9,10-olefin.

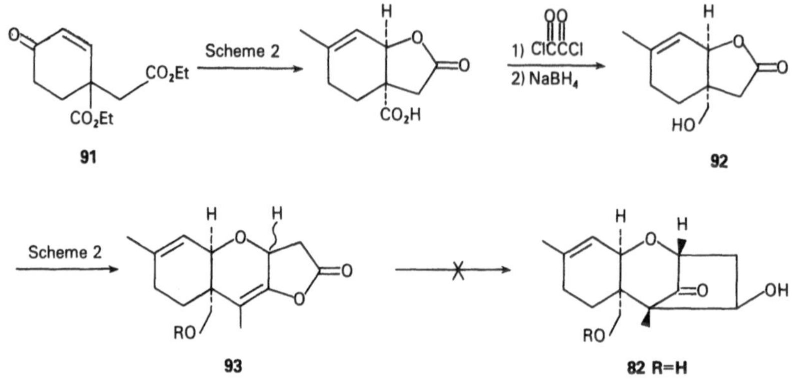

Scheme 3

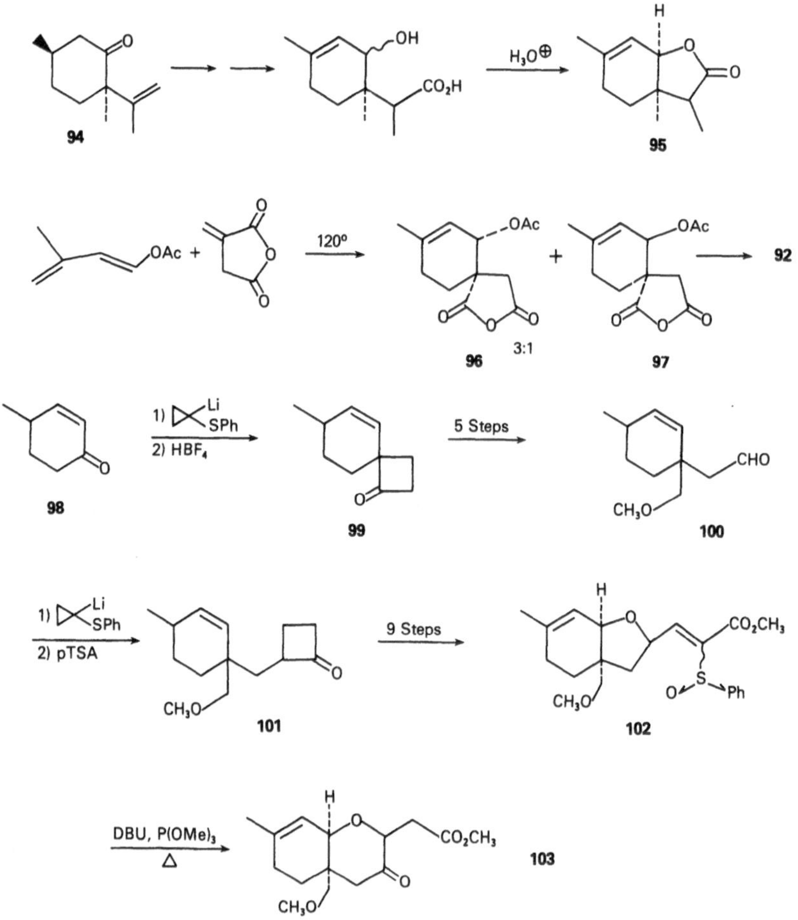

Scheme 4

RAPHAEL and COLVIN (*34*) attempted to extent this group 1 aldol strategy to the dioxygenated trichothecene verrucarol (**82**) as outlined in Scheme 3. The requisite enol ester (**93**) containing a protected angular hydroxy methyl group was synthesized from the enone (**91**) in a route similar to the earlier trichodermin synthesis (see Scheme 2). X-ray analysis has confirmed the *cis*-stereochemistry for the ring fusion in lactone (**92**). Unfortunately, all attempts to induce cyclization of (**93**) to the trichothecene skeleton (**82**) failed. Hence, the group 1 aldol approach, although providing the initial success in trichothecene synthesis, ranks as the least versatile method for construction of the trichothecene ring system.

Some alternative syntheses of intermediates related to the group 1 aldol approach are shown in Scheme 4. WELCH and WONG (*164*) prepared the optically pure lactone (**95**), an intermediate in RAPHAEL's synthesis of trichodermin (**3**), starting from (−)-methylisopulegone (**94**). In 1978, SNIDER and AMIN (*130*) synthesized the proposed verrucarol intermediate (**92**) (see Scheme 3) from the Diels-Alder adducts (**96**) and (**97**). Most significantly, this was the first example in which the cyclohexenyl A-ring was constructed by a Diels-Alder reaction, an approach which has gained favor in recent years. The advanced intermediate (**103**) was synthesized by TROST and RIGBY (*143*) in a sequence which highlights the concept of spiroannulation (**98**→**99**; **100**→**101**). Although this route was abandoned, methodology related to that embodied in the intriguing ring expansion of furan (**102**) to the pyranone (**103**) was utilized in a later synthesis of verrucarol (**82**) (*147*)

The group 2 aldol approach has proved to be a more efficient entry into the trichothecene skeleton than the just discussed group 1 approach (see Scheme 1). FUJIMOTO, TATSUNO and co-workers (*51*) were the first to utilize group 2 methodology in their 1974 synthesis of the fungal metabolite 12,13-epoxytrichothec-9-ene (**1**). Starting from the keto ester (**104**) they prepared the hydrochromanone (**105**) in eight steps as shown in Scheme 5. Intrinsically, this compound could serve as an intermediate in either one of the aldol approaches, as both approaches share the strategy of preliminary construction of the AB rings prior to attachment of ring C. The problem of establishing stereochemistry at C-5 was partially resolved by the Claisen rearrangement of enol ether (**106**) with the predominant isomer (**107**) being carried through to the nortrichothecene (**108**). Importantly, the yield of the aldol cyclization is an impressive 90%, and, as will be seen, such high yields are not limited to this example. After removal of the 3-hydroxy group, the 12,13-epoxide was installed by the method previously used by RAPHAEL for trichodermin (**3**) (see Scheme 2). That epoxidation yields the desired stereoisomer (**1**), even in the absence of hydroxy-direction, is noteworthy. However, the low yield (30% at 60% conversion) foreshadowed the problems other investigators subsequently experienced with this reaction.

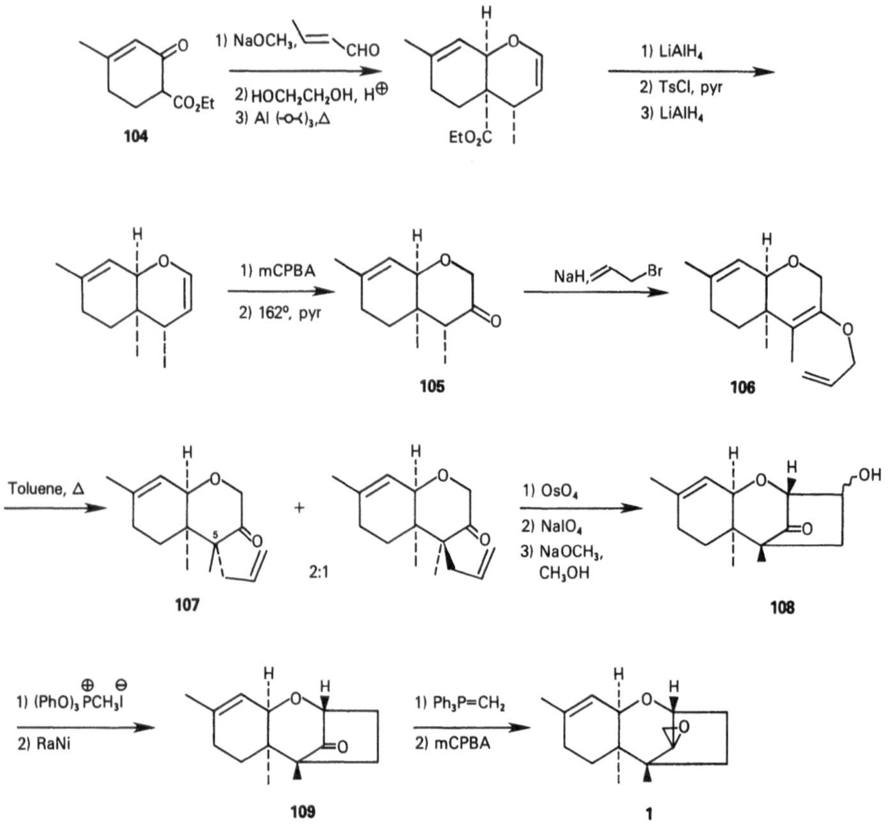

Scheme 5

Consistent with RAPHAEL's observation in the trichodermin series, reaction of dimethylsulfonium methylide with ketone (**108**) produces an epoxide epimeric with the natural product.

The synthesis of calonectrin (**5**) (*92*), a 3,14-dioxytrichothecene, was predicated on the preceding aldol condensation. KRAUS and ROTH recorded the synthesis of the requisite AB synthon (**115**) as outlined in Scheme 6. Reaction of the Diels-Alder partners (**110**) and (**111**) rapidly elaborated the cyclohexenyl A ring with adequate stereocontrol. Transformation of the major Diels-Alder isomer (**112**) to the hydrochromanone (**115**) was accomplished in 11 steps highlighted by an intramolecular Knoevenagel condensation (**113**→**114**). In a cleverly conceived process, the two carbon bridge of the C ring was installed stereoselectively. Fluoride-induced intramolecular alkylation of the enol silyl ether (**116**) generated the pivotal C-5 center with complete stereocontrol, thus improving upon FUJIMOTO's earlier solution to C-5 chirality (see Scheme 5). The aldol cyclization again

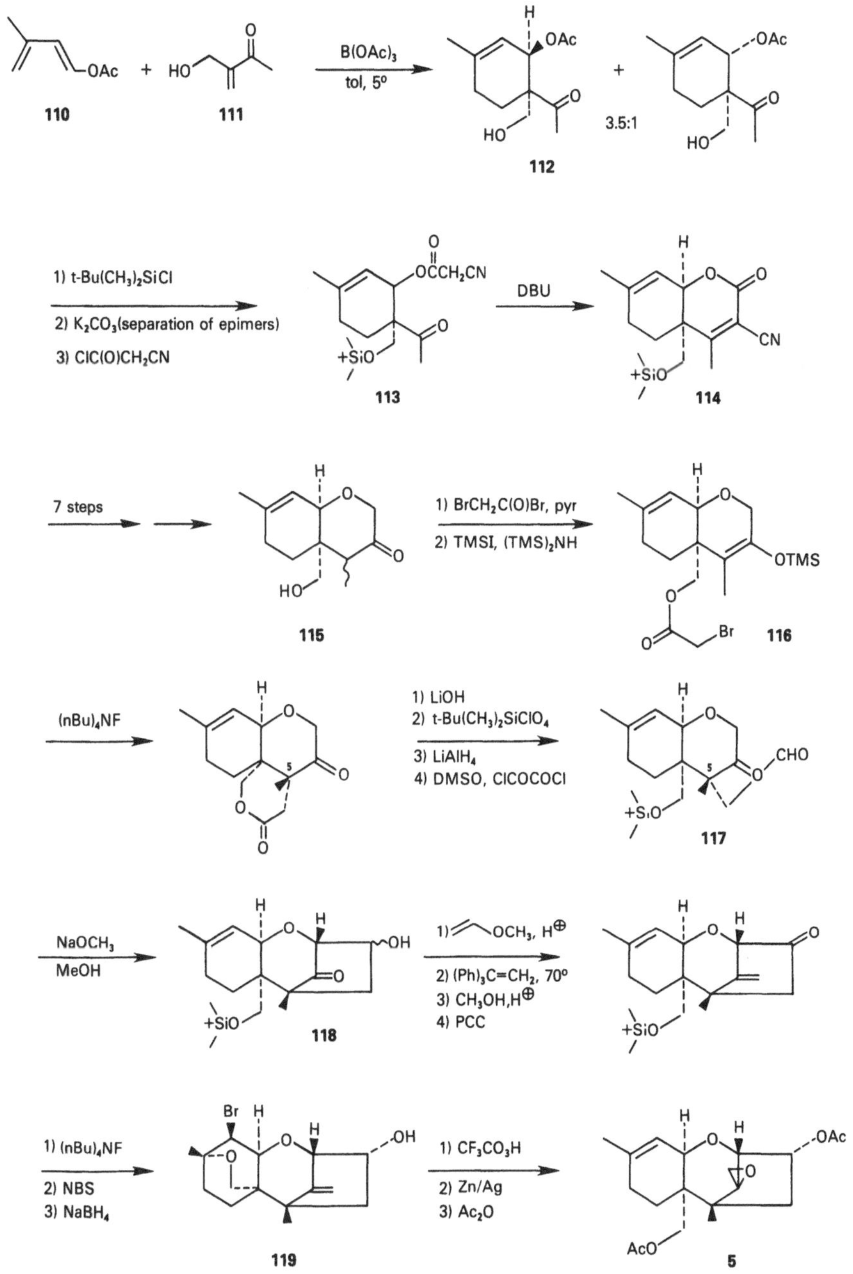

Scheme 6

proved successful as treatment of the keto aldehyde (117) with sodium methoxide yielded the alcohol (118) in 63% yield as a mixture of epimers. Realization of the final goal necessitated protection of the trisubstituted olefin as the bromo ether (119). Epoxidation of the remaining methylene group then proceeded stereoselectively, and two steps later, the synthesis of calonectrin (5) was complete. As further illustrated by Roush (117) and Schlessinger (125) in their verrucarol syntheses (see Schemes 13 and 15), protection of the 9,10-olefin as a bromo ether represents a general solution to selective incorporation of the 12,13-epoxide. However, such a solution costs the chemist an additional protection-deprotection sequence.

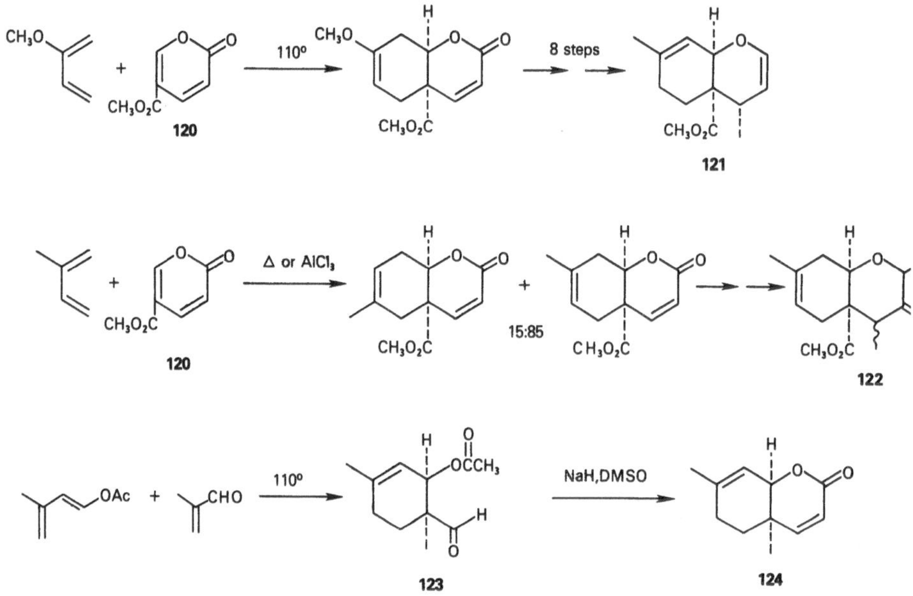

Scheme 7

A number of reports have appeared which deal specifically with the creation of the AB portion of the trichothecene skeleton. The majority of this work, as shown in Scheme 7, concerns itself with alternative Diels-Alder strategies for the construction of the A ring. When methyl coumalate (120) is used as the dienophile the AB system results immediately, although a number of manipulations are needed to convert the Diels-Alder adducts to useful synthetic intermediates. In this fashion Tatsuno and Nakahara (107) have intercepted a previous intermediate (121) in their 12,13-epoxytrichothec-9-ene (1) synthesis (see Scheme 5), while Kraus and Frazier (91) have prepared ketone (122), an olefin isomer of the calonectrin intermediate (115) (see Scheme 6). Most recently, Banks and co-workers (12) have prepared lactone (124) from the Diels-Alder adduct (123). Scheme 8 summarizes some early work from Goldsmith's laboratory (57) in which the *cis* fusion of the AB system was generated from a cuprate addition to enone (125).

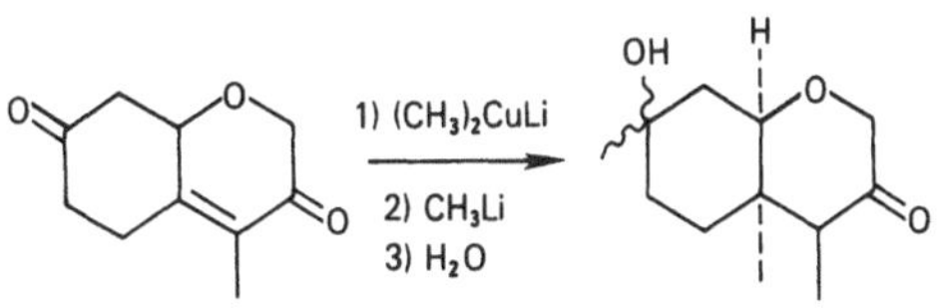

125

Scheme 8

3. The Biomimetic Approach

As mentioned in the Introduction, the group 3 biomimetic approach (see Scheme 1) has been the most popular route to the trichothecene skeleton. Two different moieties have served as the electrophilic site for biomimetic cyclization. When the cyclization proceeds *via* an allylic carbonium ion (**127**) (Path A, Scheme 9), the desired trisubstituted olefin (**126**) is obtained directly. On the other hand, the Michael acceptor (**128**) (Path B) yields, upon cyclization, a ketone (**129**) which must then be transformed into the olefin (**126**), a process which shows good but not complete regioselectivity. Hence, Path A, which can also be entered from the enone (**128**), is the superior route. Further analysis of Scheme 9 reveals that the primary stereochemical challenge of the biomimetic approach is to control the relative stereochemistry at the two quaternary centers C-5 and C-6. Within the context of trichothecene synthesis, a number of useful protocols have been devised for this purpose and include photocyclization (*99*), selective ring contraction (*134*), Diels-Alder cycloaddition (*117, 125*) conjugate addition (*27, 120*), and interconversion of dienyl iron complexes (*114*).

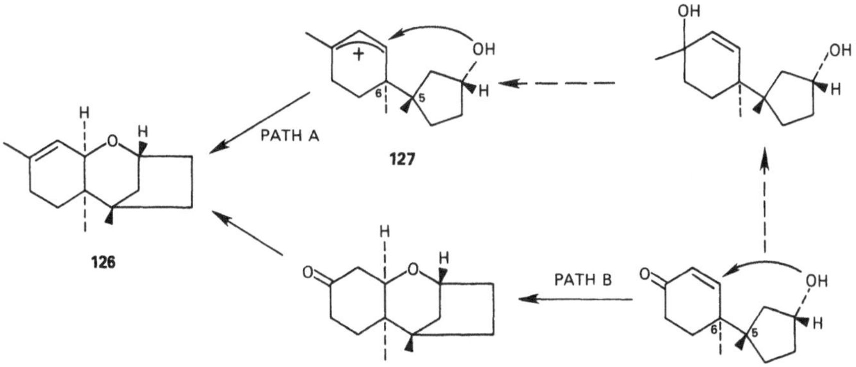

Scheme 9

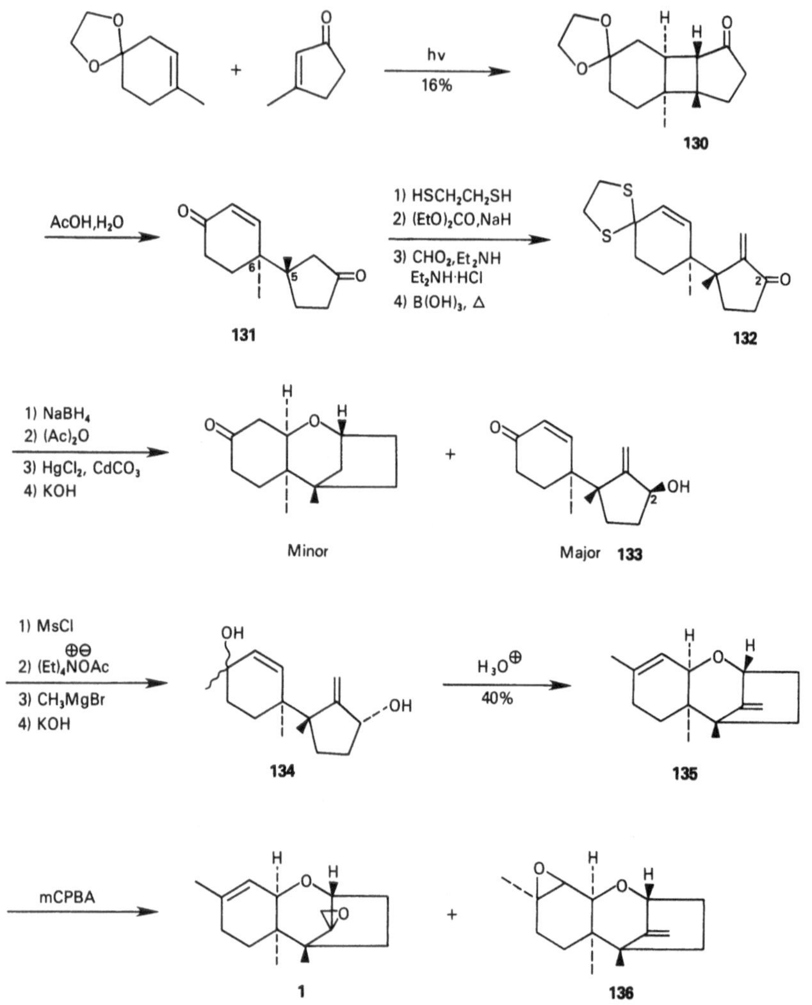

Scheme 10

Work on the group 3 biomimetic strategy was initiated in 1974 by
Masuoka, Kamikawa, and Kubota (*99*) with the synthesis of a simple
trichothecanoid model. Two years later, the same laboratory reported the
synthesis of the mycotoxin 12,13-epoxytrichothec-9-ene (**1**), shown in
Scheme 10, based on their previous model study (*98*). The pivotal relay for
both syntheses is the bicyclodione (**131**). Control of the C-5, C-6
stereochemistry was achieved by a [2 + 2] photoaddition in which the major
crystalline isomer was the *cis, anti, cis* tricycle (**130**). This was readily
unraveled to the dione (**131**) by simple acid hydrolysis. Reduction of the
enone (**132**) introduced the stereochemical center at C-2. Unfortunately,

References, pp. 211—219

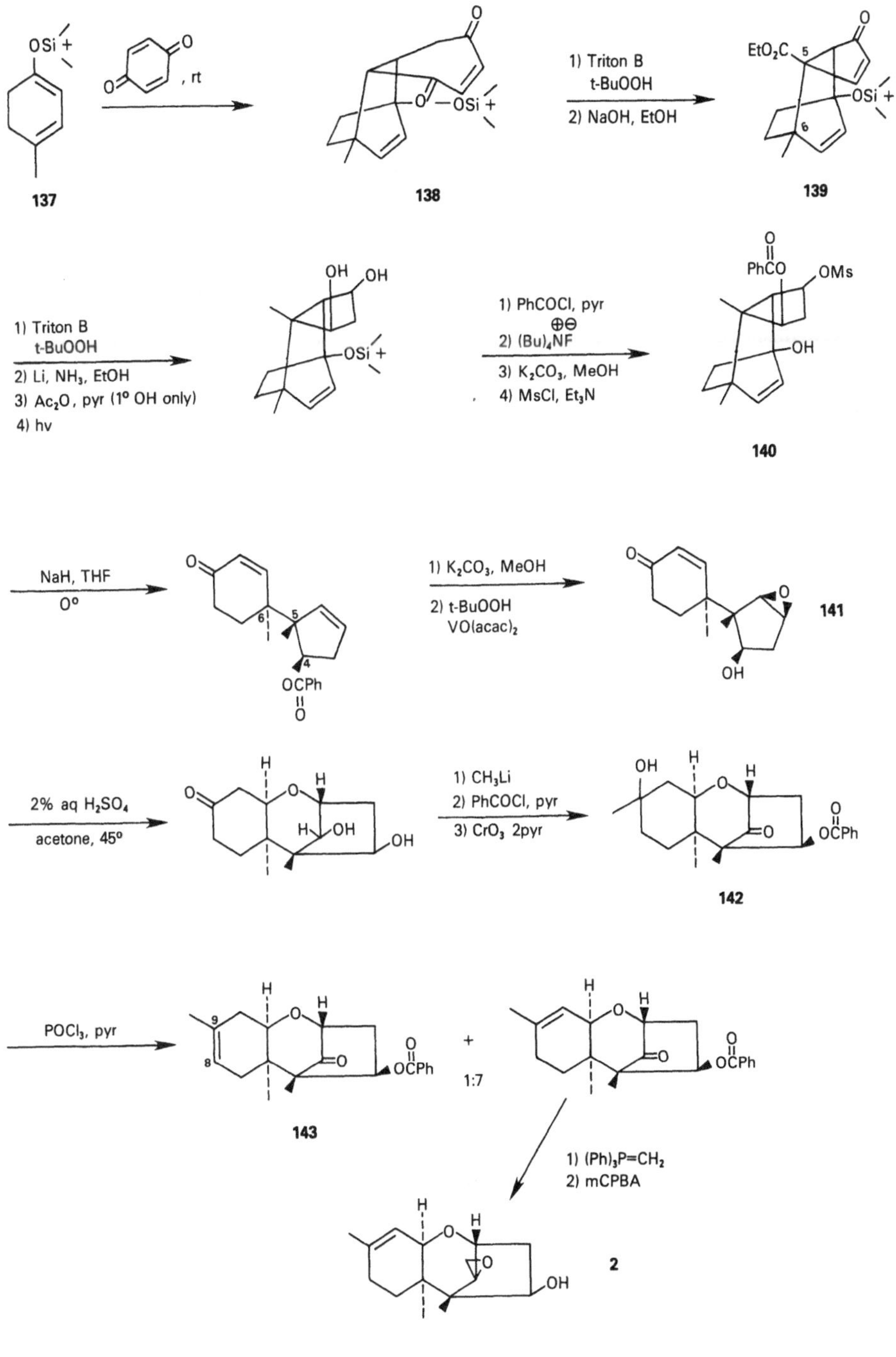
137
138
139
140
141
142
143
2
, rt
1) Triton B
t-BuOOH
2) NaOH, EtOH
1) Triton B
t-BuOOH
2) Li, NH3, EtOH
3) Ac2O, pyr (1° OH only)
4) hv
1) PhCOCl, pyr
2) (Bu)4NF
3) K2CO3, MeOH
4) MsCl, Et3N
NaH, THF
0°
1) K2CO3, MeOH
2) t-BuOOH
VO(acac)2
2% aq H2SO4
acetone, 45°
1) CH3Li
2) PhCOCl, pyr
3) CrO3 2pyr
POCl3, pyr
1:7
1) (Ph)3P=CH2
2) mCPBA

Scheme 11

subsequent manipulations revealed that the major reduction product (133) possessed the wrong stereochemistry to undergo cyclization. This problem was circumvented by an alcohol inversion sequence furnishing diol (134) which, upon exposure to acid, yielded the trichothecene (135) in moderate yield. Oxidation of diene (135) with m-CBPA resulted in the isolation of two epoxides (1) and (136), thereby substantiating the non-selective nature of peracid epoxidation on certain trichothecadienes. This early study of the group 3 biomimetic cyclization set the stage for the recent surge of related methodology which has culminated in four successful total syntheses.

STILL and TSAI were the next to utilize the group 3 approach in their synthesis of trichodermol (2) (134). A unique aspect of this work is the selective Herz-Favorski ring contraction of the Diels-Alder adduct (138) to the cyclopentenone (139). This reaction, in tandem with the well-precedented stereoselectivity exhibited by the Diels-Alder reaction (137→138), established the desired stereochemistry between C-5 and C-6. This stereochemistry, as well as that at C-4, obtained by a selective dissolving metal reduction, was unmasked by a Grob fragmentation of the hydroxy mesylate (140). Hydroxy-assisted epoxidation across the cyclopentene ring followed by acid hydrolysis of epoxide (141) furnished the trichothecene skeleton directly in 60% yield. Introduction of the 9,10-double bond by dehydration of alcohol (142) produced a small amount of the 8,9-regioisomer (143). Incorporation of the 12,13-epoxide followed the familar script of peracid epoxidation, thus concluding a novel synthesis of trichodermol (2).

PEARSON and ONG (114) had equally good success with the tandem epoxide opening-cyclization methodology (150→151) in a synthesis of the model trichothecene (152). A novel aspect of their work lies in the control of the C-5, C-6, stereochemistry as shown in Scheme 12. Reaction of the dienylium iron complex (144) with the keto ester (145) produced a 1:1 mixture of diene diastereoisomers (146) and (147). It is important to note that double bond migration in the cyclohexadienyl ring serves to invert the stereochemistry at C-6, thereby acting as a mechamism to interchange the relative stereochemistry of C-5 and C-6. Advantage of this fact was taken with the alcohols (148) and (149) which were found to equilibrate to a mixture in which the diastereoisomer (148), desired for trichothecene synthesis, predominated. After separation of this isomer, the "wrong" isomer (149) was reequilibrated, thus eventually providing an 80% yield of diastereomer (148) even though the original alkylation yielded only 50% of (146).

SCHLESSINGER and NUGENT (125) also employed the group 3 biomimetic cyclization in their benchmark synthesis of verrucarol (82). The key intermediate in their synthetic plan was the bicyclic compound (154) obtained in 10 steps from the readily available dione (153). With the C-ring

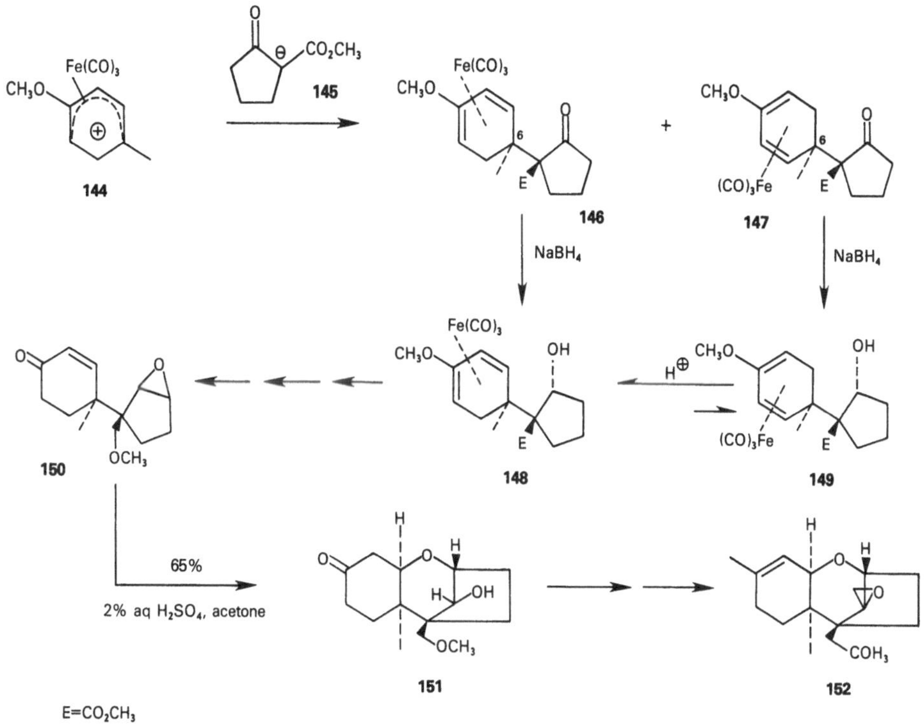

E=CO₂CH₃

nearly fully elaborated in this intermediate, SCHLESSINGER and NUGENT next addressed the problem of stereocontrol at C-5 and C-6. Not surprisingly, the molecular topography of the bicyclic system resulted in exclusive attack of diene (155) on the top face of the enone component yielding, after acid hydrolysis, the cyclohexenone (156) as a single diastereomer. Selective addition of methyl lithium to the enone carbonyl followed by reduction of the lactone then gave triol (157) which readily cyclized to the trichothecene (158). As epoxidation of the exocyclic olefin in the presence of the trisubstituted double bond proved impossible, formation of the bromo ether (159) prior to peracid oxidation was necessitated. Deprotection of the bromo ether then completed this first total synthesis of verrucarol (82). A particularly attractive feature of this sequence is the availability of the starting dione (153) in optically pure form, via an enantioselective Robinson annulation, thereby establishing the first chiral synthesis of verrucarol (82) (124).

In a series of three publications, ROUSH and D'AMBRA developed another group 3 biomimetic sequence culminating in a synthesis of verrucarol (82). Their initial report, appearing in 1980, focused on the

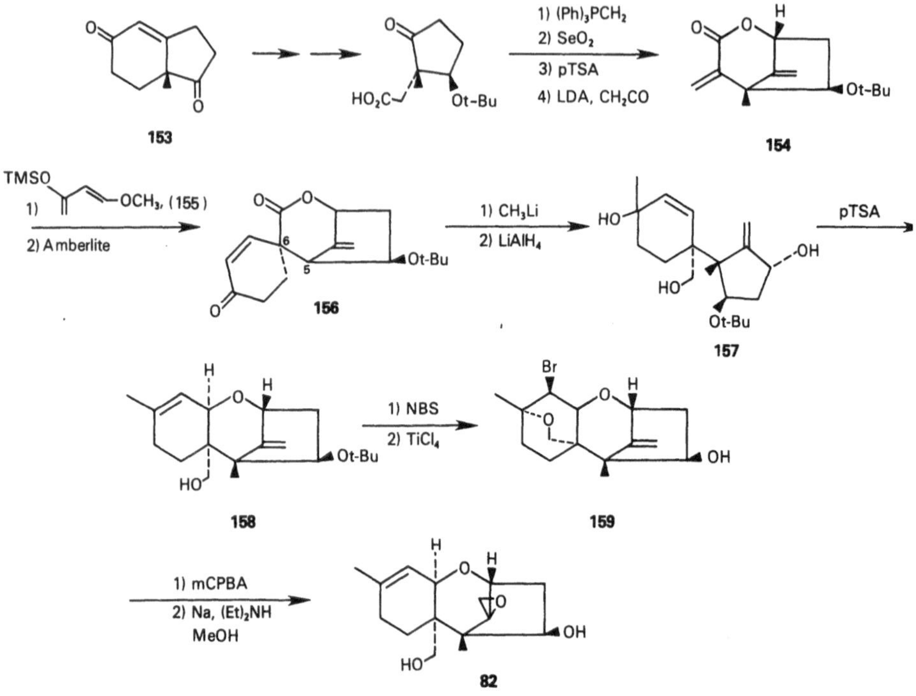

Scheme 13

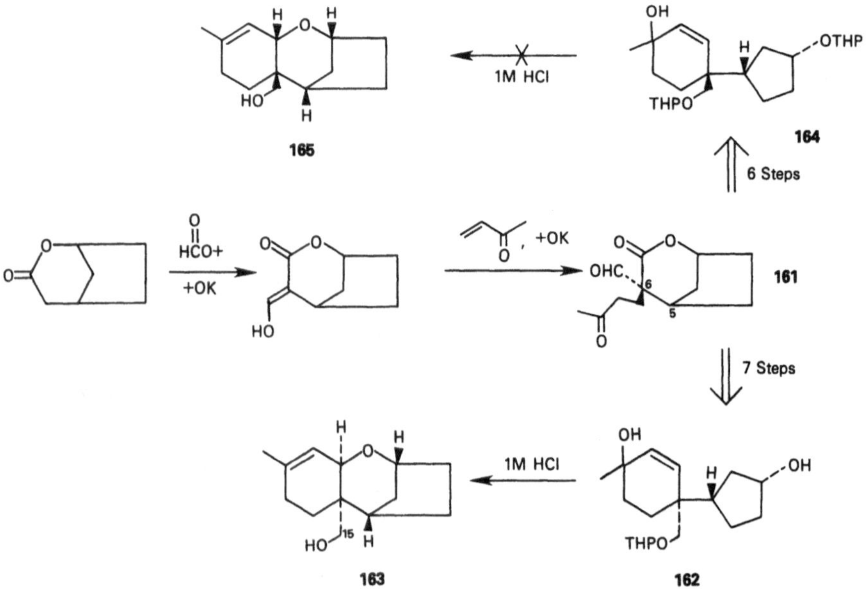

Scheme 14

model system (**163**) shown in Scheme 14 (*120*). The critical C-5, C-6 stereochemistry was established by a Michael addition which, as expected, occurred on the less hindered *exo*-face of the oxabicyclo[3.2.1]octane. Subsequent manipulations yielded diol (**162**) which cyclized smoothly to give the dinortrichothecene (**163**). At the time this represented the first synthesis of a trichothecene derivative possessing a 15-hydroxymethyl group. In an ancillary study, the authors attempted to prepare the trichothecene analog (**165**) having the novel *cis, syn, cis* ring stereochemistry. Unfortunately, attempted cyclization of allylic alcohol (**164**), prepared from the aforementioned Michael adduct (**161**), gave none of the desired ring system. As we shall see, this ring system can be created through an intramolecular Diels-Alder reaction (*89*).

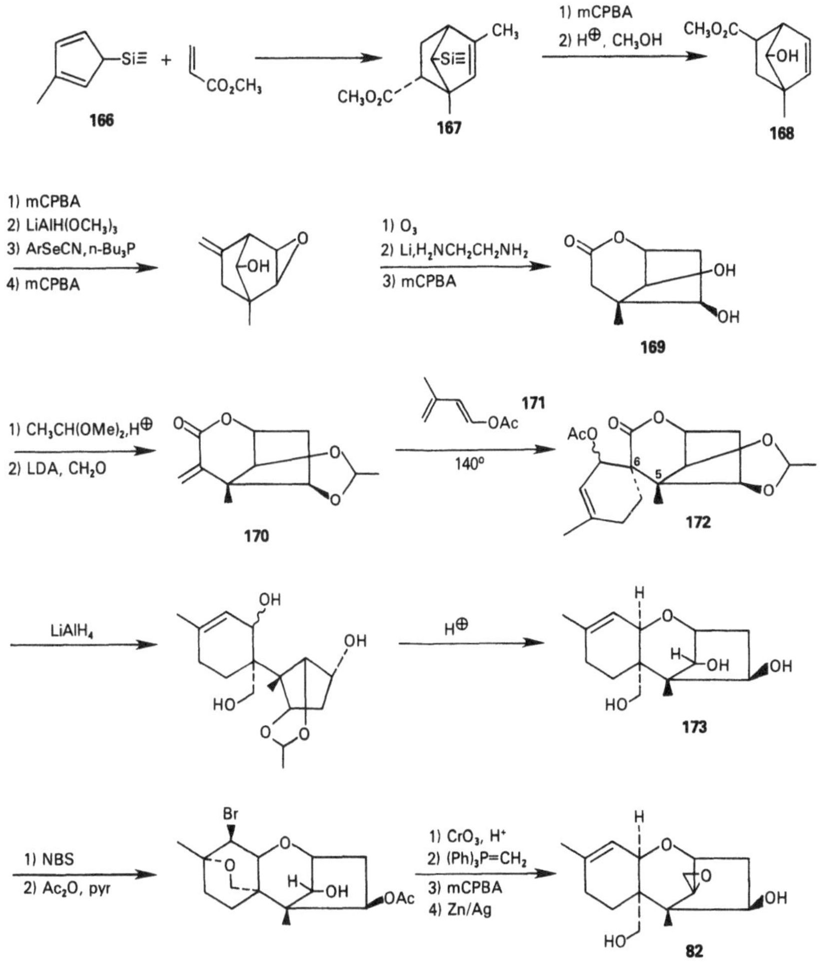

Scheme 15

With their basic strategy substantiated, Roush and D'Ambra next turned to the construction of the substituted bicyclo[3.2.1]octane (169) as outlined in Scheme 15 (116). Diels-Alder reaction of methyl acrylate and the silylcyclopentadiene (166) formed the bicyclo[2.2.1]heptene (167) as the major product. The propensity of silicon to stabilize β-carbocations was then demonstrated by a unique silyl-controlled Wagner-Meerwein rearrangement which furnished the bicyclo[2.2.1]heptenol (168) with a high degree of stereocontrol. After a number of standard transformations, the key BC-ring synthon (169) was in hand. At this point, the authors deviated slightly from their model studies and chose to install the A-ring *via* a Diels-Alder reaction (117). Consistent with the results of Schlessinger and Nugent (see Scheme 13) the topology of the bicyclic system directs the attack of the acetoxydiene (171) on the α-methylene lactone (170) so as to establish the requisite C-5, C-6 stereochemistry. After reduction of the Diels-Alder adduct (172), acid catalyzed cyclization again proceeded smoothly to give the triol (173). This was transformed without event into the trichothecene verrucarol (82).

The most recent foray into the group 3 biomimetic approach launched by Brooks, Grothaus and Mazdiyasni (27) has resulted in the first synthesis of the cytostatic agent anguidine (9). Importantly, this work also represented the first chiral synthesis of a trichothecene metabolite, as their sequence opened with an enantioselective microbial reduction (26) of dione (174) shown in Scheme 16. The fact that this reduction yielded the unnatural configuration at the carbon destined to be C-4 in the trichothecene skeleton is easily rectified by an alcohol inversion sequence furnishing, after suitable protection, the optically pure cyclopentane (175). Further elaboration of this chiral synthon, as reported in a preliminary communication (28), led to the heavily substituted oxabicyclo[3.2.1]octane (176).

In analogy to Roush's earlier studies (see Scheme 14) the A-ring was incorporated by a Robinson annulation (176→177) which provided the desired stereochemistry at C-5 and C-6. Reduction then unraveled the bicyclic system to yield the expected precursor for cyclization (178). Although attempts to cyclize this tetraol were unsuccessful, simple conversion of the 15-hydroxy group to the acetate (179) allowed for smooth cyclization and provided the 13-nortrichothecene (180). After standard elaboration to the hydroxy diene (181) it was found that, in sharp contrast to other oxygenated trichothecenes, peracid oxidation selectively epoxidized the 12,13-double bond. This result allowed for rapid completion of the synthesis yielding the natural antipode of anguidine (9).

The alternative biomimetic cyclization, the group 4 approach (see Scheme 1), has not enjoyed the popularity of the aforementioned group 3 approach. This preference might relate to the innate ability of the group 3 methodology to control C-11 stereochemistry; that is, a group 3 cyclization

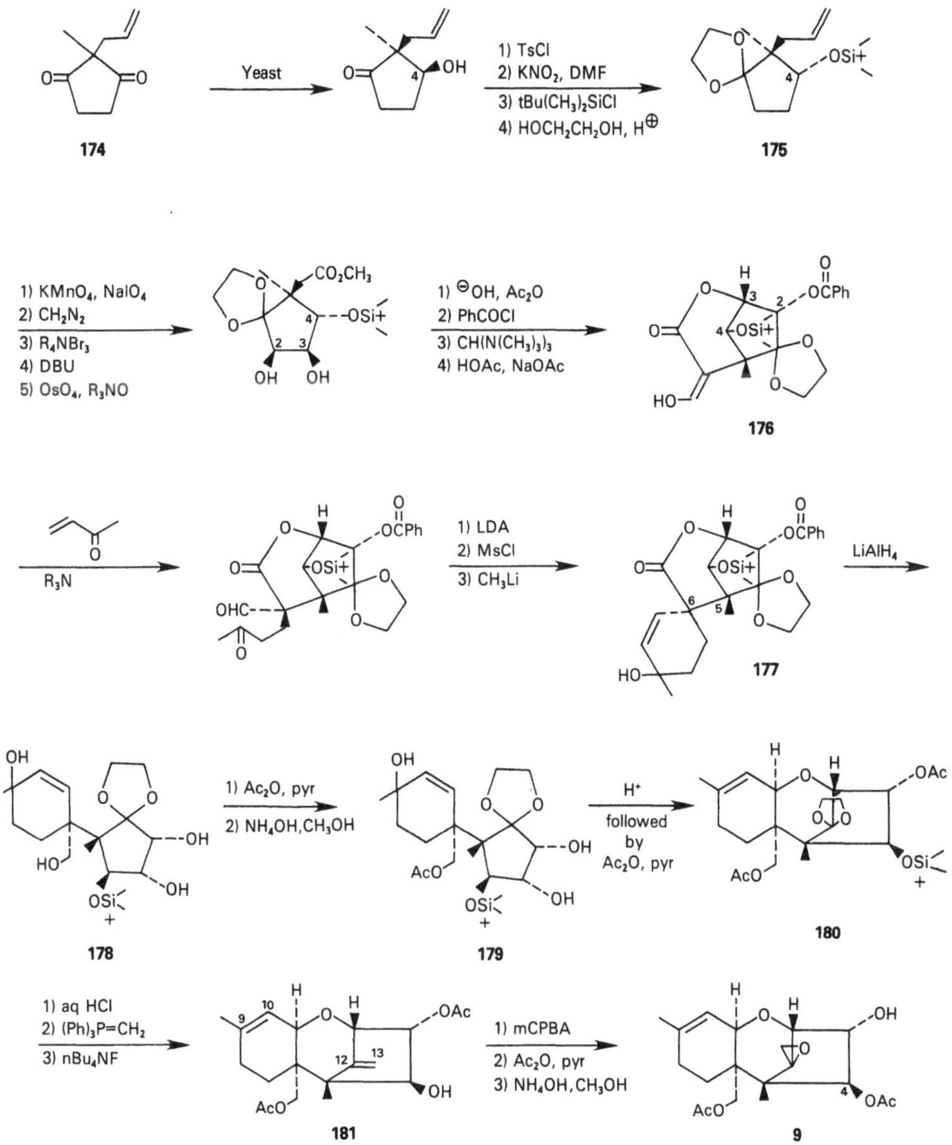

Scheme 16

always yields the *cis* AB-ring junction. At first glance, employment of the group 4 cyclization would necessite prior control of the C-11 center, thereby introducing additional synthetic complications. Within the context of this evaluation, it is not surprising that the earliest work on the group 4 approach involved model trichothecanoids possessing an aromatic A-ring, thus circumventing the question of C-11 stereochemistry.

ANDERSON and co-workers (5, 6) first exploited a group 4 cyclization in route to the construction of two aromatic trichothecene model systems, one of which is depicted in Scheme 17. The group 4 approach is embodied in the base catalyzed cyclization of keto bromide (182) to the modified trichothecene skeleton (184). Cyclization could be envisaged to proceed through the hemiketal intermediate (183). Due to the possibility of C-2 epimerization both epimers of the keto bromide (182) can be converted to product. A similar base catalyzed cyclization was peformed by GOLDSMITH and co-workers (58) in their preparation of ketone (187). In this case, epimerization of the starting keto chloride (186) was impossible; consequently, only one of the two epimers underwent ring closure.

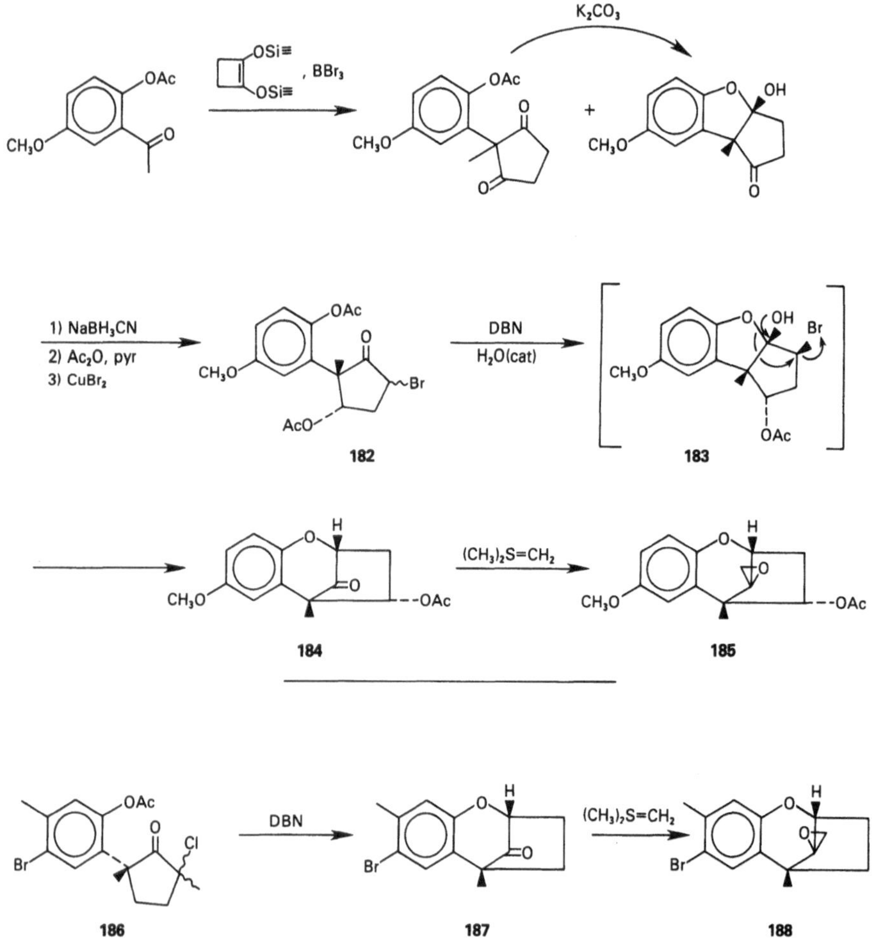

Scheme 17

Within this series of compounds there has been some disagreement as to the stereochemical outcome of epoxide formation *via* sulfonium ylides. While ANDERSON implied that the *anti*-epoxide (185) was formed from ketone (184), GOLDSMITH has presented strong evidence that this reaction actually yields the *syn*-stereoisomer (188), the wrong stereoisomer for trichothecene synthesis. This latter result is consistent with earlier studies in the trichothecene field (*33, 51*).

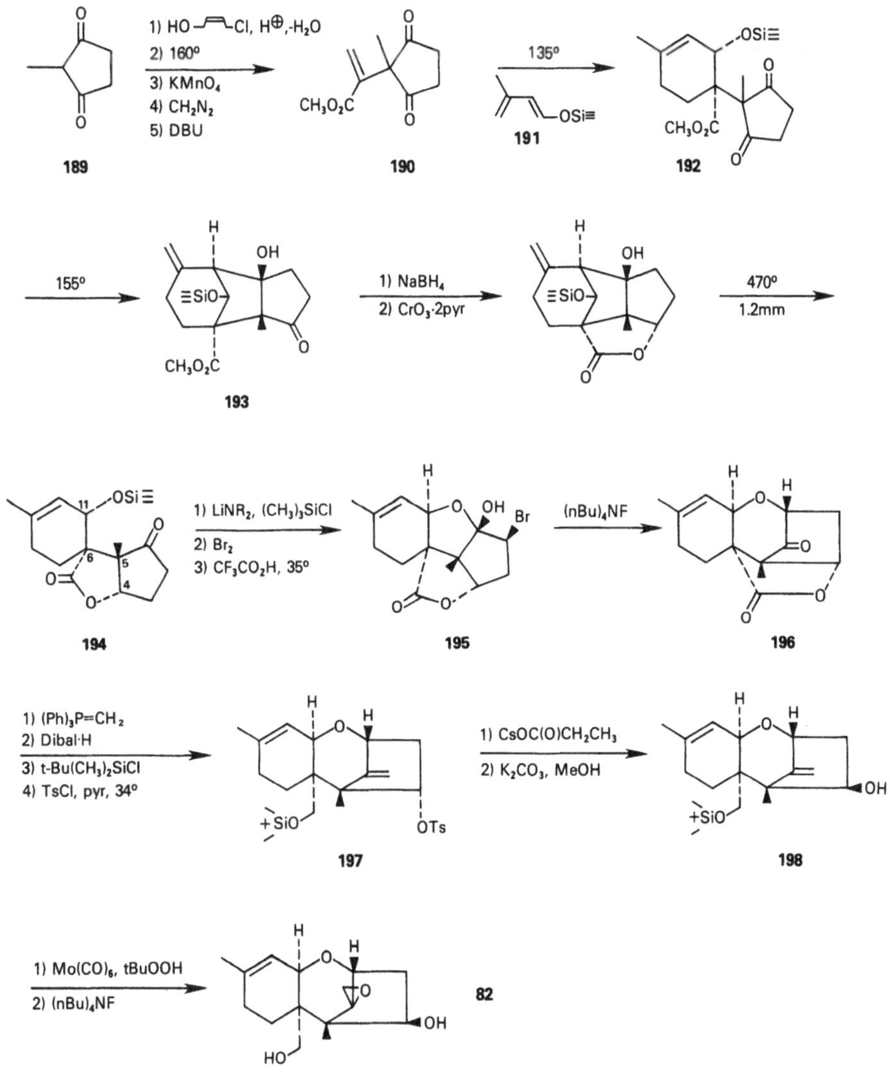

Scheme 18

A novel application of the group 4 biomimetic approach can be found in the synthesis of verrucarol (**82**) by Trost and McDougal (*145, 146*). Importantly, their work demonstrates that, contrary to fears raised in the earlier analysis, complete control of C-11 stereochemistry is possible with the group 4 approach, thus allowing for the synthesis of both the trichothecene and 11-epitrichothecene skeletons. As outlined in Scheme 18, their synthetic sequence commenced with the conversion of 2-methyl-1,3-cyclopentadione (**189**) to the methyl acrylate (**190**) in which the quaternary carbon was assembled by a Claisen rearrangement. Interestingly, the dione (**189**) serves as the starting material and C-ring synthon for three of the reported trichothecene syntheses (*27, 125, 146*). Heating of the dienophile in the presence of the siloxydiene (**191**) at 130° gave the expected Diels-Alder adduct (**192**). Unexpectedly, additional heating of this initial adduct to 155° triggered an intramolecular ene reaction, yielding the tricyclic compound (**193**). Analysis of this ene process revealed that only one of the two diastereotopic keto groups in dione (**192**) could align itself to undergo the ene reaction, a situation which serves as a mechanism to differentiate the two ketones. Maintenance of this differentiation was realized upon reduction and pyrolysis; the second step effects a *retro* ene reaction, which furnishes the modified Diels-Alder adduct (**194**). Inspection of the stereocenters in this compound reveals that while it possesses the desired relative stereochemistry at C-5 and C-6, an important consideration in the group 4 approach as it was in the group 3 methodology, inversion of both C-4 and C-11 is required if it is to serve for the eventual synthesis of verrucarol. After introduction of bromide at C-2, the C-11 stereochemistry was easily inverted by treatment with trifluoroacetic acid. Presumably, inversion occurs by intramolecular trapping of the cyclohexenyl cation by a hydrated form of the ketone which produces hemiketal (**195**) in a process reminiscent of the group 3 biomimetic cyclization. The key cyclization was then accomplished in approximately 70% by using fluoride ion as the catalyst. Employment of the more common amine or oxygen bases failed to produce any of the cyclized product (**196**). Upon formation of the tosylate (**197**), the stereochemistry at C-4 was inverted with cesium propionate. While peracid epoxidation of the resulting diene (**198**) again exhibited poor selectivity, molybdenum based epoxidation, pioneered by Sharpless, did result in exclusive formation of the desired monoepoxide. The generality of this latter oxidation within the trichothecene field remains to be explored. Simple desilylation yielded the target molecule verrucarol (**82**).

As mentioned above, the ability to synthesize the 11-epitrichothecenes is an added bonus of this methodology. Scheme 19 outlines how this was accomplished from the intermediary ketone (**194**). After introduction of bromine, treatment with fluoride ion yielded the novel 11-epitrichothecene (**199**), the structure of which was confirmed by X-ray analysis. In an effort

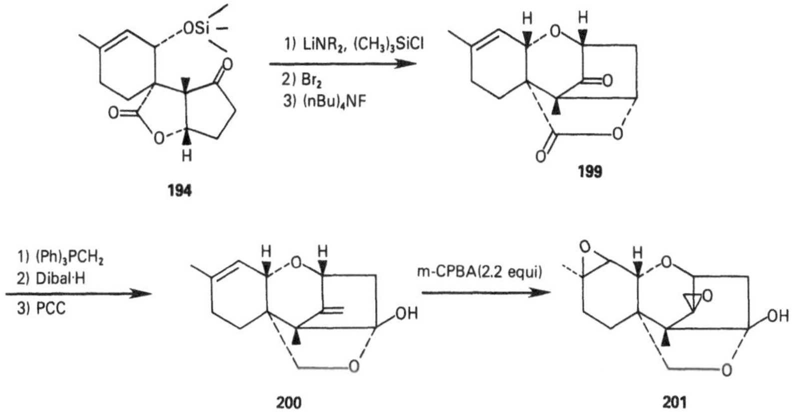

Scheme 19

to prepare analogs for biological testing, the epoxidation of the hemiketal (**200**), synthesized in 3 steps from the initial cyclization product, was studied. Unfortunately, in contrast to the verrucarol series, the molybenum catalyzed reaction resulted in slow decomposition of starting material. The diepoxide (**201**) could be formed by peracid oxidation and is currently undergoing biological screening.

4. Miscellaneous Approaches

Although to date no naturally occurring trichothecene has been synthesized by other than an aldol or biomimetic sequence, three model trichothecanoid systems have been constructed by alternative methodologies. As illustrated in Scheme 20, WHITE and co-workers (*165*) were the

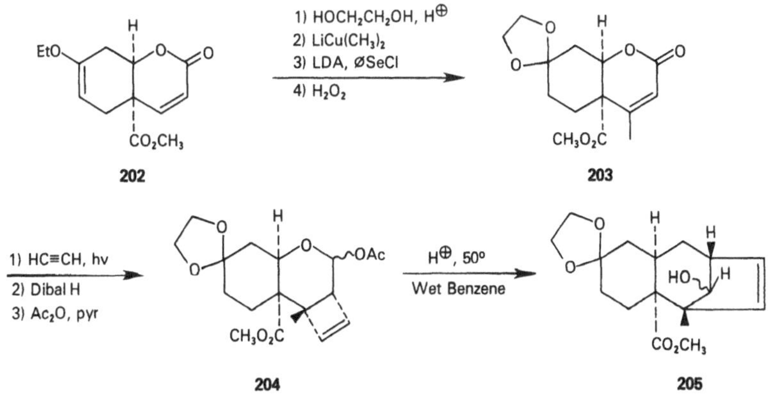

Scheme 20

first to publish an alternative strategy to the trichothecene skeleton in their synthesis of the trichothecene (**205**) which possesses all the necessary functionalities for eventual production of verrucarol. The trichothecene skeleton was formed by a cationic ring expansion of cyclobutene (**204**), itself readily prepared from the previously described Diels-Alder adduct (**202**) (see Scheme 7). The critical C-5, C-6 stereochemistry was established by the photoaddition of acetylene to enone (**203**).

Recently, Fraser-Reid and Tsang have extended their studies on the use of carbohydrates as chiral starting materials into the trichothecene field (*47, 48*). The triacetate (**206**), available from D-glucose, has been transformed into the modified trichothecene skeleton (**208**) as outlined in Scheme 21. The key ring-forming reaction in this sequence was an intramolecular alkylation of the amide (**207**). A number of transformations, including the transposition of the 8,9-olefin, still remain to convert product (**208**) into a naturally occurring trichothecene.

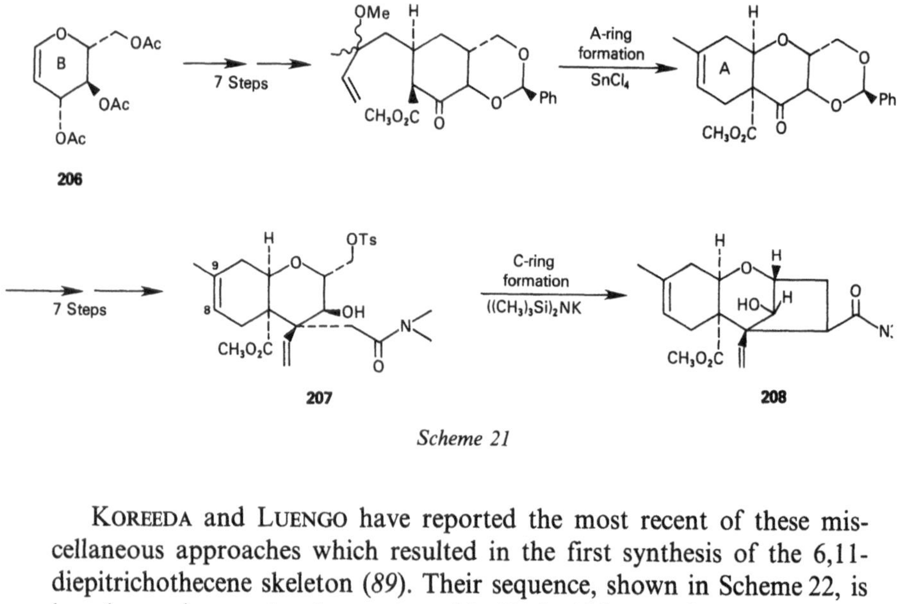

Scheme 21

Koreeda and Luengo have reported the most recent of these miscellaneous approaches which resulted in the first synthesis of the 6,11-diepitrichothecene skeleton (*89*). Their sequence, shown in Scheme 22, is based on an intramolecular version of the Diels-Alder reaction. Preparation of the two dienol ethers (**210**) and (**211**) was carried out from the common intermediate (**209**). While heating the *E*-isomer (**210**) produced no cyclized materials, the *Z*-isomer, upon heating, produced the 6,11-diepitrichothecene (**212**) in an 80% yield. It should be recalled that this trichothecene isomer could not be prepared *via* a group 3 type cyclization (see Scheme 14). Obviously, the extension of this approach to more complex trichothecene skeletons remains to be explored.

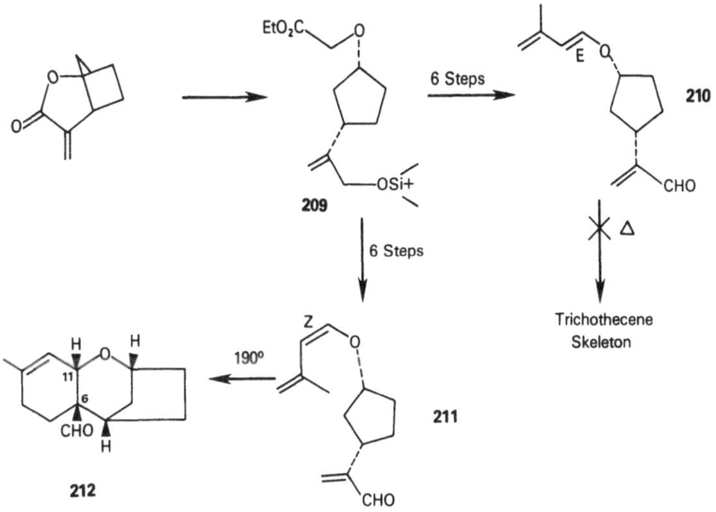

Scheme 22

V. Synthesis of the Trichoverroids

1. Introduction

The isolation of trichodermadiene (**72**) marked the discovery of the newest class of trichothecene metabolites, the trichoverroids. As noted previously, these are derivatives of the simple sesquiterpenes which possess complex ester side chains, structurally related to those found in the macrocyclic trichothecenes, at C-4 or both C-4 and C-15 (see Table VIII). Evidence for their role as biosynthetic intermediates to the macrocyclic trichothecenes (*42, 78*) has spurred interest in their synthesis.

Initially, the most demanding challenge for the synthetic chemist was assigning both the relative and absolute stereochemistry at C-6' and C-7' (see Scheme 23) of the ester sidechain. This was first accomplished by TULSHIAN and FRASER-REID (*149*) in 1981. Their methodology was used later to synthesize trichoverrin B (**71**), the first trichoverroid to be obtained by total synthesis (*42*). ROUSH and SPADA reported an alternative synthesis of the chiral trichoverroid side chains (*118*) and have most recently completed a synthesis of trichoverrol B (**69**) (*119*).

The synthetic work on both the trichoverroids and the macrocyclic trichothecenes demands a good source of the sesquiterpene verrucarol (**82**). At first, this presented a problem as the main source of verrucarol (**82**) was the hydrolysis of verrucarin A (**37**), a macrocyclic trichothecene whose supply was somewhat limited. This problem has been rectified by TULSHIAN and Fraser-Reid (*150*) who have developed a three-step synthesis of verrucarol from the readily available mycotoxin anguidine (**9**). Consequently, a surge of research involving the conversion of verrucarol (**82**) to other naturally occurring metabolites has occurred.

2. Total Syntheses

Tulshian and Fraser-Reid have established the relative and absolute stereochemistry at C-6′, C-7′ of the ester sidechain through a chiral synthesis originating from simple sugars (*149*). As outlined in Scheme 23, D-glucose was converted into the diacetate (**213**) which contained the two chiral centers destined to become C-6′ and C-7′ of the trichoverroids. Acid catalyzed opening of the dihydropyran gave the hydroxy aldehyde (**214**) which was condensed with the silyl acetate (**216**) to yield a 50:50 mixture of olefin isomers from which the desired Z diene (**217**) was isolated. Upon treatment with sodium methoxide the resulting dihydroxy diene (**218**) was found to be identical with the hydrolysis products from the B group of the trichoverroids (see Table VIII), thereby establishing their stereochemistry as D-erythro. Stereoselective formation of the (−)-epoxide (**219**) estab-

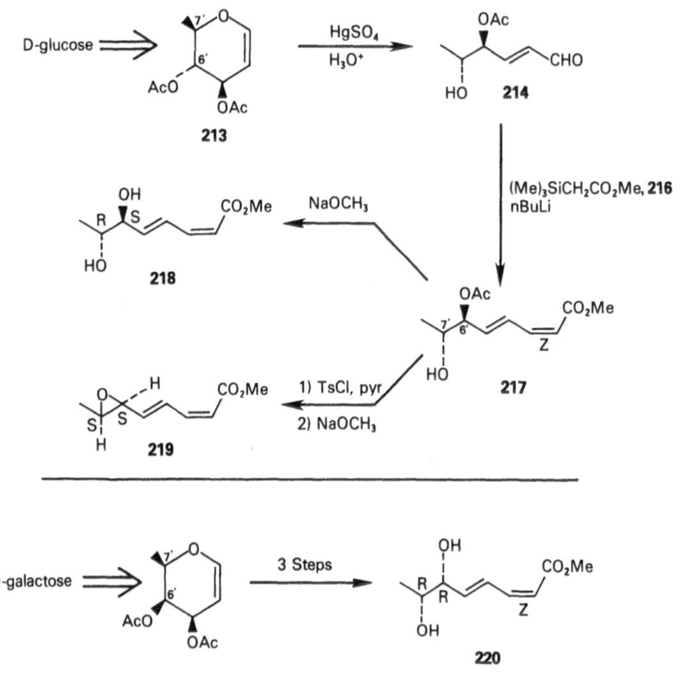

Scheme 23

lished the absolute configuration of trichodermadiene (**72**) as 6′-*R*, 7′-*R*. A similar sequence of reactions converted D-galactose into the *R,R*-diene (**220**) which was shown to be the enantiomer of the hydrolysis products from the A group trichoverroids. Hence, the A group trichoverroids possess the L-threo or 6′-*S*, 7′-*S* stereochemistry.

With the answer to these important stereochemical questions in hand, ESMOND, FRASER-REID and JARVIS turned their attention to the synthesis of trichoverrin B (**71**), a possible biogenic precursor of the macrocyclic trichothecenes (*42*). The successful synthesis, shown in Scheme 24, followed a convergent strategy which first necessitated the construction of the two side chain pieces. The acyl-activated chiral diene (**222**) was prepared from aldehyde (**214**), a product of D-glucose (see Scheme 23). As before, a Peterson olefination introduced the second ene unit and produced a mixture

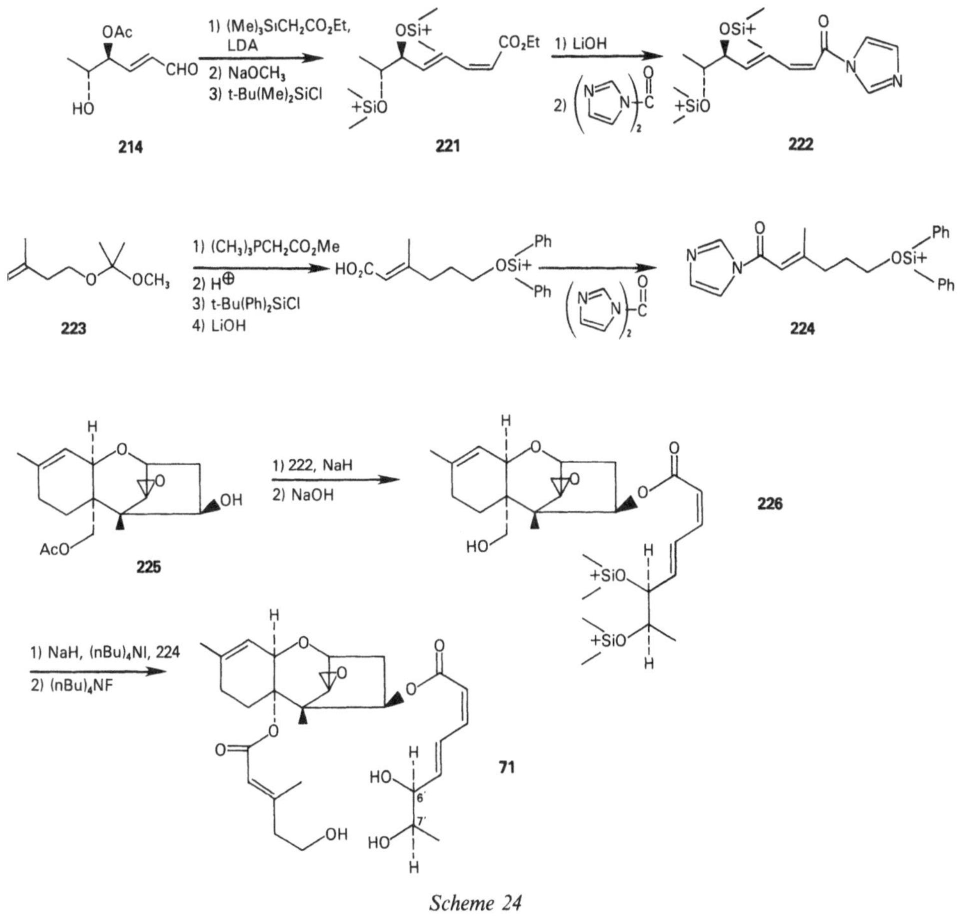

Scheme 24

of *E*- and *Z* (**221**)-isomers from which the pure *Z*-isomer (**222**) was ultimately obtained. The second side chain (**224**) was prepared in 5 steps from ketone (**223**). In both fragments the acyl groups were activated as an acyl imidazole.

Attachment of the dienoic ester (**222**) onto 15-acetoxyverrucarol (**225**) was achieved *via* the sodium alkoxide. This esterification was accompanied by some diene isomerization which required that the desired *Z*,*E*-isomer (**226**) be purified by HPLC. Esterification of the second acyl piece (**224**) onto the C-15 hydroxy group was sluggish although inclusion of am-monium iodide allowed realization of a 60% yield of diester. Desilylation then gave a pure sample of trichoverrin B (**71**).

Roush and Spada have reported an alternative synthesis of the optically pure trichoverroid ribbons based on an enantioselective reduction (*118*). The chiral propargyl alcohol (**227**) (see Scheme 25) was prepared from the corresponding ketone by reduction with lithium aluminium hydride in the presence of the chiral ligand Darvon alcohol. Subsequent reduction of the propargyl alcohol to a *trans*-allylic alcohol and epoxidation gave the epoxy alcohol (**228**) as a single enantiomer. In accord with Fraser-Reid's observations, control of the 2',3'-double bond geometry during olefination reactions was difficult. Roush and Spada have nicely circumvented this problem by introducing the olefin into the cyclohexane (**229**) thus assuring

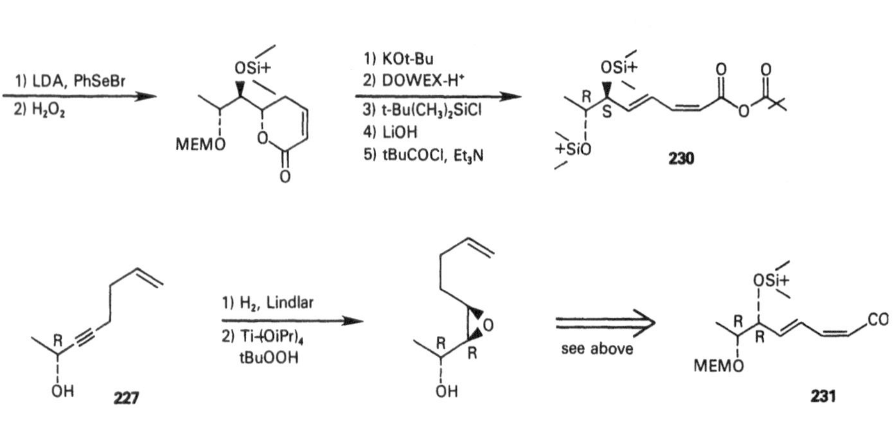

Scheme 25

formation of the Z-isomer. A series of standard transformations then yielded the diene component (230) with the acyl group activated as a mixed anhydride. The diastereomeric diene (231) could be constructed from the identical chiral alcohol (227) by first forming the cis-allylic alcohol via catalytic hydrogenation. Hydroxy-assisted epoxidation followed by the same sequence of reactions then yielded the 6'-R, 7'-R dienoic acid (231).

In a separate communication ROUSH and SPADA (119) utilized their chiral ribbons in a synthesis of trichoverrol B (69) as shown in Scheme 26. Verrucarol (82) was selectively esterified at C-15 to yield the monoester (232). Formation of the sodium alkoxide and acylation with the mixed anhydride (230) successfully introduced the diene fragment with no observable diene isomerization. The authors hypothesized that the lack of a nucleophilic species capable of undergoing reversible Michael additions with the diene unit accounts for the absence of olefin isomerization. Simple treatment with fluoride then yielded the monoester trichoverrol B (69).

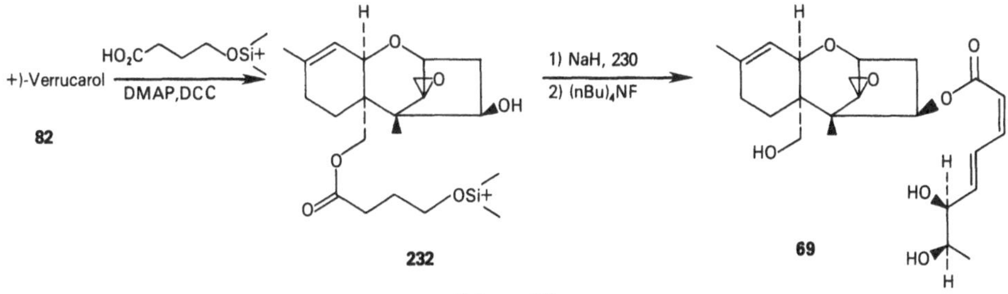

Scheme 26

VI. Synthesis of Macrocycles

1. Introduction

The first report dealing with the synthesis of a macrocyclic trichothecene, albeit a non-natural product, originated from TAMM's laboratory in 1978 (22). Three years later, STILL and OHMIZU synthesized the first naturally occurring macrocyclic trichothecene verrucarin A (37) (133). Since 1981, the increased availability of verrucarol (82) (150), the sesquiterpenoid backbone for most of the macrocycles, has stimulated work in this area and culminated in STILL's recent syntheses (135) of baccharin B5 (59) and roridin E (47), the most complex trichothecenes synthesized to date. Sandwiched between these landmark syntheses have been equally exciting studies initiated by TAMM (verrucarin A and 3α-hydroxyverrucarin A (106), FRASER-REID and JARVIS (verrucarin J) (42) and ROUSH (verrucarin J and verrucarin B) (*115, 121*). Besides the initial model studies of Tamm and Breitenstein (*22*), a number of other model systems have been synthesized (*109, 148*). Finally, a good deal of literature exists on the construction of the acyclic ribbons without detailing their attachment to the verrucarol nucleus. Most noticeable in this latter category is Schlessinger's work (*124*) on the "southern half" of vertisporin (**55**), perhaps the most complex member of the trichothecene family.

Many of the existing syntheses share two common strategies. Firstly, the syntheses are convergent with either one or most often two acyclic units assembled prior to their attachment onto verrucarol. Such a strategy assures the efficient use of the most precious commodity, verrucarol. Secondly, a common retrosynthetic disconnection has been at the ester linkages for both the attachment of the acyclic pieces and for the formation of the macrocycle. This allows chemists to avail themselves of the extensive methodology developed for the formation of esters and, in particular, for macrolactonizations. A notable exception to this latter strategy has been Still's recent synthesis of the dilactide macrocycles in which an intramolecular carbon bond forming reaction serves to close the macrocycle (*135*).

As dictated by the two aforementioned strategies, the selective functionalization of verrucarol (**82**) at either the C-4 or C-15 hydroxy group has been studied in detail. While initial reports in this area seemed to be contradictory, a clearer picture has now emerged. Esterifications which involve acyl chlorides or acyl imidazoles tend to a favor reaction at the secondary hydroxy group of C-4 (see Scheme 27). Conversely, esterifications which utilize acid anhydrides or acyl groups activated by dicyclohexylcarbodimide (DCC) tend to favor reaction at the C-15 primary hydroxy function (see Scheme 30). In general the selectivity for C-15 monoesterification is higher than for C-4 monoesterification, although a number of highly selective reactions at C-4 do exist (*109*).

2. Model Systems

As model studies represented most of the early activity in the synthesis of macrocyclic trichothecenes and, in addition, contributed a number of important precedents for future synthetic endeavors, the syntheses of these systems will be outlined first.

Historically, it is no coincidence that Tamm, who has had a long involvement with all aspects of trichothecene chemistry and, hence, had access to an ample supply of verrucarol, initially lead the field in the synthesis of the macrocycles. In 1978, Breitenstein and Tamm synthesized the model macrocycle tetrahydroverrucarin J (**236**) as shown in Scheme 27

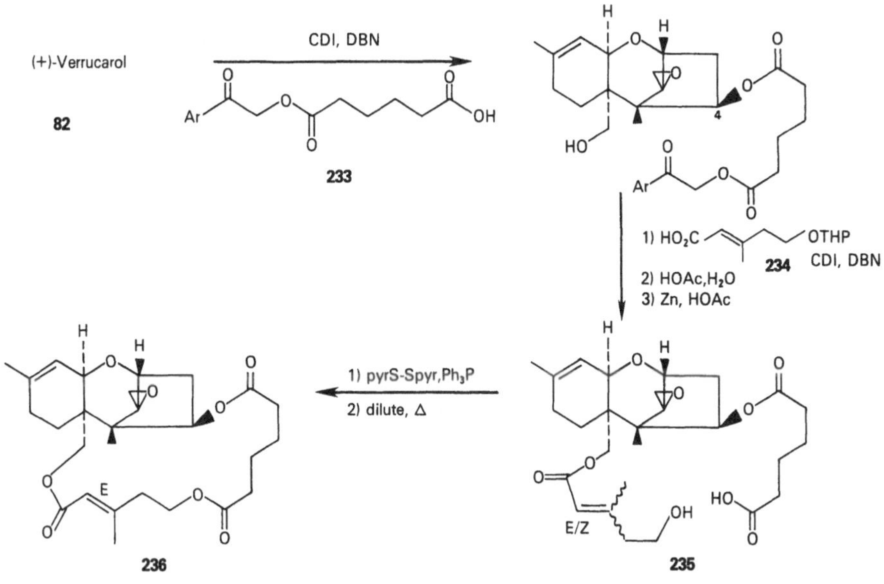

Scheme 27

(*22*). This report first noted the different reactivities of acyl chlorides and imidazoles *vs.* acid anhydrides towards esterification at C-4 or C-15. Based on this difference, the monoacid (**233**) was selectively attached to the secondary hydroxy group (C-4) *via* an acyl imidazole in modest yield. Attachment of the second arm (**234**) by a similar procedure followed by the removal of all protecting groups gave the *seco*-acid (**235**) as a mixture of *E,Z*-isomers. Macrolactonization of this mixture, through the intermediacy of a thioester, yielded the macrocycle (**236**) as the pure *E*-isomer. Evidently only the *E*-isomer, which is the olefin geometry found in verrucarin J, can undergo cyclization. This selective cyclization of a naturally occurring olefin isomer from isomeric mixtures has been observed in other trichothecanoid systems (*106, 133, 148*).

The synthesis of another macrocyclic model compound was communicated by NOTEGEN, TORI and TAMM in 1981 and is summarized in Scheme 28 (*109*). Selective protection (72%) of the C-4 hydroxy moiety as the acetal (**237**) allowed esterification with acid (**238**) to take place at C-15. The hydroxy acid (**239**), devoid of protecting groups, was then lactonized, this time onto the C-4 hydroxy group of verrucarol, to yield the model macrocycle (**240**). These early examples established that macrolactonization, either at C-4 (Scheme 28) or at the primary hydroxy group of the side chain (Scheme 27) would be equally valid approaches to the naturally occurring verrucarins, a fact which has since been substantiated by synthesis of the natural products (*121, 133*).

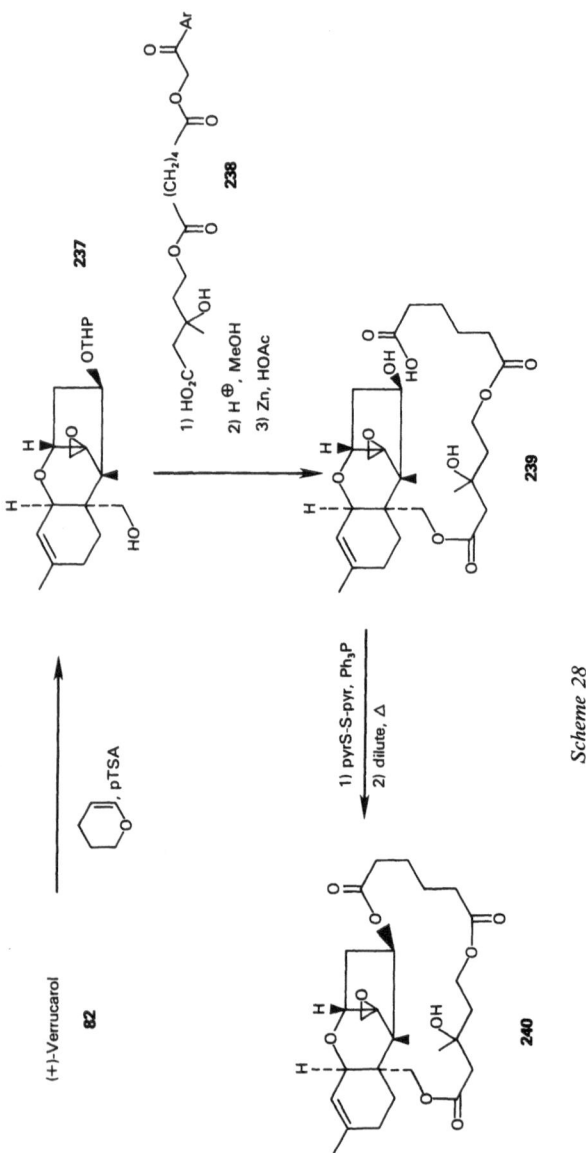

Scheme 28

The most recent model study, undertaken by TROST and McDOUGAL, has explored an unusual approach to verrucarin-type trilactides (*148*). Their supposition was that thermal isomerization of the *cis*-cyclobutene (**242**), which could produce two possible E,Z-isomers by competing conrotatory processes, would exhibit selectivity due to the conformational constraint imposed by the macrocycle. The synthesis of the target molecule is outlined in Scheme 29. Importantly, the key macrolactonization (**241**→**242**) proceeded without isomerization of the *cis*-cyclobutene moiety. Upon thermolysis a mixture of E,Z-isomers (**243**) and (**244**) was obtained. Comparison with an authentic sample revealed that the "natural" isomer (**243**) was indeed the major compound present (2:1). The applicability of such an approach to the verrucarins must yet be explored. Interestingly, these relatively simple models inhibit protein synthesis in a fashion reminiscent of the natural trichothecenes.

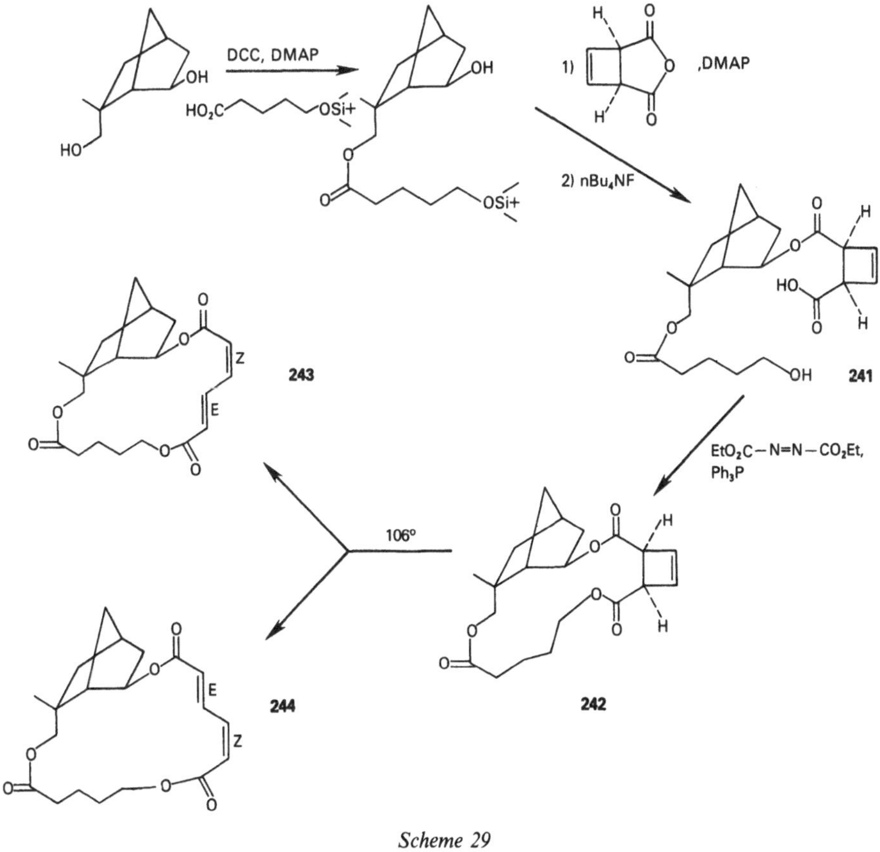

Scheme 29

3. Verrucarin A

Most synthetic work directed toward the macrocyclic trichothecenes has focused on the verrucarins, particularly verrucarin A (37) and verrucarin J (40). For verrucarin A there now exist two total syntheses (106, 133), starting from verrucarol (82), as well as innumerable reports on the synthesis of verrucarinic acid derivatives, a principal component of the macrocyclic ribbon (see Scheme 32).

The synthesis of verrucarin A by Still and Ohmizu in 1981 (133) was the first total synthesis of a naturally occurring macrocyclic trichothecene. As outlined in Scheme 30, their strategy, as that of others, first necessitated

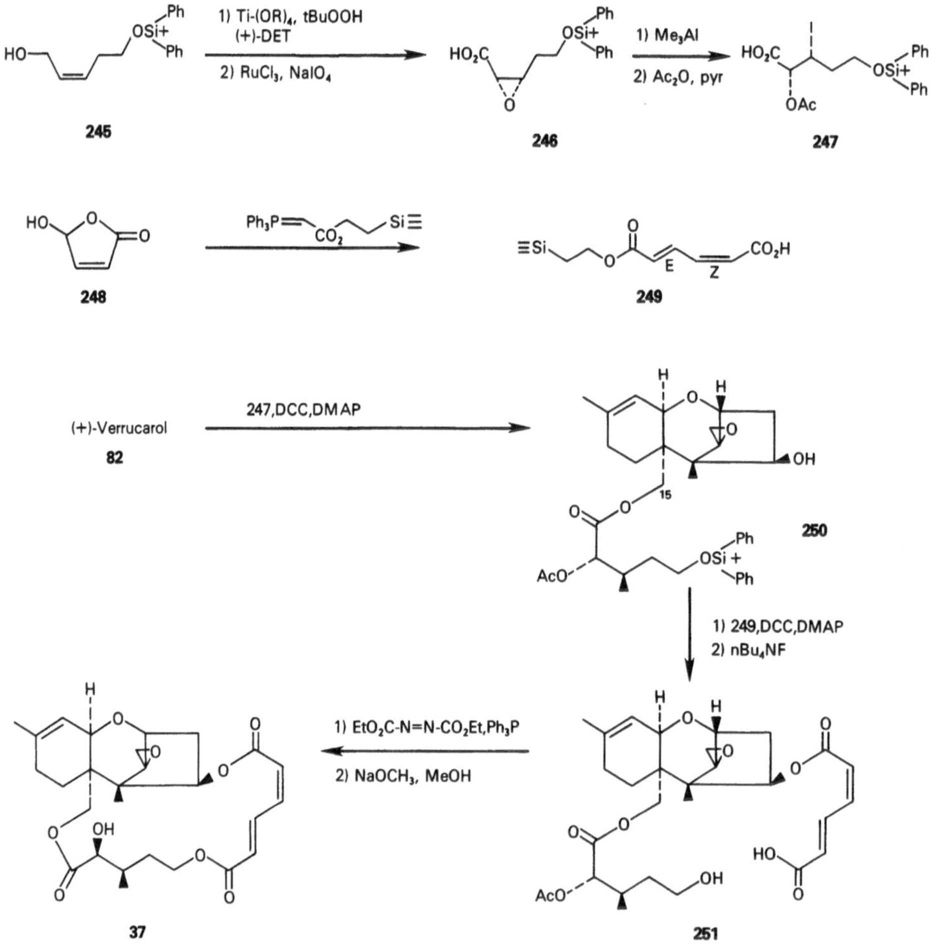

Scheme 30

the construction of the acyclic precursors (**247**) and (**249**), the latter in optically pure form. It should be emphasized that whenever the macrocyclic ribbon contains a chiral center the absolute stereochemistry of that center must be controlled in order to avoid the formation of diastereomers when the center is attached to chiral verrucarol. The enantioselective synthesis of acid (**247**) was accomplished by chiral epoxidation of olefin (**245**). Regioselective opening of the epoxy acid (**246**) with trimethylaluminium inverted the stereochemistry at C-3 to yield a protected form of verrucarinic acid (**247**). The second acyclic component, diene (**249**), was prepared in one step from malealdehydic acid (**248**). Two independent syntheses of diene (**249**) exist (*106, 148*) as well as a synthesis of its related half ester (**287**) and *E,E*-isomer (**288**) (see Scheme 35) (*115*).

Reaction of the verrucarinic acid derivative (**247**) with verrucarol (**82**), mediated by DCC and 4-N,N-dimethylaminopyridine (DMAP), yielded the monoester (**250**) by exclusive acylation at C-15. A similar esterification procedure attached the muconic acid (**249**) to C-4. Following desilylation the *seco* acid (**251**) was cyclized (52%) and deacetylated to yield verrucarin A (**37**). When alternative conditions for the esterification of acid (**249**) were employed, partial isomerization of the *E,Z*-isomer (**251**) to an *E,E*-isomer was observed. However, lactonization of this mixture produced only the natural *E,Z*-macrocyclic (**37**) as the *E,E*-isomer failed to cyclize. Such reactivity differences have been observed with other trichothecanoid macrocycles (*22, 148*), although, as shall be seen, *E,E*-macrocycles can be prepared if so desired (*115, 121*).

An alternative synthesis of verrucarin A (**37**) was achieved by TAMM and co-workers in 1982 (*106*). Strategically, the approach (shown in Scheme 31) was identical with that of STILL and OHMIZU. The difference of the two syntheses resides principally in the construction of the acyclic synthons. TAMM controlled the absolute stereochemistry of the verrucarinic acid piece (**254**) through the chiral acid (**252**), itself obtained by an enzyme mediated hydrolysis of the corresponding prochiral dimethylester. Introduction of the second chiral center by hydroxylation of the ester (**253**) was only partially stereoselective (2:1), but ultimately yielded the desired chiral acid (**254**). Base catalyzed elimination of the lactonic ester (**255**) produced an alternative synthesis of the dienoic acid (**249**). Sequential acylation of the chiral acid (**254**) and the dienoic acid (**249**) onto verrucarol (**82**) gave diester (**256**) contaminated with substantial amounts of the *E,E*-isomer. In general, the *E,Z*-muconate moiety is prone to isomerization, both during its incorporation and, in some cases, during macrolactonization. However, consonant with STILL's observations, only the natural *E,Z*-isomer underwent cyclization yielding verrucarin A (**37**) upon removal of the THP protecting group. A similar series of reactions originating from the anguidine derivative (**257**) produced the unnatural macrocycle (**258**).

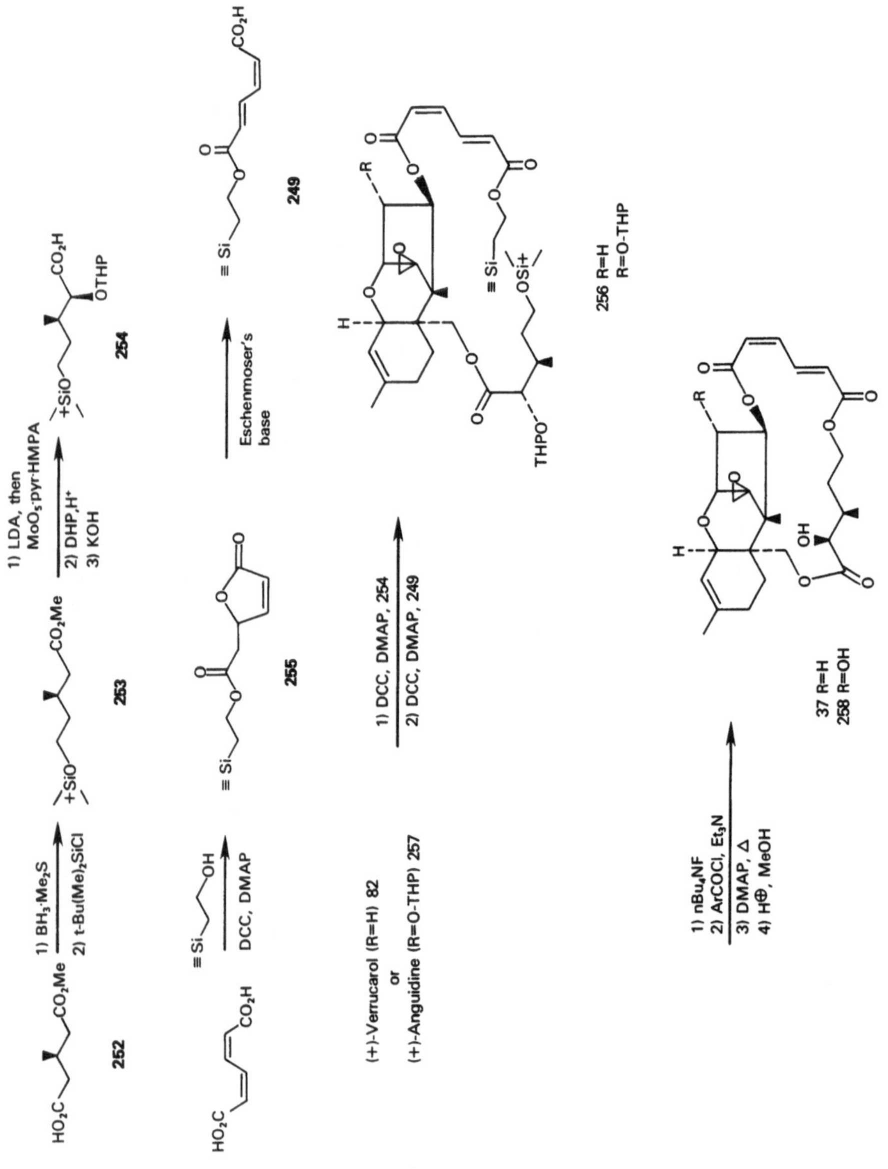

252
1) BH3·Me2S
2) t-Bu(Me)2SiCl
253
1) LDA, then MoO5·pyr·HMPA
2) DHP,H+
3) KOH
254
DCC, DMAP
255
Eschenmoser's base
249
(+)-Verrucarol (R=H) 82
or
(+)-Anguidine (R=O-THP) 257
1) DCC, DMAP, 254
2) DCC, DMAP, 249
256 R=H
R=O-THP
1) nBu4NF
2) ArCOCl, Et3N
3) DMAP, Δ
4) H⊕, MeOH
37 R=H
258 R=OH

Scheme 31

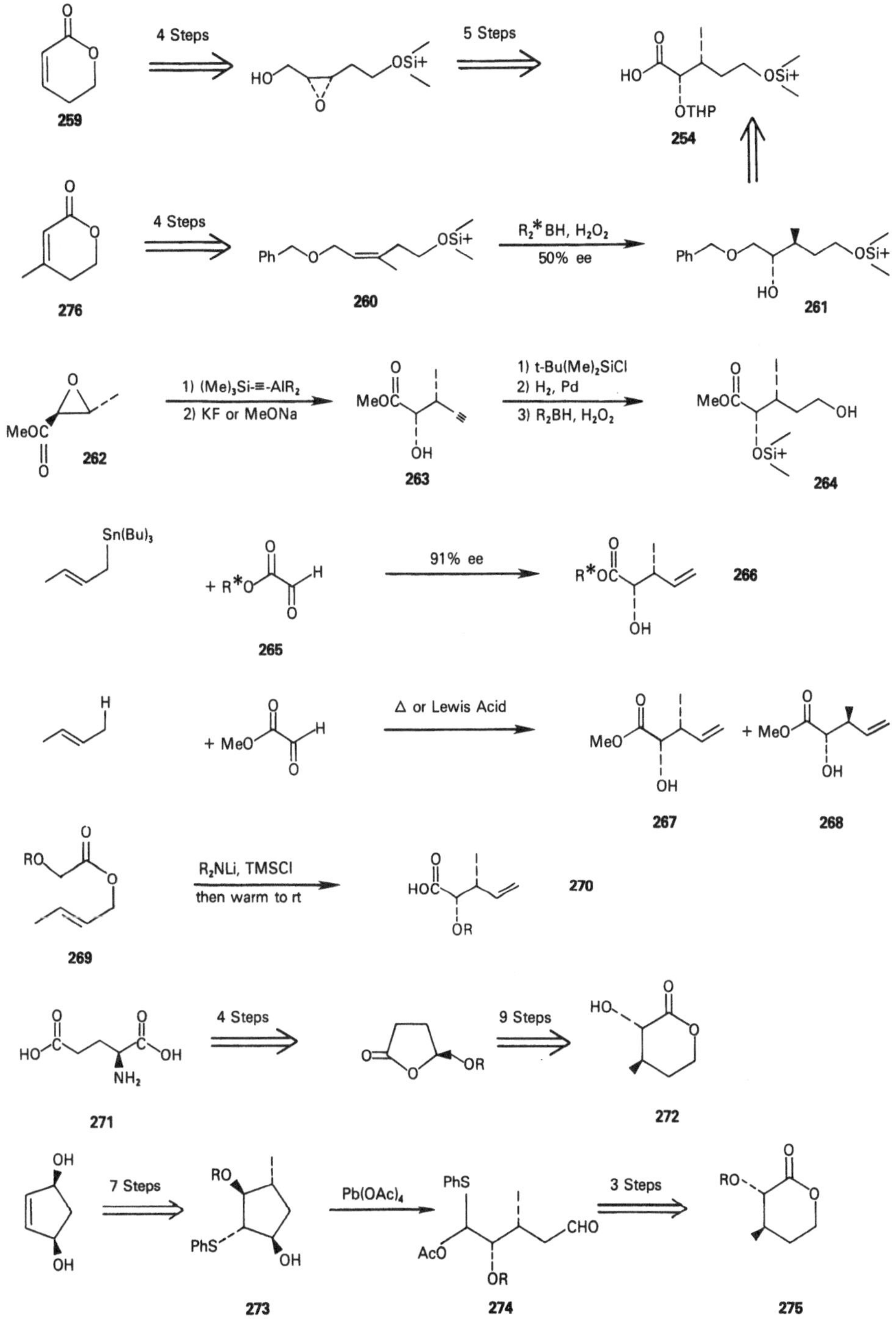

4 Steps
5 Steps
OTHP
4 Steps
R₂*BH, H₂O₂
50% ee
1) (Me)₃Si-≡-AlR₂
2) KF or MeONa
1) t-Bu(Me)₂SiCl
2) H₂, Pd
3) R₂BH, H₂O₂
Sn(Bu)₃
91% ee
Δ or Lewis Acid
R₂NLi, TMSCl
then warm to rt
4 Steps
9 Steps
7 Steps
Pb(OAc)₄
3 Steps

Scheme 32

As part of the recent surge in methodology related to control of acyclic stereochemistry, numerous reports have appeared on the synthesis of racemic and chiral verrucarinic acid derivatives. These are summarized in Scheme 32. In most instances, the eventual synthesis of verrucarin A is not the primary concern of these efforts.

In addition to the route discussed in Scheme 31, TAMM and co-workers have developed two other syntheses of chiral verrucarinic acid derivatives (67). One of their newer routes (**259→254**) closely mimics STILL's original work in the area (see Scheme 30) by utilizing a chiral epoxidation to establish the desired absolute stereochemistry. Their second alternative involves an enantioselective hydroboration (**260→261**) which proceeded with only mediocre optical induction.

Nearly identical routes to the hydroxy ester (**264**) have been developed independently by ROUSH and by TROST (122, 144). Selective opening of the trans-epoxide (**262**) gave alcohol (**263**) with correct relative stereochemistry at C-2 and C-3. Subsequent conversion to the hydroxy ester (**264**) was straightforward. Successful resolution of alcohol (**263**) allows this route to be used for the synthesis of chiral material (144).

A tin-mediated ene reaction on the chiral glyoxylate ester (**265**, R* = chiral group) has yielded the verrucarinic acid precursor (**266**) as virtually a single enantiomer (168). The selectivity of this metallo-ene process stands in sharp contrast to the proto-ene reaction (129) which produces a mixture of diastereomers (**267**) and (**268**) in low yield.

In connection with studies on the Ireland-Claisen rearrangement of α-oxyacetates, groups led by BARTLETT (13), BURKE (30) and FUJISAWA (123) have synthesized various derivatives of acid (**270**). In general, these workers have found that derivatives of ester (**269**) where R was not a proton gave the highest yields and diastereoselectivities. BURKE and co-workers converted their product to racemic verrucarinolactone (**272**).

Finally, two relatively long syntheses to verrucarinolactone derivatives exist. The first by KOGA and co-workers (140) accounts for the synthesis of chiral material as their sequence originates with L-glutamic acid (**271**). TROST and McDOUGAL (144) have prepared lactone (**275**) via a bond cleavage reaction (**273→274**) developed in TROST's laboratories.

4. Verrucarin J

Two total syntheses of verrucarin J (**40**) have been recorded. The first by ESMOND, FRASER-REID and JARVIS (see Scheme 34) is an extension of their work in the trichoverroid area (see Scheme 24) (42). ROUSH and BLIZZARD have published a second synthesis of verrucarin J and have developed routes

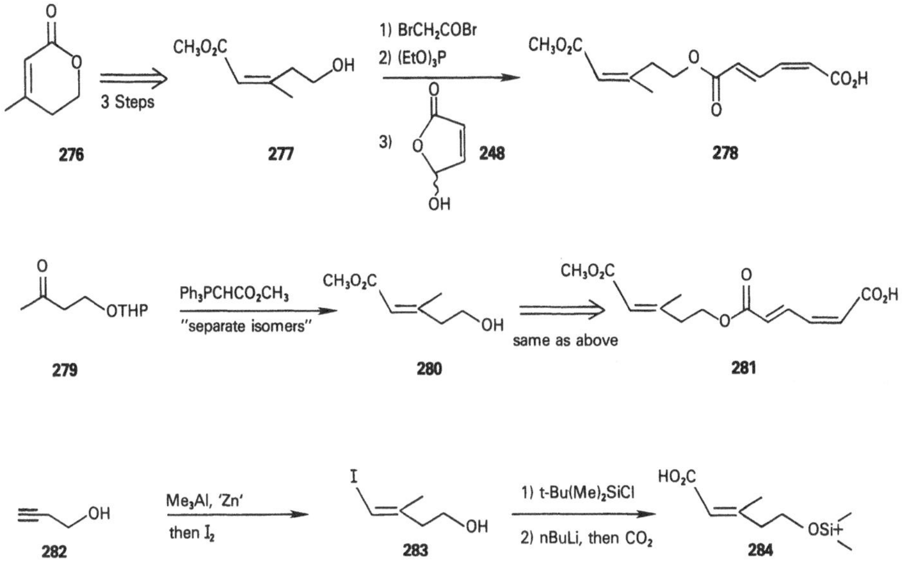

Scheme 33

to most of the olefin isomers related to the naturally occurring compound
(*115, 121*).

As was the case with verrucarin A, a number of routes to the acyclic
precursors have been described and are summarized in Scheme 33.
Although the macrocyclic ribbon of verrucarin J (**40**) contains no chiral
centers, it does contain three olefins whose geometries must be controlled.
The acrylic ester portion of the side chain, originally assigned the *Z*
configuration, has subsequently been shown to possess the *E* configuration
as a result of n.O.e. experiments (*102*) as well as by synthetic correlation (*22,
42, 121*). Due to the uncertainty associated with this geometry stereospecific
syntheses of both isomers (**277** and **280**, Scheme 33) have been reported.

Three groups, lead by TAMM (*22*), WHITE (*166*) and ROUSH (*122*) have
synthesized the unnatural *Z*-isomer (**277**) starting from lactone (**276**). The
ROUSH and WHITE groups further elaborated their compounds into the
complete side chain (**278**) of 2′,3′-*Z*-verrucarin J *via* a Horner-Emmons
reaction on malealdehydic acid (**248**). The *E*-acrylates, corresponding to the
natural configuration of verrucarin J, have been synthesized by two routes.
The first, utilized by two groups (*42, 166*), involves a nonstereoselective
Wittig-type condensation with the butanone (**279**). WHITE and co-workers
again elaborated their acrylate piece (**280**) into the complete acyclic ribbon
(**281**). ROUSH provided a stereoselective synthesis based on Negishi's
zirconium-catalyzed addition of organometallics across acetylenes
(**282**→**283**). This ultimately yielded the 5-carbon piece (**284**) which was

both transformed into the complete ribbon (**281**) (*122*) and, more importantly, used directly in a successful synthesis of verrucarin J (see Scheme 35) (*121*).

The synthesis of verrucarin J (**40**) by Esmond, Fraser-Reid, and Jarvis (*42*) involved an oxidative ring closure of trichoverrin B (**71**) as shown in Scheme 34. This two-step oxidation, initiated by the chromium-based oxidant PDC, compares favorably with standard macrolactonizations as a 50% yield of verrucarin J (**40**) can be realized. Presumably, the hemiacetal (**285**) is the crucial intermediate for the second oxidation step. The ability to form the macrocycle without isomerization of the olefin bonds is a significant feature of this reaction.

Extensive efforts toward the synthesis of the macrocyclic trichothecenes have been initiated by Roush and co-workers and culminated in the synthesis of verrucarin J (**40**) and related olefin isomers (*121*). The successful approach, outlined in Scheme 35, was one in which a growing side chain was ultimately cyclized onto the hydroxy group at C-4 of verrucarol (**82**). Initial efforts to attach an intact ribbon such as (**281**) (see Scheme 33) to verrucarol were frustrated by concomitant isomerization of

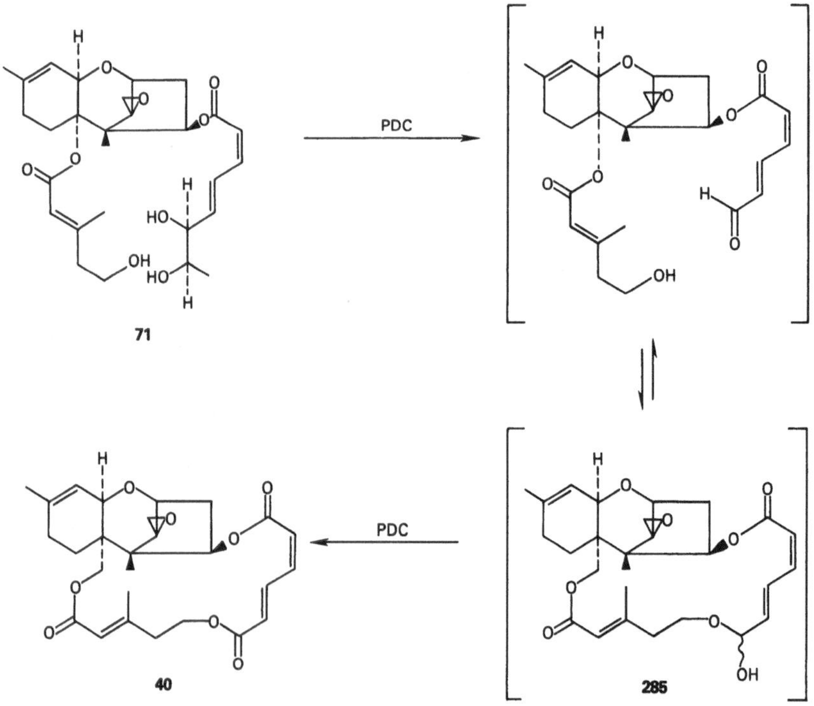

Scheme 34

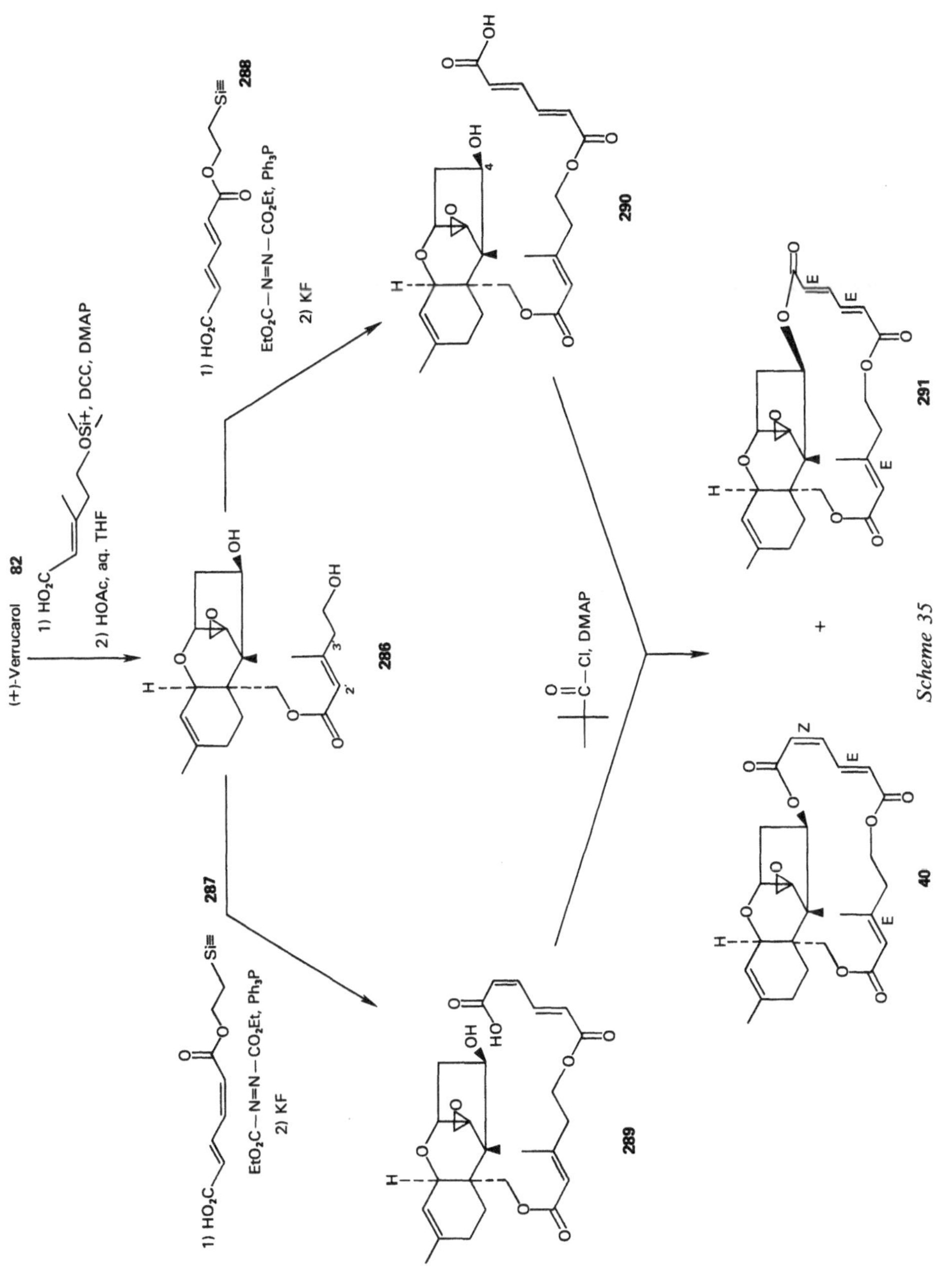
(+)-Verrucarol
82
1) HO2C
OSi+, DCC, DMAP
2) HOAc, aq. THF
286
1) HO2C
Si≡
288
EtO2C—N=N—CO2Et, Ph3P
2) KF
287
290
289
C—Cl, DMAP
291
40
+
E
Z

Scheme 35

the E,Z-muconate moiety to the thermodynamically more stable E,E-
isomer. This type of isomerization is a continuing problem with all
compounds containing the muconate side chain.

The Roush approach to the verrucarin J system began with the well-
precedented DCC-mediated acylation of the primary alcohol function of
verrucarol (82) to give the flexible intermediate (286). At that point either
one of the two muconate isomers (287) or (288) could be attached to the
primary alcohol of diol (286). After suitable deprotection, the E,E,Z-seco
acid (289) and the E,E,E-seco acid (290) were obtained. Lactonization of
either pure isomer via a mixed anhydride produced mixtures of the isomeric
macrocycles (40) and (291). In both cases the major isomer obtained has the
configuration of the starting seco acid. Isomer crossover most likely occurs
prior to macrolactonization and is dependent on the reaction conditions
chosen. For example cyclization of the E,E,Z-isomer (289) with DCC yields
only the E,E,Z-macrocycle (40), albeit in poor yield (115). Two additional
macrocycles were created starting from the acrylate (286) having the 2',3'-Z-
configuration (only the E geometry is depicted in Scheme 35). Again
lactonization of the Z,Z,E-macrocycle corresponding to (289) yielded a

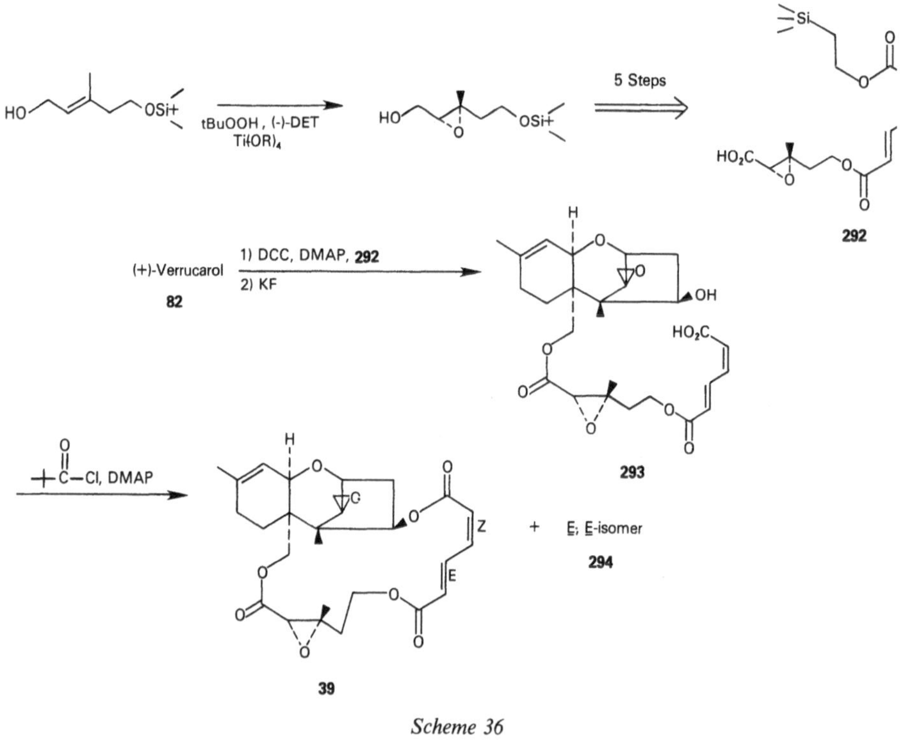

Scheme 36

mixture of Z,Z,E- and Z,E,E-macrocycles from muconate isomerization. Given the current interest in the biological activity of modified trichothecenes (*87*), these isomeric macrocycles, easily separated by chromatography, are ideal candidates for biological study.

5. Verrucarin B

ROUSH and BLIZZARD have also synthesized the macrocyclic trichothecene verrucarin B (**39**) (*115*). Their approach is shown in Scheme 36. In this case the intact ribbon (**292**), containing a chiral epoxide, was attached onto the primary hydroxyl group of verrucarol (**82**). After unmasking of the acid with fluoride, the resulting *seco* compound (**293**) was cyclized *via* a mixed anhydride. Again isomerization accompanied macrolactonization so that a 2:1 mixture of the E,Z- (**39**) and E,E-isomers (**294**) were obtained. Although the mixed anhydride method is haunted by isomerization, the overall yields from this procedure $(70-90\%)$ are among the highest reported for macrolactonization of a trichothecene. A careful study of the lactonization step has shown that isomerization can only be avoided at the expense of overall yield (*115*).

6. Roridin E and Baccharin B5

The crowning achievement so far of synthetic efforts toward the macrocyclic trichothecenes has been STILL's synthesis of the roridins, roridin E (**47**) and baccharin B5 (**59**) (*135*). The roridins are structurally more complex than the verrucarins since they possess additional chiral centers at C-6' and C-13' [see (**47**) and (**59**), Scheme 37]. In general it is difficult to ascertain the stereochemistry at C-6' and C-13' by standard spectroscopic methods and, in fact, STILL's synthesis of roridin E (**47**) served to establish its configuration as 6'-R, 13'-R. Baccharin B5, a plant metabolite of the roridins, owns the same 6',13'-configuration, which enabled STILL to construct both compounds from a single advanced intermediate. For this reason the syntheses of roridin E (**47**) and baccharin B5 (**59**) are discussed together.

The synthetic strategy utilized by STILL and coworkers, detailed in Scheme 37, contained two aspects unique within the macrocyclic trichothecene field. First, the macrocycle was constructed by a carbon-carbon bond forming reaction (**299**→**300**) in contrast to the usual cyclization by way of a carbon-oxygen bond. Second, additional chiral centers were introduced onto an already formed macrocycle. Previous endeavors had accounted for all the chiral centers prior to their incorporation into the

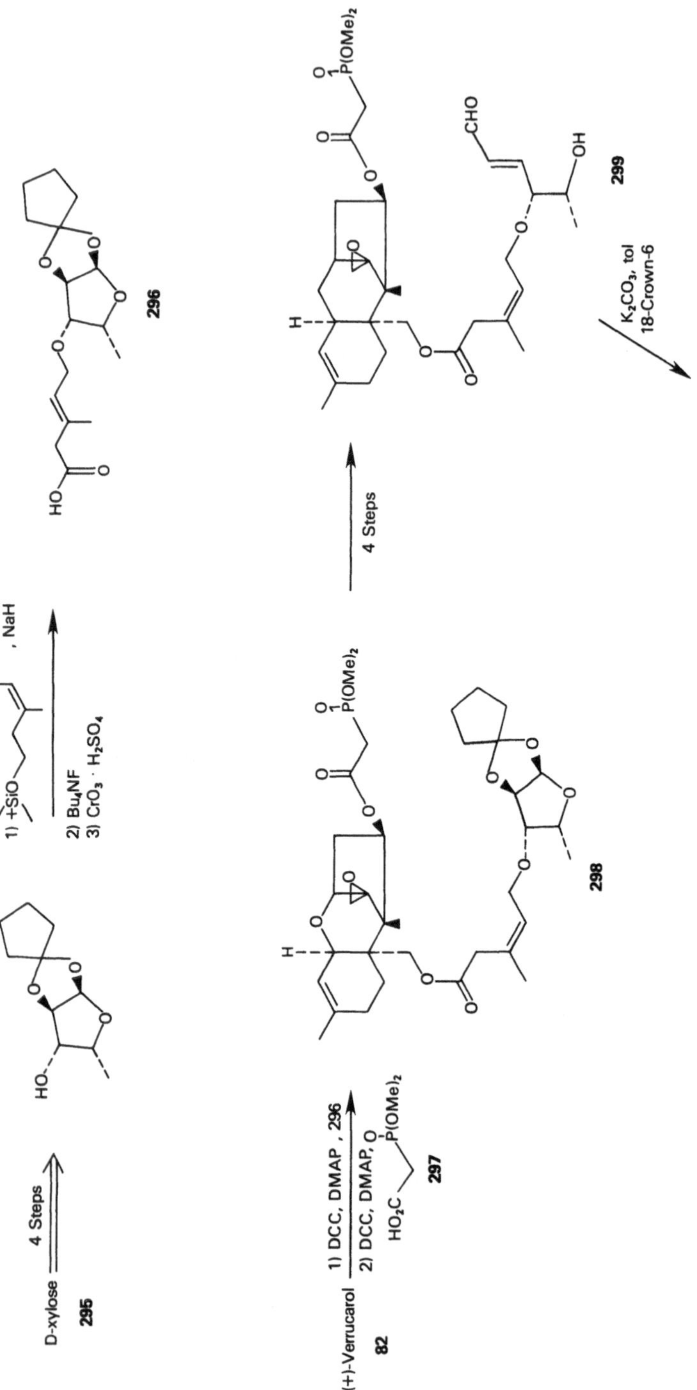
D-xylose
295
4 Steps
1) +SiO / X , NaH
2) Bu4NF
3) CrO3 · H2SO4
296
(+)-Verrucarol
82
1) DCC, DMAP, 296
2) DCC, DMAP, HO2C P(OMe)2
297
298
4 Steps
299
K2CO3, tol
18-Crown-6

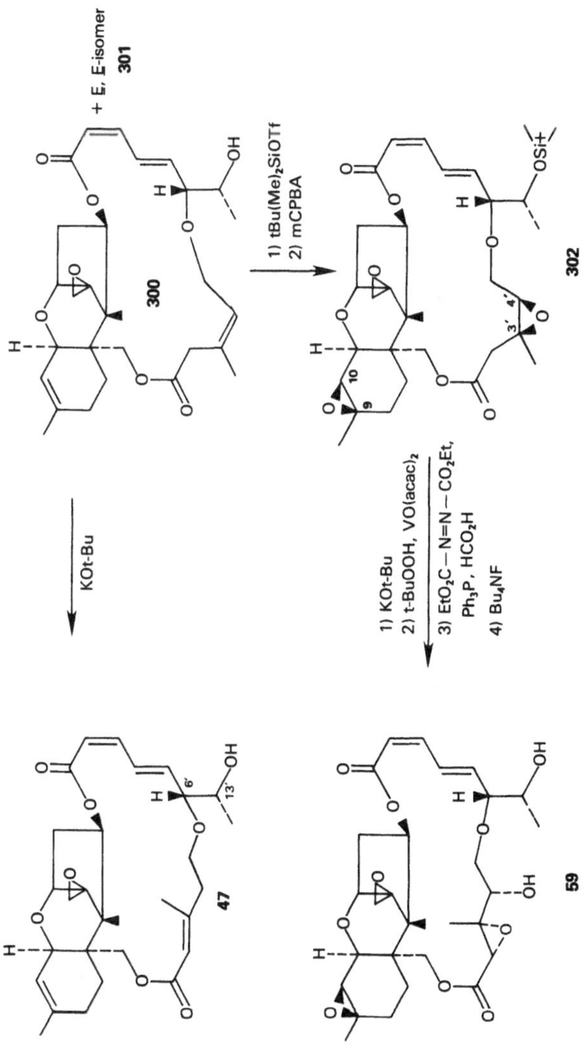
+ E, E-isomer
301
300
1) tBu(Me)2SiOTf
2) mCPBA
302
KOt-Bu
1) KOt-Bu
2) t-BuOOH, VO(acac)2
3) EtO2C—N=N—CO2Et,
Ph3P, HCO2H
4) Bu4NF
47
59

Scheme 37

macrocycle. This novel strategy permitted utilization of the conformational bias of the macrocycle to obtain stereocontrol.

STILL chose to establish the aboulte stereochemistry at C-6′ and C-13′ by use of the chiral starting material, D-xylose (295). In seven steps this readily available sugar was transformed into the chiral ether (296) which represented a major portion of the macrocyclic chain. Selective attachment of the acid (296) to the primary hydroxyl group of verrucarol (82) proceeded smoothly and, following incorporation of the phosphonate (297), the diacyl verrucarol derivative (298) was obtained. Modification of the sugar nucleus followed by intramolecular condensation yielded the E,Z-macrocycle (300) together with nearly equivalent amounts of the E,E-isomer (301). It is this advanced intermediate (300) which has served as a precursor to both roridin E (47) and baccharin B5 (59).

The synthesis of roridin E (47) was completed by simple base catalyzed conjugation of the 3′,4′-double bond (300→47). The exclusive formation of the desired olefin geometry $(2′,3′-E)$ demonstrated that selective reactions could be performed on the intact macrocycle. An even more impressive display of stereocontrol was witnessed in the synthesis of baccharin B5 (59). Of the five additional chiral centers necessary to complete the synthesis of baccharin B5, four were introduced by simple peracid oxidation yielding the triepoxide (302) as a single diastereomer ($> 15:1$). While formation of the 9,10-β-epoxide had ample precedent (87), the selective epoxidation of the 3′,4′-double bond must result from conformational constraints imposed by the macrocyclic ring. This stereocontrol is maintained and elaborated in a four-step sequence, featuring the familiar hydroxy-directed epoxidation of an allylic alcohol, to yield baccharin B5 (59) with its 14 asymmetric centers.

7. Vertisporin

While the complete synthesis of vertisporin (55, see Table V) has yet to be achieved, SCHLESSINGER and coworkers have constructed the acyclic portion or "southern half" of the macrocycle (124). As seen in Scheme 38, even this acyclic portion (305) represents a significant synthetic challenge. The strategy employed by SCHLESSINGER is an especially efficient example of the use of carbohydrates for constructing chiral substances, as all four chiral centers present in D-xylose (295) are retained in the final molecule (305). The key step in the sequence was the acid catalyzed dehydration of enol (303) to yield the bicyclic ether (304). This ether contains the basic ring system of the side chain and following modest modifications the target acid (305) was obtained. Completion of the synthesis calls for attachment of the acid (305) to the primary hydroxyl group of verrucarol with the final ring closure to take place at the C-4 oxygen of verrucarol. This plan awaits laboratory verification.

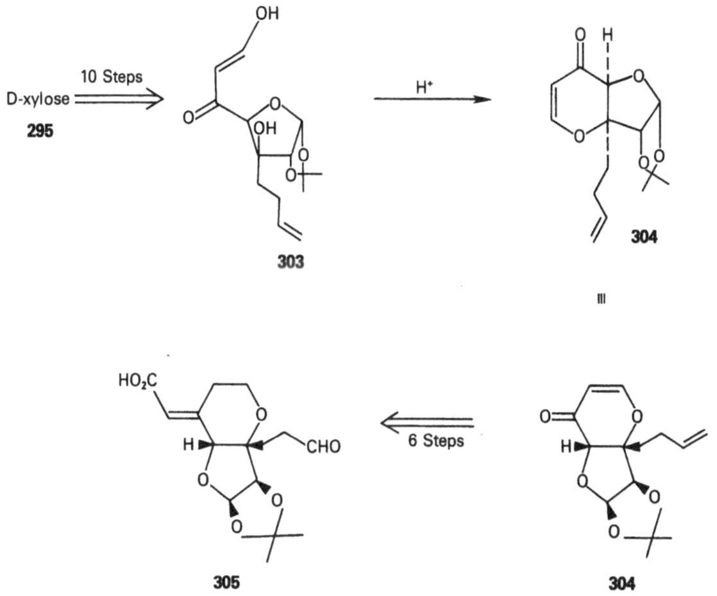

Scheme 38

References

1. ABRAHAMSSON, S., and B. NILSSON: The molecular structure of trichodermin. Acta Chem. Scand. **20,** 1044 (1966).

2. — — Direct determination of the molecular structure of trichodermin. Proc. Chem. Soc. (London) **1964,** 188.

3. ACHILLADELIS, B., and J. R. HANSON: Minor terpenoids of *Trichothecium.* Phytochem. **8,** 765 (1969).

4. ADLER, S. S., S. LOWENBRAUN, B. BIRCH, R. JARRELL, and J. GARRARD: Anguidine: a broad phase II study of the Southeastern Cancer Study Group. Cancer Treat. Rep. **68,** 423 (1984).

5. ANDERSON, W. K., E. J. LAVOIE, and G. E. LEE: Synthesis of 6,9-bisnormethyl-8-methoxy-6,8,10-trichothecatriene. J. Organ. Chem. (USA) **42,** 1045 (1977).

6. ANDERSON, W. K., and G. E. LEE: Synthesis of C-ring-functionalized A-ring-aromatic trichothecane analogs. J. Organ. Chem. (USA) **45,** 501 (1980).

7. ASAKE, Y., and S. TAKITANI: Thin-layer chromatographic analysis of trichothecene mycotoxins. Dev. Food Sci. **4,** 113 (1983).

8. BAMBURG, J. R., and F. M. STRONG: 12,13-Epoxytrichothecenes. Microbial Toxins **7,** 207 (1971).

9. BAMBURG, J. R.: Chemical and biochemical studies of the trichothecene mycotoxins. In: Mycotoxins and other fungal related food problems (RODRICKS, J. V., ed.), p. 144. Washington, D.C.: American Chemical Society. 1976.

10. Bamburg, J. R., and F. M. Strong: Mycotoxins of the trichothecene family produced by *Fusarium tricinctum* and *Trichoderma lignorum*. Phytochem. **12**, 2405 (1969).

11. Bamburg, J. R.: Biological and biochemical actions of trichothecene mycotoxins. Proc. Mol. Subcell. Biol. **8**, 41 (1983).

12. Banks, R. E., J. A. Miller, M. J. Nunn, P. Stanley, T. J. R. Weakley, and Z. Ullah: Diels-Alder route to potential trichothecene precursors. J. Chem. Soc., Perkin Trans. I **1981**, 1096.

13. Bartlett, P. A., D. J. Tanzella, and J. F. Barstow: Ester-enolate Claisen rearrangement of lactic acid derivatives. J. Organ. Chem. (USA) **47**, 3941 (1982).

14. Bata, A., A. Vanyl, and R. Lasztity: Study of mycotoxins in foods. V. Parallel detection of some fusariotoxins by capillary gas chromatography. Elelmiszervizsgalati Kozl. **28**, 189 (1982).

15. Bennett, G. A., R. E. Peterson, R. D. Plattner, and O. L. Shotwell: Isolation and purification of deoxynivalenol and a new trichothecene by high-pressure liquid chromatography. J. Amer. Oil Chem. Soc. **58**, 1002 (1981).

16. Betina, V., and M. Vankova: Trichothecin − an antibiotic morphogenic factor. Mycotoxin and bitter substance of apples. Biologia (Bratislava) **32**, 943 (1977).

17. Blight, M. M., and J. F. Grove: New metabolic products of *Fusarium culmorum*. Toxic trichothec-9-en-8-ones and 2-acetylquinazolin-4(3H)-one. J. Chem. Soc., Perkin Trans. I **1974**, 1691.

18. Bloem, R. J., T. A. Smitka, R. H. Bunge, J. C. French, and E. P. Mazzola: Roridin L-2, a new trichothecene. Tetrahedron Letters, **24**, 249 (1983).

19. Bohner, B., and Ch. Tamm: Die Konstitution von Roridin A. Helv. Chim. Acta **49**, 2527 (1966).

20. − − Die Konstitution von Roridin D. Helv. Chim. Acta **49**, 2547 (1966).

21. Breitenstein, W., and Ch. Tamm: Carbon-13 NMR spectroscopy of the trichothecane derivatives verrucarol, verrucarins A and B, and roridins A, D, and H. Helv. Chim. Acta **58**, 1172 (1975).

22. − − Partial synthesis of tetrahydroverrucarin J. Helv. Chim. Acta **61**, 1975 (1978).

23. Breitenstein, W., Ch. Tamm, E. V. Arnold, and J. Clardy: The absolute configuration of the fungal metabolite Verrucarin B. Biosynthetic consequences. Helv. Chim. Acta **62**, 2699 (1979).

24. Breitenstein, W., and Ch. Tamm: Verrucarin K, the first natural trichothecane derivative lacking the 12,13-epoxy group. Helv. Chim. Acta **60**, 1522 (1977).

25. Brian, P. W., and J. G. McGowan: Biologically active metabolic products of the mould *Metarrhizium glutinosum* S. Pope. Nature **157**, 334 (1946).

26. Brooks, D. W., P. G. Grothaus, and W. L. Irwin: Chiral cyclopentanoid synthetic intermediates *via* asymmetric microbial reduction of prochiral 2,2-disubstituted cyclopentanediones. J. Organ. Chem. (USA), **47**, 2820 (1982).

27. Brooks, D. W., P. G. Grothaus, and H. Mazdiyasni: Total synthesis of the trichothecene mycotoxin anguidine. J. Amer. Chem. Soc. **105**, 4472 (1983).

28. Brooks, D. W., P. G. Grothaus, and J. T. Palmer: Synthetic studies of trichothecenes. An enantioselective synthesis of a C-ring precursor of anguidine. Tetrahedron Letters **23**, 4187 (1982).

29. Buening, G. M., D. D. Mann, B. Hook, and G. D. Osweiler: The effect of T-2 toxin on the bovine immune system: Cellular factors. Vet. Immunol. Immunopathol. **3**, 411 (1982).

30. Burke, S. D., W. F. Fobare, and G. J. Pacofsky: Chelation control of enolate geometry. Acyclic diastereoselection *via* the enolate Claisen rearrangement. J. Organ. Chem. (USA) **48**, 5221 (1983).

31. Busam, L., and G. G. Habermehl: Accumulation of mycotoxins by *Baccharis coridifolia*: A reason for livestock poisoning. Naturwissenschaften **69**, 392 (1982).

32. COLE, R. J., J. W. DORNER, R. H. COX, B. M. CUNFER, H. G. CUTLER, and B. J. STUART: The isolation and identification of several trichothecene mycotoxins from *Fusarium heterosporum*. J. Nat. Prod. **44**, 324 (1981).

33. COLVIN, E. W., S. MALCHENKO, R. A. RAPHAEL, and J. S. ROBERTS: Total synthesis of (±)-trichodermin. J. Chem. Soc. Perkins Trans. I **1973**, 1989.

34. — — — — Synthetic studies on the sesquiterpene antibiotic verrucarol. J. Chem. Soc. Perkin Trans. I, **1978**, 658.

35. CORDELL, G. A.: Biosynthesis of sesquiterpenes. Chem. Rev. **76**, 426 (1976).

36. CUNDLIFFE, E., and J. E. DAVIES: Inhibition of initiation, elongation and termination of eukaryotic protein synthesis by trichothecene fungal toxins. Antimicrob. Agents Chemother. **11**, 491 (1977).

37. DAWKINS, A. W., J. F. GROVE, and B. K. TIDD: Diacetoxyscirpenol and some related compounds. Chem. Commun. **1965**, 27.

38. DAWKINS, A. W.: Phytotoxic compounds produced by *Fusarium equiseti* II. The chemistry of diacetoxyscirpenol. J. Chem. Soc. (C) **1966**, 116.

39. DOYLE, T. W., and W. T. BRADNER: Trichothecenes. In: Anticancer agents based on natural product models (CASSIDY, J. M., and J. D. DOUROS, eds.), p. 43. New York: Academic Press. 1980.

40. EPPLEY, R. M., E. P. MAZZOLA, M. E. STACK, and P. A. DREIFUSS: Structure of satratoxin F and satratoxin G, metabolites of *Stachybotrys atra*: Application of proton and carbon-13 nuclear magnetic resonance spectroscopy. J. Organ. Chem. (USA) **45**, 2522 (1980).

41. EPPLEY, R. M., E. P. MAZZOLA, R. J. HIGHET, and W. J. BAILEY: Structure of satratoxin H, a metabolite of *Stachybotrys atra*. Application of proton and carbon-13 nuclear magnetic resonance. J. Organ. Chem. (USA) **42**, 240 (1977).

42. ESMOND, R., B. FRASER-REID, and B. B. JARVIS: Synthesis of trichoverrin B and its conversion to verrucarin J. J. Organ. Chem. (USA) **47**, 3358 (1982).

43. EVANS, R., J. R. HANSON, and T. MARTEN: Terpenoid biosynthesis XI. Stereochemistry of some stages in trichothecane biosynthesis. J. Chem. Soc. Perkins I. **1974**, 857.

44. FETZ, E., B. BOHNER, and CH. TAMM: Die Konstitution von Verrucarin J. Helv. Chim. Acta **48**, 1669 (1965).

45. FISHMAN, J., E. R. H. JONES, G. LOWE, and M. C. WHITING: The chemistry and stereochemistry of trichothecin. J. Chem. Soc. **1960**, 3948.

46. FLURY, E., R. MAULI, and H. P. SIGG: Constitution of diacetoxyscirpenol. Chem. Commun. **1965**, 26.

47. FRASER-REID, B.: Synthetic approach to trichothecenes from glucose. Kagaku. Zokan (Kyoto) **1983**, 29.

48. FRASER-REID, B., and R. TSANG: Personal Communication.

49. FREEMAN, G. G., J. E. GILL, and W. S. WARING: The structure of trichothecin and its hydrolysis products. J. Chem. Soc. **1959**, 1105.

50. FREEMAN, G. G., and R. I. MORRISON: The isolation and chemical properties of trichothecin, an antifungal substance from *Trichothecium roseum* Link. Biochemistry **44**, 1 (1949).

51. FUJIMOTO, Y., S. YOKURA, T. NAKAMURA, T. MORIKAWA, and T. TATSUNO: Total synthesis of (±)-12,13-epoxytrichothec-9-ene. Tetrahedron Letters **1974**, 2523.

52. GARDNER, D., A. T. GLEN, and W. B. TURNER: Calonectrin and 15-deacetylcalonectrin, new trichothecanes from Calonectria nivalis. J. Chem. Soc. Perkin Trans. I **1972**, 2576.

53. GHOSAL, S., D. K. CHAKRABARTI, A. K. SRIVASTAVA, and R. S. SRIVASTAVA: Toxic 12,13-epoxytrichothecenes from anise fruits infected with *Trichothecium roseum*. J. Agric. Food Chem. **30**, 106 (1982).

54. GODTFREDSEN, W. O., J. F. GROVE, and CH. TAMM: Zur Nomenklatur einer neuen Klasse von Sesquiterpenen. Helv. Chim. Acta **50**, 1666 (1967).

55. Godtfredsen, W. O., and S. Vangedal: Trichodermin, a new antibiotic related to trichothecin. Proc. Chem. Soc. (London) **1964**, 188.
56. — — Trichodermin, a new sesquiterpene antibitiotic. Acta Chem. Scand. **19**, 1088 (1965).
57. Goldsmith, D. J., A. J. Lewis, and W. C. Still, Jr.: Bicyclic intermediates for trichothecane synthesis. Exploitation of an enolate as a protecting group. Tetrahedron Letters 4807 (1973).
58. Goldsmith, D. J., T. K. John, C. D. Kwong, and G. R. Painter, III: Preparation and rearrangement of trichothecane-like compounds. Synthesis of aplysin and filiformin. J. Organ. Chem. (USA) **45**, 3989 (1980).
59. Gutzwiller, J., and Ch. Tamm: Über die Struktur von Verrucarin A. Helv. Chim. Acta **48**, 157 (1965).
60. — — Über die Struktur von Verrucarin B. Helv. Chim. Acta **48**, 177 (1965).
61. Grove, J. F.: Phytotoxic compounds produced by *Fusarium equiseti*. VI. 4β,8α,15-triacetoxy-12,13-epoxytrichothec-9-ene-3α,7α-diol. J. Chem. Soc. (C) **1970**, 378.
62. — Phytotoxic compounds produced by *Fusarium equiseti*. V. Transformation products of 4β,15-diacetoxy-3α,7α-dihydroxy-12,13-epoxytrichothec-9-en-8-one and the structures of nivalenol and fusarenone. J. Chem. Soc. (C) **1970**, 375.
63. — The Constituents of Glutinosin. J. Chem. Soc. **1968**, 810.
64. Harrach, B., G. Danko, G. Cseh, and M. Benko: Isolation of macrocyclic trichothecenes from straw associated with death of calves (Stachybotryotoxicosis). Magy. Allatorv. Lapja **37**, 808 (1982).
65. Härri, E., W. Loeffler, H. P. Sigg, H. Stahelin, Ch. Stoll, Ch. Tamm, and D. Wiesinger: Über die Verrucarine und Roridine, eine Gruppe von cytostatisch hochwirksamen Antibiotica aus *Myrothecium*-Arten. Helv. Chim. Acta **45**, 839 (1962).
66. Hayakawa, S., E. Kondo, Y. Wakisaka, H. Minato, and K. Katagiri: Vertisporin, a new antibiotic from *Verticimonosporium diffractum*. J. Antibiot. **28**, 550 (1975).
67. Herold, P., P. Mohr, and Ch. Tamm: Syntheses of optically active verrucarinic acid. Helv. Chim. Acta **66**, 744 (1983).
68. Ilus, T., P. J. Ward, M. Nummi, H. Adlercreutz, and J. Gripenberg: A new mycotoxin from *Fusarium*. Phytochemistry **16**, 1839 (1977).
69. Ishii, K., and Y. Ueno: Isolation and characterization of two new trichothecenes from *Fusarium sporotrichioides* strain M-1-1. Appl. Environ. Microbiol. **42**, 541 (1981).
70. Ishii, K.: Chemistry and bioproduction of nonmacrocyclic trichothecenes. Dev. Food Sci. **4**, 7 (1983).
71. — Two new trichothecenes produced by *Fusarium* species. Phytochemistry **14**, 2469 (1975).
72. Ishii, K., K. Sakai, Y. Ueno, H. Tsunoda, and M. Enomoto: Solaniol, a toxic metabolite of *Fusarium solani*. Appl. Microbiol. **22**, 718 (1971).
73. Ishii, K., S. V. Pathre, and C. J. Mirocha: Two new trichothecenes produced by *Fusarium roseum*. J. Agric. Food Chem. **26**, 649 (1978).
74. Jarvis, B. B., J. O. Midiwo, D. Tuthill, and G. A. Bean: Interaction between the antibiotic trichothecenes and the higher plant *Baccharis megapotamica*. Science **214**, 460 (1981).
75. Jarvis, B. B., R. M. Eppley, and E. P. Mazzola: Chemistry and bioproduction of macrocyclic trichothecenes. Dev. Food Sci. **4**, 20 (1983).
76. Jarvis, B. B., J. O. Midiwo, T. Desilva, and E. P. Mazzola: Verrucarin L, a new macrocyclic trichothecene. J. Antibiot. **34**, 120 (1981).
77. Jarvis, B. B., J. O. Midiwo, G. P. Stahly, G. Pavanasasivam, and E. P. Mazzola: Trichodermadiene: a new trichothecene. Tetrahedron Letters 787 (1980).
78. Jarvis, B. B., G. Pavanasasivam, C. E. Holmlund, T. Desilva, G. P. Stahly, and E. P. Mazzola: Biosynthetic intermediates to the macrocyclic trichothecenes. J. Amer. Chem. Soc. **103**, 472 (1981).

79. JARVIS, B. B., G. P. STAHLY, G. PAVANASASIVAM, and E. P. MAZZOLA: Antileukemic compounds derived from the chemical modification of macrocyclic trichothenecenes. 1. Derivatives of verrucarin A. J. Med. Chem. **23**, 1054 (1980).

80. — — — — Structure of roridin J, a new macrocyclic trichothecene from *Myrothecium verrucaria*. J. Antibiot. **33**, 256 (1980).

81. JARVIS, B. B., and E. P. MAZZOLA: Macrocyclic and other novel trichothecenes: Their structure, synthesis, and biological significance. Accounts Chem. Res. **15**, 388 (1982).

82. JARVIS, B. B., J. O. MIDIWO, J. L. FLIPPEN-ANDERSON, and E. P. MAZZOLA: Stereochemistry of the roridins. J. Nat. Prod. **45**, 440 (1982).

83. JARVIS, B. B., G. P. STAHLY, G. PAVANASASIVAM, J. O. MIDIWO, T. DESILVA, C. E. HOLMLUND, E. P. MAZZOLA, and R. F. GEOGHEGAN, JR.: Isolation and characterization of the trichoverroids and new roridins and verrucins: J. Organ. Chem. (USA) **47**, 1117 (1982).

84. JARVIS, B. B., and V. M. VRUDHULA: New trichoverroids from *Myrothecium verrucaria*: 16-hydroxytrichodermadienediols. J. Antibiot. **36**, 459 (1983).

85. JARVIS, B. B., V. M. VRUDHULA, J. O. MIDIWO, and E. P. MAZZOLA: New trichoverroids from *Myrothecium verrucaria*: verrol and 12,13-deoxytrichodermadiene. J. Organ. Chem. (USA) **48**, 2576 (1983).

86. JARVIS, B. B., V. M. VRUDHULA, and G. PAVANASASIVAM: Trichoverritone and 16-hydroxyroridin L-2, new trichothecenes from *Myrothecium roridum*. Tetrahedron Letters **24**, 3539 (1983).

87. JARVIS, B. B., J. O. MIDIWO, and E. P. MAZZOLA: Antileukemic compounds derived from the chemical modification of macrocyclic trichothecenes. 2. Derivatives of roridins A and H and verrucarins A and J. J. Med. Chem. **27**, 239 (1984).

88. KANEKO, T., H. SCHMITZ, J. M. ESSERY, W. ROSE, H. G. HOWELL, F. A. O'HERRON, S. NACHFOLGER, J. HUFTALEN, W. T. BRADNER, R. A. PARTYKA, T. W. DOYLE, J. E. DAVIES, and E. CUNDLIFFE: Structural modifications of anguidine and antitumor activities of its analogs. J. Med. Chem. **25**, 579 (1982).

89. KOREEDA, M., and J. I. LUENGO: A novel type of intramolecular Diels-Alder reaction involving dienol ethers: An unusual preference for a boat transition state in the incipient ring formation. J. Organ. Chem. (USA) **49**, 2079 (1984).

90. KOTSONIS, F. N., R. A. ELLISON, and E. B. SMALLEY: Isolation of acetyl T-2 toxin from *Fusarium poae*. Appl. Microbiol. **30**, 493 (1975).

91. KRAUS, G. A., and K. FRAZIER: Synthetic studies toward verrucarol. 1. Synthesis of the AB ring system. J. Organ. Chem. (USA) **45**, 4820 (1980).

92. KRAUS, G. A., B. ROTH, K. FRAZIER, and M. SHIMAGAKI: Stereoselective synthesis of calonectrin. J. Amer. Chem. Soc. **104**, 1114 (1982).

93. KUPCHAN, S. M., B. B. JARVIS, R. G. DAILEY, JR., W. BRIGHT, R. F. BRYAN, and Y. SHIZURI: Baccharin, a novel potent antileukemic trichothecene triepoxide from *Baccharis megapotamica*. J. Amer. Chem. Soc. **98**, 7092 (1976).

94. KUPCHAN, S. M., D. R. STEELMAN, B. B. JARVIS, R. G. DAILEY, JR., and A. T. SNEDEN: Isolation of potent new antileukemic trichothecenes from *Baccharis megapotamica*. J. Organ. Chem. (USA) **42**, 4221 (1977).

95. KUPCHAN, S. M., B. B. JARVIS, M. S. KUPCHAN, and R. G. DAILEY: Antileukemic trichothecin epoxides. Germany Offen. DE 2846210 (1979).

96. LANSDEN, J. A., R. J. COLE, J. W. DORNER, R. H. COX, H. G. CUTLER, and J. D. CLARK: A new trichothecene mycotoxin isolated from *Fusarium tricinctum*. J. Agric. Food Chem. **26**, 246 (1978).

97. MACHIDA, Y., and S. NOZOE: Biosynthesis of trichothecin and related compounds. Tetrahedron **28**, 5113 (1972).

98. MASUOKA, N., and T. KAMIKAWA: A synthesis of 12,13-epoxytrichothec-9-ene. Tetrahedron Letters **1976**, 1691.

99. Masuoka, N., T. Kamikawa, and T. Kubota: Synthesis of 13-nortrichothec-9(10)-ene. Model reaction toward the total synthesis of trichodermin. Chem. Letters 1974, 751.

100. Matsumoto, M.: Structures of isororidin E, epoxyisororidin E, epoxyroridin H, and diepoxyroridin H, new metabolites isolated from species of Cyclindrocarpon. J. Sci. Hiroshima Univ. 43, 107 (1979).

101. Matsumoto, M., H. Minati, N. Uotani, K. Matsumoto, and E. Kondo: New antibiotics from Cylindrocarpon sp. J. Antibiot. 30, 618 (1977).

102. Matsumoto, M., H. Minato, K. Tori, and M. Ueyama: Structures of isororidin E, epoxyisororidin E, and epoxy- and diepoxyroridin H, new metabolites isolated from Cylindrocarpon species determined by carbon-13 and hydrogen-1 NMR spectroscopy. Revision of C-2':C-3' double bond configuration of the roridin group. Tetrahedron Letters 1977, 4093.

103. McPhail, A. T., and G. A. Sim: The structure of Verrucarin A; X-ray analysis of Verrucarin A p-iodobenzene sulfonate. J. Chem. Soc. 1966, 1394.

104. Minato, H., T. Katayama, and K. Tori: Vertisporin, a new antibiotic from Verticimonosporium diffractum. Tetrahedron Letters 1975, 2579.

105. Mirocha, C. J., and S. Pathre: Identification of the toxic principle in a sample of poaefusarin. Appl. Microbiol. 26, 719 (1973).

105a. Mohr, P., Ch. Tamm, W. Zürcher, and M. Zehnder: Sambucinol and Sambucoin, two new metabolites of Fusarium sambucinum possessing modified trichothecane structures. Helv. Chim. Acta 67, 406 (1984).

106. Mohr, P., M. Tori, P. Grossen, P. Herold, and Ch. Tamm: Synthesis of verrucarin A and 3α-hydroxyverrucarin A from verrucarol and diacetoxyscripenol (anguidine). Helv. Chim. Acta 65, 1412 (1982).

107. Nakahara, Y., and T. Tatsuno: Toxicological research on substances from Fusarium nivale; an alternative synthesis of 12,13-epoxy-trichothec-9-ene. Chem. Pharm. Bull. 28, 1981 (1980).

108. Naoi, Y.: Clean-up procedures and GLC analysis (of trichothecene mycotoxins). Dev. Food Sci. 4, 121 (1983).

109. Notegen, E. A., M. Tori, and Ch. Tamm: Partial synthesis of 3'-hydroxy-2'-deoxy-2'',3'',4'',5''-tetrahydroverrucarin A. Helv. Chim. Acta 64, 316 (1981).

110. Ohtsubo, K.: Chronic toxicity of trichothecenes. Dev. Food Sci. 4, 171 (1983).

111. Okuchi, M., M. Itoh, Y. Kaneko, and S. Dio: A new antifungal substance produced by Myrothecium. Agr. Biol. Chem. (Tokyo) 32, 394 (1968).

112. Ong, C. W.: Trichothecanes. – A review. Heterocycles 19, 1685 (1982).

113. Pathre, S. V., C. J. Mirocha, C. M. Christensen, and J. Behrens: Monoacetoxyscirpenol. New mycotoxin produced by Fusarium roseum Gibbosum. J. Agric. Food Chem. 24, 97 (1976).

114. Pearson, A. J., and C. W. Ong: Trichothecene analogs. Total synthesis of 12,13-epoxy-14-methoxytrichothecene via organoiron complexes. J. Amer. Chem. Soc. 103, 6686 (1981).

115. Roush, W. R., and T. A. Blizzard: Synthesis of verrucarin B. J. Organ. Chem. (USA). To be published.

116. Roush, W. R., and T. E. D'Ambra: Synthesis of a bicyclic precursor to verrucarol: Application of a trimethylsily-controlled Diels-Alder reaction and Wagner-Meerwein rearrangement sequence. J. Organ. Chem. (USA) 46, 5045 (1981).

117. — — Total synthesis of (±)-verrucarol. J. Amer. Chem. Soc. 105, 1058 (1983).

118. Roush, W. R., and A. P. Spada: Enantio- and stereoselective syntheses of the dihydroxyoctadienoic acid fragments of the roridins and trichoverrins. Tetrahedron Letters 23, 3773 (1982).

119. — — Synthesis of trichoverrol B. Tetrahedron Letters 24, 3693 (1983).

120. ROUSH, W. R., and T. E. D'AMBRA: Total synthesis of verrucarol: A stereoselective synthesis of 13,14-dinor-15-hydroxytrichothec-9-ene. J. Organ. Chem. (USA) **45**, 3927 (1980).

121. ROUSH, W. R., and T. A. BLIZZARD: Synthesis of verrucarin J. J. Organ. Chem. (USA) **48**, 758 (1983).

122. ROUSH, W. R., T. A. BLIZZARD, and F. Z. BASHA: Methodology for the synthesis of the acyclic portions of verrucarins A and J. Tetrahedron Letters **23**, 2331 (1982).

123. SATO, T., K. TAJIMA, and T. FUJISAWA: Diastereoselective synthesis of erythro- and threo-2-hydroxy-3-methyl-4-pentenoic acids by the ester enolate Claisen rearrangement of 2-butenyl 2-hydroxyacetate. Tetrahedron Letters **24**, 729 (1983).

124. SCHLESSINGER, R. H.: Personal communication.

125. SCHLESSINGER, R. H., and R. A. NUGENT: Total synthesis of racemic verrucarol. J. Amer. Chem. Soc. **104**, 1116 (1982).

126. SCOTT, P. M.: Assessment of quantitative methods for determination of trichothecenes in grains and grain products. J. Assoc. Off. Anal. Chem. **65**, 876 (1982).

127. SIEGFRIED, R., and H. K. FRANK: Contribution to the analysis of *Fusarium* toxins (Trichothecene toxins). S. Lebensm.-Unters. Forsch. **166**, 363 (1978).

128. SIGG, H. P., R. MAULI, E. FLURY, and D. HAUSER: Die Konstitution von Diacetoxyscirpenol: Helv. Chim. Acta **48**, 962 (1965).

129. SNIDER, B. B., and J. W. VAN STRATEN: Stereochemistry of ene reactions of glyoxylate esters. J. Organ. Chem. (USA) **44**, 3567 (1979).

130. SNIDER, B. B., and S. G. AMIN: A synthetic precursor of verrucarin A. Syn. Commun. **8**, 117 (1978).

131. STEKOL'NIKOV, L. I.: New antibiotic roridin E. Priroda (Moscow) **1971**, 104.

132. STEYN, P. S., R. VLEGGAAR, C. J. RABIE, N. P. KRIEK, and J. S. HARINGTON: Trichothecene mycotoxins from *Fusarium sulphureum*. Phytochemistry **17**, 949 (1978).

133. STILL, W. C., and H. OHMIZÜ: Synthesis of verrucarin A. J. Organ. Chem. (USA) **46**, 5242 (1981).

134. STILL, W. C., and M.-Y. TSAI: Total synthesis of (±)-trichodermol. J. Amer. Chem. Soc. **102**, 3654 (1980).

135. STILL, W. C., C. GENNARI, J. A. NOGUEZ, and D. A. PEARSON: Synthesis of macrocyclic trichothecanoids: Baccharin B5 and Roridin E. J. Amer. Chem. Soc. **106**, 260 (1984).

136. TAMM, CH.: The antibiotic complex of the verrucarins and roridins. Fortschr. Chem. Org. Naturst. **31**, 63 (1974).

137. TAMM, CH., and W. BREITENSTEIN: The biosynthesis of trichothecene mycotoxins. In: Biosynthesis of Mycotoxins (STEYN, P. S., ed.), p. 69. New York: Academic Press. 1980.

138. TATSUNO, T.: Chemical synthesis of trichothecenes. Dev. Food Sci. **4**, 47 (1983).

139. TATSUNO, T., Y. FUJIMOTO, and Y. MORITA: Toxicological research on substances from *Fusarium nivale*. III. Structure of nivalenol and its monoacetate. Tetrahedron Letters **1969**, 2823.

140. TOMIOKA, K., F. SATO, and K. KOGA: Synthetic approaches toward verrucarin A. Chiral synthesis of (−)-verrucarinolactone. Heterocycles **17**, 311 (1982).

141. TORI, M.: Syntheses of trichothecane-type sesquiterpenes and verrucarins. Yuki Gosei Kagaku Kyokaishi **39**, 642 (1981).

142. TRAXLER, P., W. ZÜRCHER, and CH. TAMM: Die Struktur des Antibioticums Roridin E. Helv. Chim. Acta **53**, 2071 (1970).

143. TROST, B. M., and J. H. RIGBY: Synthetic strategy toward verrucarins. An approach toward verrucarol. J. Organ. Chem. (USA) **43**, 2938 (1978).

144. TROST, B. M., and P. G. MCDOUGAL: Synthesis of optically active verrucarinic acid derivatives. Tetrahedron Letters **23**, 5497 (1982).

145. — — Total synthesis of verrucarol. J. Amer. Chem. Soc. **104**, 6110 (1982).

146. TROST, B. M., P. G. MCDOUGAL, and K. J. HALLER: A tandem cycloaddition-ene

strategy for the synthesis of (±)-verrucarol and (±)-4,11-diepi-12,13-deoxyverrucarol. J. Amer. Chem. Soc. **106**, 383 (1984).

147. Trost, B. M., P. G. McDougal, and J. H. Rigby: Studies directed towards verrucarins: A synthesis of verrucarol. An approach to verrucarin A. In: Current Trends in Organic Synthesis (Nozaki, H., ed.), p. 45. Oxford: Pergamon Press. 1983.

148. Trost, B. M., and P. G. McDougal: Rotational selectivity in cyclobutene ring openings. Model studies directed toward a synthesis of verrucarin A. J. Organ. Chem. (USA) **49**, 458 (1984).

149. Tulshian, D. B., and B. Fraser-Reid: A synthetic route to the C-4 octadienic esters of trichothecenes from D-glucose. J. Amer. Chem. Soc. **103**, 474 (1981).

150. — — The ready conversion of anguidine into verrucarol and trichodermol. Tetrahedron Letters **21**, 4549 (1980).

151. Ueno, Y. (ed.): Developments in Food Science 4. Trichothecenes — Chemical, Biological and Toxicological Aspects. Amsterdam-Oxford-London: Elsevier. 1983.

152. Ueno, Y., I. Ueno, T. Tatsuno, K. Ohokubo, and H. Tsunoda: Fusarenon-X, a toxic principles of *Fusarium nivale* culture filtrate. Experientia **25**, 1062 (1969).

153. Ueno, Y.: Biological detection of trichothecenes. Dev. Food Sci. **4**, 125 (1983).

154. — General toxicology of trichothecene mycotoxins. Dev. Food Sci. **4**, 135 (1983).

155. — Mode of action of trichothecenes. Pure Appl. Chem. **49**, 1737 (1977).

156. — Toxicological properties of trichothecenes. Maikotokishin (Tokyo) **13**, 11 (1981).

157. Ueno, Y., K. Ishii, K. Sakai, S. Kanaeda, H. Tsunoda, T. Toshitsugu, and M. Enomoto: Microbial survey on bean hulls poisoning of horses with the isolation of toxic trichothecenes, neosolaniol, and T-2 toxin of *Fusarium solani*. Jpn. J. Exp. Med. **42**, 187 (1972).

158. Ueno, Y., K. Nakayama, K. Ishii, F. Tashiro, Y. Minoda, Y. Omori, and K. Komagata: Metabolism of T-2 toxin in *Curtobacterium* sp. strain 114-2. Appl. Environ. Microbiol. **46**, 120 (1983).

159. Ueno, Y., M. Hosoya, and Y. Ishikawa: Inhibitory effects of mycotoxins on protein synthesis in rabbit reticulocytes. J. Biochem. (Tokyo) **66**, 419 (1969).

160. Ueno, Y., K. Ishii, M. Sawan, K. Ohtsubo, Y. Matsuda, T. Tanaki, H. Kurata, and M. Ichinoe: Toxicological approaches to the metabolites of *Fusaria*. XI. Trichothecenes and zearalenone from *Fusarium* species isolated from river sediments. Jpn. J. Exp. Med. **47**, 177 (1977).

161. Vesonder, R. F., A. Ciegler, and A. H. Jensen: Isolation of the emetic principle from *Fusarium*-infected corn. Appl. Microbiol. **26**, 1008 (1973).

162. Vesonder, R. F., A. Ciegler, A. H. Jensen, W. K. Rohwedder, and D. Weisleder: Co-identity of the refusal and emetic principle from *Fusarium*-infected corn. Appl. Environ. Microbiol. **31**, 280 (1976).

163. Wei, R.-D., F. M. Strong, E. B. Smalley, and H. K. Schnoes: Chemical interconversion of T-2 and HT-2 toxins and related compounds. Biochem. Biophys. Res. Commun. **45**, 396 (1971).

164. Welch, S. C., and R. V. Wong: Synthetic intermediate for trichothecane phytotoxic sesquiterpenoids. Syn. Commun. **2**, 291 (1972).

165. White, J. D., T. Matsui, and J. A. Thomas: Novel synthesis of the tricyclic nucleus of verrucarol. J. Organ. Chem. (USA) **46**, 3376 (1981).

166. White, J. D., J. P. Carter, and H. S. Kezar, III: Stereoselective synthesis of the macrocycle segment of verrucarin J. J. Organ. Chem. (USA) **47**, 929 (1982).

167. Xu, Y., X. Huang, and Y. Cai: Isolation and structure of CBD2 — a new trichothecene toxin. Weishengwu Xuebao **22**, 35 (1982).

168. Yamamato, Y., N. Maeda, and K. Maruyama: Enantio- and diastereo-selective reaction of but-2-enylstannane with glyoxylate esters and its application to a short synthesis of verrucarinolactone. Chem. Commun. **1983**, 774.

169. YATES, S. G., H. L. TOOKEY, J. J. ELLIS, and H. J. BURKHARDT: Mycotoxins produced by *Fusarium nivale* isolated from tall fescue: (Festuca arundinacea) Phytochem. **7,** 139 (1968).

170. YOSHIZAWA, T., and M. MOROOKA: Trichothecenes from mold-infested cereals in Japan. In: Mycotoxins in Human and Animal Health, Conference Proceedings (RODRICKS, J. V., C. W. HESSELTINE, M. A. MEHLMAN, eds.), p. 309. Park Forest South: Pathotox Publ., Inc. 1977.

171. — — Deoxynivalenol and its monoacetate. New mycotoxins from *roseum* and moldy barley. Agr. Biol. Chem. **37,** 2933 (1973).

172. ZÜRCHER, W., and CH. TAMM: Isolierung von 2′-Dehydroverrucarin A als Metabolit von *Myrothecium roridum* Tode *ex* Fr. Gattungstyp bei Fries. Helv. Chim. Acta **49,** 2594 (1966).

(Received July 19, 1984)

Quassinoid Bitter Principles II

By Judith Polonsky, Institut de Chimie des Substances Naturelles, C. N. R. S, Gif-Sur-Yvette, France

Contents

I. Introduction

More than ten years have elapsed since the publication of a comprehensive review on the quassinoids, the bitter principles of the Simaroubaceae family (80). Interest in these terpenoids has increased enormously in recent years due in part to the finding of the American National Cancer Institute in the early 1970s that these compounds display marked antileukemic activity. Furthermore, a wide spectrum of other biological properties for the quassinoids has been discovered and studies on chemical modifications of inactive members to yield biologically active ones were undertaken. New structures have been established also and numerous synthetic approaches have been developed which include the total synthesis of the parent compound, quassin (p. 250) and also that of castelanolide (p. 253).

It is intended that this present chapter will be an extension of my first review in this series and will contain references up to September 1984.

A short article on some aspects of this subject was published recently (81).

II. Quassinoid General Features

In reviewing the essential features of the quassinoids, the new structural types discovered during the last decade will be emphasised.

The quassinoids can be divided into distinct groups according to their basic skeletons. The five skeletons observed are presented on Chart 1. Skeletal types C and E are new. The C_{18} skeleton C differs from B by loss of one carbon atom in ring A, probably due to a benzylic rearrangement resulting in ring contraction. The second new skeletal system found in the last decade was type E which differs from type D by the mode of formation of the γ-lactone. In type D this function is formed by linkage of C-23 to C-21 whereas in type E, the γ-lactone is formed by linkage of C-23 to C-17.

The number and the positions of the methyl groups are the same on these five basic skeletons and all the quassinoids so far known have only one methyl group at C-4.

By far the greatest majority of the numerous quassinoids known have the (C_{20}) basic skeleton (also named picrasane), seven are C_{19}-compounds with the basic skeleton B, three quassinoids with skeleton C, three with D and three with E have been isolated to date. The quassinoids are heavily oxygenated lactones (δ-lactones in the C_{20}-quassinoids, γ-lactones in the C_{19} and C_{18}, δ- and γ-lactones in the C_{25}-compounds). They have varying numbers of different oxygen-containing groups; with the exception of carbons C-5, C-9 and the methyl groups at C-4 and C-10, these oxygenated functions have been found on all the other carbon atoms.

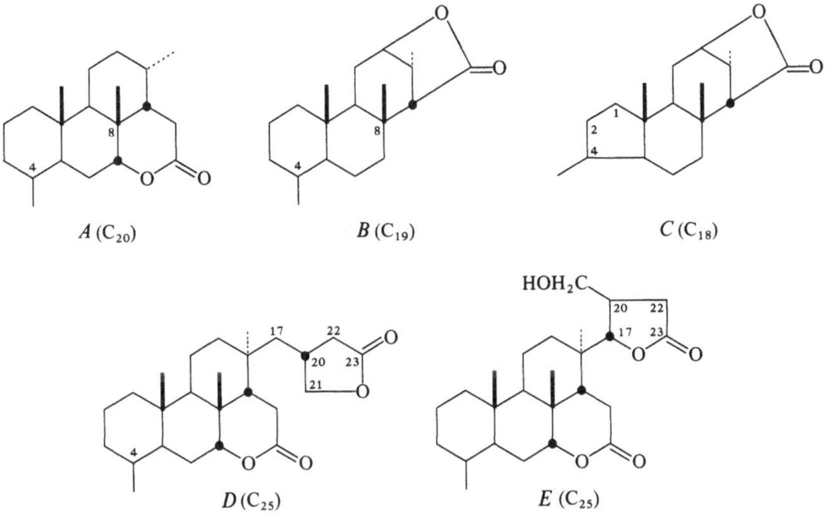

Chart 1

III. Structure Determination of Quassinoids

The recently found quassinoids will be presented according to their basic skeleton and when relevant comment will be made on the specific biological activity of the individual compounds.

1. C_{18}-Quassinoids

Samaderine A (**1**) isolated from *Samadera indica* (*111*), laurycolactone A (**2**) and B (**3**) isolated from the Vietnamese Simaroubaceae *Eurycoma longifolia* Jack (*67*) are the only three C_{18}-quassinoids so far known. The structures of (**1**) and (**2**) were established by X-ray analysis and that of (**3**) by spectral evidence.

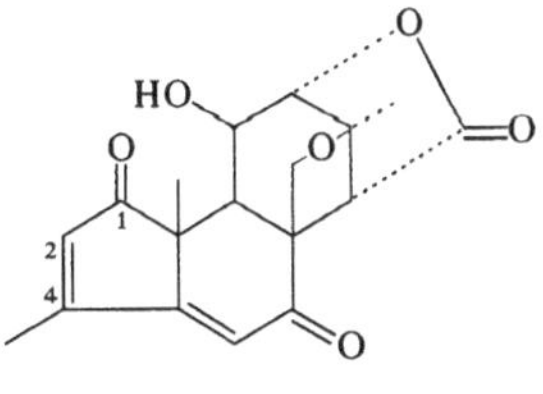

(**1**) Samaderine A 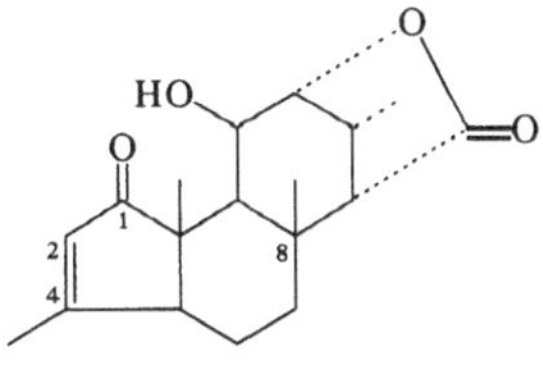   (**2**) Laurycolactone A 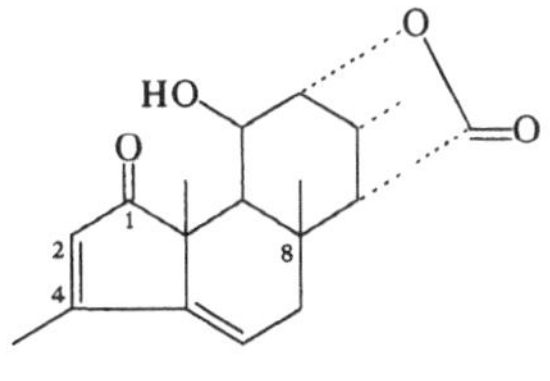   (**3**) Laurycolactone B

2. C₁₉-Quassinoids

Only one new C_{19}-quassinoid was isolated during the last decade, namely shinjulactone B (**4**) which was extracted from *Ailanthus altissima* Swingle (*22*). Its structure was determined by X-ray diffraction method. Shinjulactone B (**4**) has a quite original structure. It possesses a free primary alcohol at C-8 and contracted A and B rings. It may derive biogenetically from a C_{20}-quassinoid, such as ailanthone (**19**) *via* a 1,2-dioxo derivative (**5**). Under oxidative conditions, (**5**) might undergo successively a $C_{(1)}$-$C_{(2)}$ bond cleavage, decarboxylation, a contraction of the B ring and formation of the α,β-unsaturated γ-lactone to give (**4**).

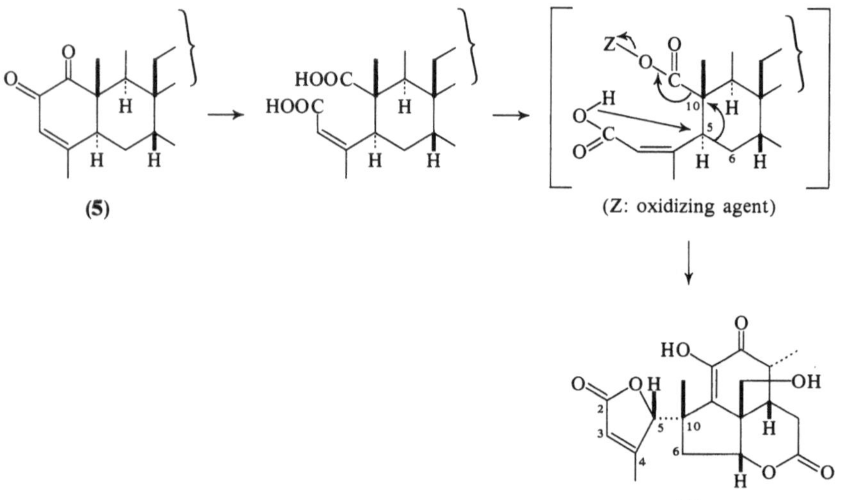

(5)

(Z: oxidizing agent)

(4)

Eurycomalactone, isolated previously from *Eurycoma longifolia* Jack, which was assigned structure (**6**) (*53, 70*) had its structure revised to (**7**) as a result of a X-ray analysis (*67*).

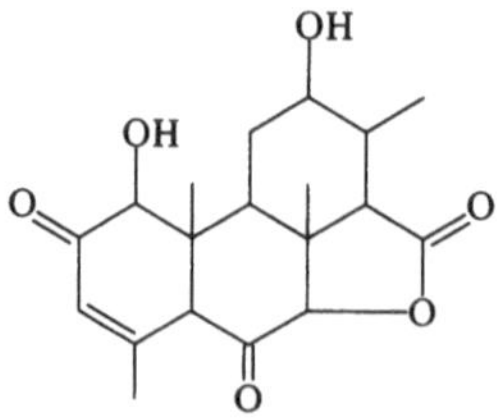

(6) Eurycomalactone

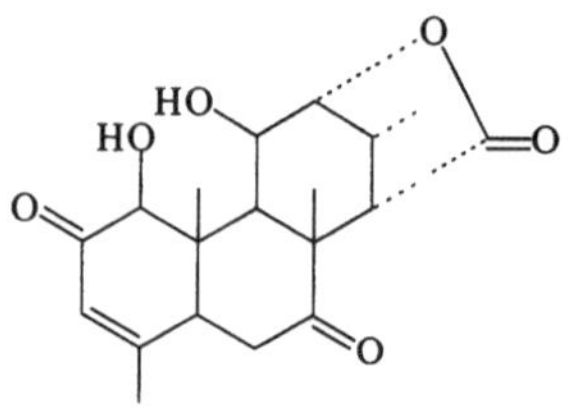

(7) Revised structure

3. C_{20}-Quassinoids

The principal variations in the structure of those quassinoids which are basically C_{20}-compounds are as follows:

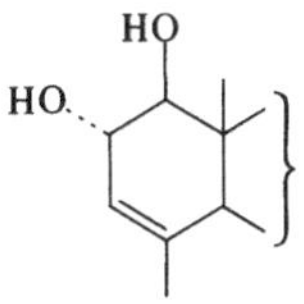

(a) 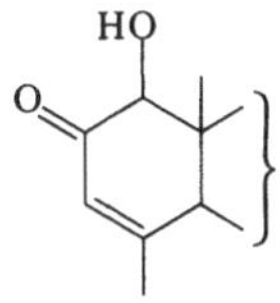   (b) 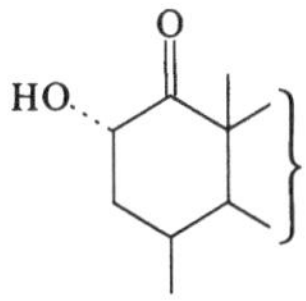   (c) 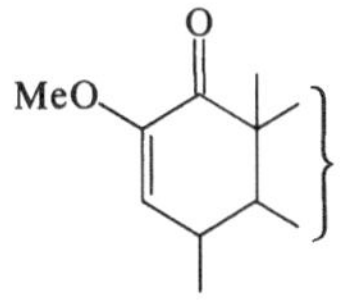   (d)

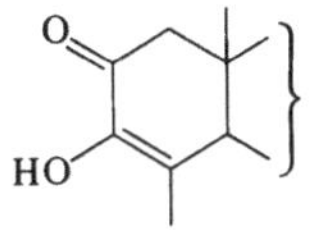

(e) 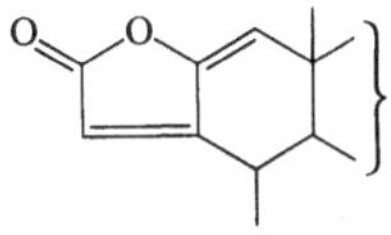   (f)

Ring A may have the structure (a), (b), (c), (d), (e) or (f).

Structure (e) has so far been found only in certain quassinoids isolated from the genus *Brucea* whilst structure (f) occurs so far only in two quassinoids, sergeolide (**62**) and 15-deacetylsergeolide (**63**) (see below).

In ring C at position 8 one finds either a methyl group or a primary alcohol. The latter generally is involved in a hemiketal bridge to C-11 or an oxide bridge to C-13. The δ-lactone ring may have hydroxyl groups and in particular a β-oriented hydroxyl group at C-15 which is often found esterified with various small fatty acids. Ring B may possess an oxygenated group at position 6.

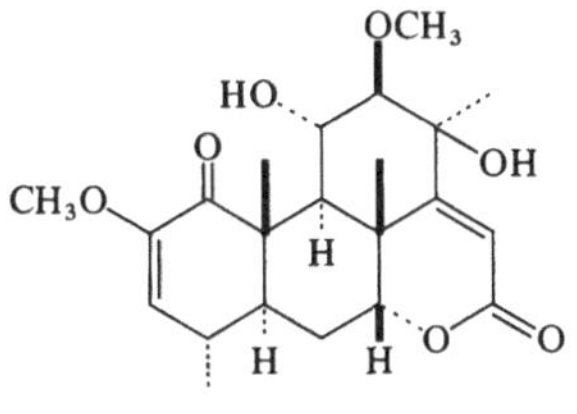

(**8**) 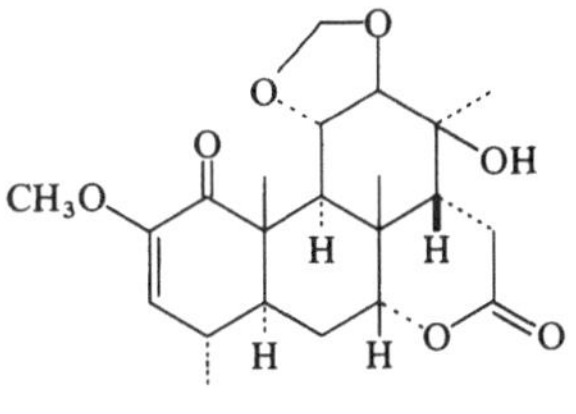   (**9**)

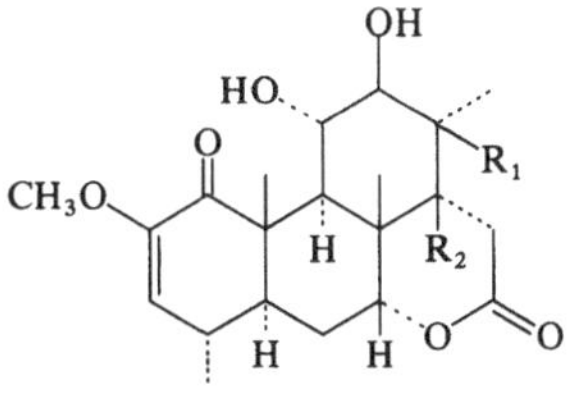

(**10**) R_1=H, R_2=OH
(**11**) R_1=OH. R_2=H

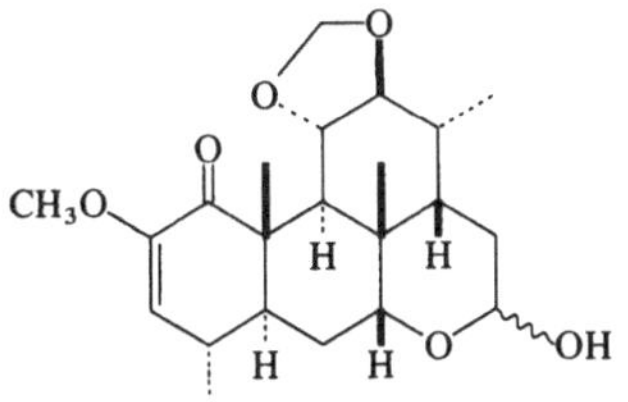

(**13**) 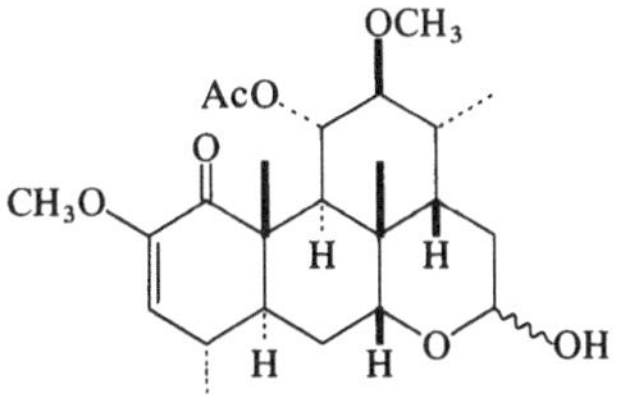   (**12**)

Investigation of the minor bitter constituents of *Picrasma ailanthoides* Planchon led to the isolation and structural elucidation of the following new quassinoids: Nigakilactone K (**8**), L (**9**), M (**10**) and N (**11**) (*62*). Quite recently two quassinoids hemiacetals, picrasinol-A (**12**) and -B (**13**) have been isolated from the bark of the same plant (*71*); they all have a methyl group at C-8.

Soulameanone (**14**) and 1,12-di-O-acetylsoulameanone (**15**), isolated from the New Caledonian species *Soulamea muelleri* Brongn. *et* Gris (*89*), have also a methyl group at position 8. They have the α,β-unsaturated ketol group in ring A, a structural feature common to a number of quassinoids, and β-hydroxyls at C-13 and C-15. The structure of (**14**) has been confirmed by X-ray analysis. The ^{1}H-n.m.r. spectrum of (**14**) displays a singlet for H-15. The quite unusual absence of spin coupling with H-14 can be explained by a boat conformation of the lactone ring, caused mainly by a repulsive interaction between the hydroxyls at C-13 and C-15. Soulameanone (**14**) did not show significant inhibitory *in vivo* activity against P-388 lymphocytic leukemia in the mouse.

Δ^2-Picrasin B (**16**) was also isolated from the same Simaroubaceae (*89*). This quassinoid had already been prepared from picrasin B (*30*) but was not reported as a naturally occurring compound.

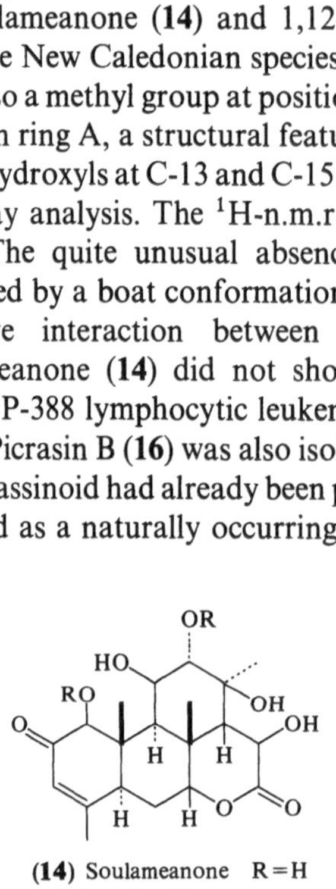

(**14**) Soulameanone R=H
(**15**) R=CH$_3$CO

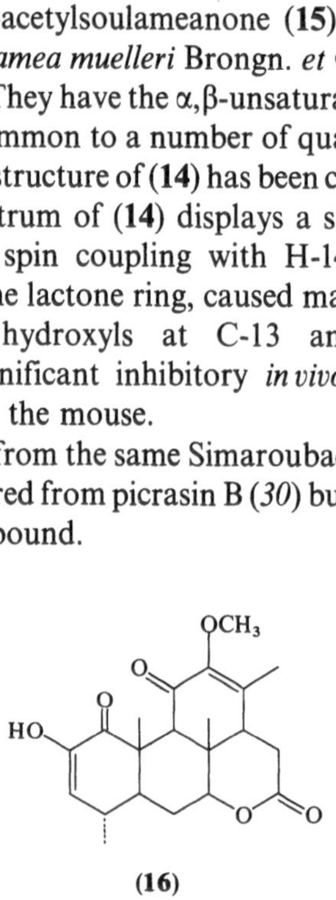

(**16**)

Karinolide (**17**), isolated from the stem bark of the French Guyanan Simaroubaceae *Simaba multiflora* A. Juss. (*87*), shinjulactone C (**18a**) (*36*) and shinjulactone F (**18b**) (*37*), isolated from the Japanese *Ailanthus altissima* Swingle (= *A. glandulosa* Desf.), are the only three C$_{20}$-quassinoids so far known which have a primary alcohol group at C-8. Their structures have been established by X-ray diffraction analysis. Karinolide (**17**) is the first example of a natural quassinoid to have a double bond in ring B and a bis-hemiketal function between C-1 and C-11. Shinjulactone C (**18a**), a non-bitter quassinoid, has a quite original structure with inversed configuration at C-9 and C$_{(1)}$-C$_{(12)}$ and C$_{(5)}$-C$_{(13)}$ linkages. Shinjulactone F (**18b**) has a 5 βH-picrasane skeleton with all *cis*-configurations at the A/B, B/D and C/D ring junctions.

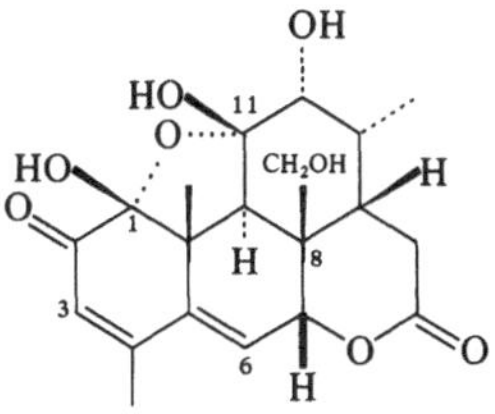

(17) Karinolide 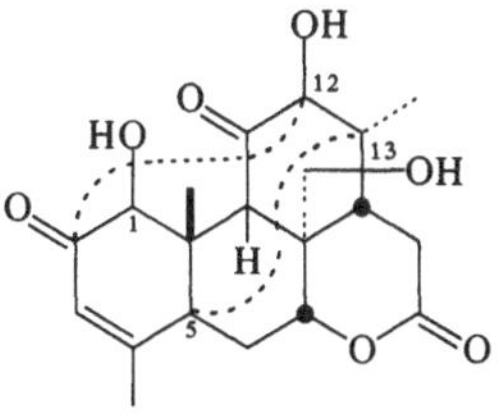 **(18a)** Shinjulactone C 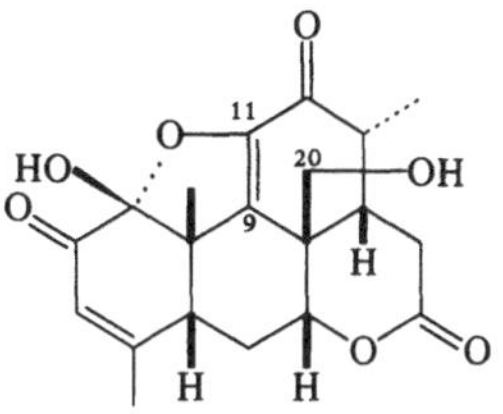 **(18b)** Shinjulactone F

The configuration of the 12-hydroxyl group of ailanthone (**19**), the major constituent of *Ailanthus altissima,* has recently been revised and found to be 12 α (*64,65*).

Ailanthone (**19**) has been converted into shinjulactone C (**18a**) in the following manner (*35*): The monosilyl derivative (**20**) was acetylated to give the monosilyl diacetate (**21**) which isomerised to a monosilyl enolacetate (**22**) on heating in pyridine. Desilylation of (**22**) afforded (**23**) which was oxidised to the α-diketone (**24**). The latter was transformed by refluxing in pyridine, into di-O-acetylshinjulactone C (**25**) and then hydrolysed to shinjulactone C (**18a**).

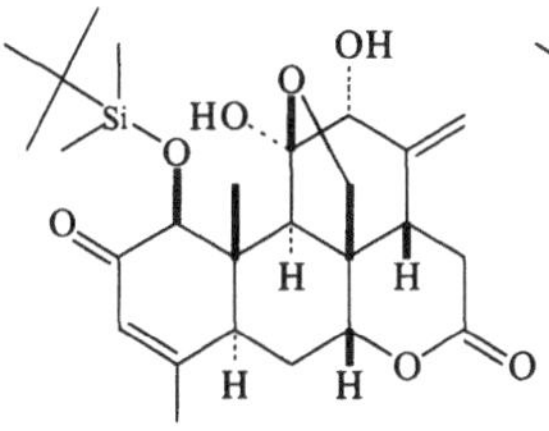

(20) 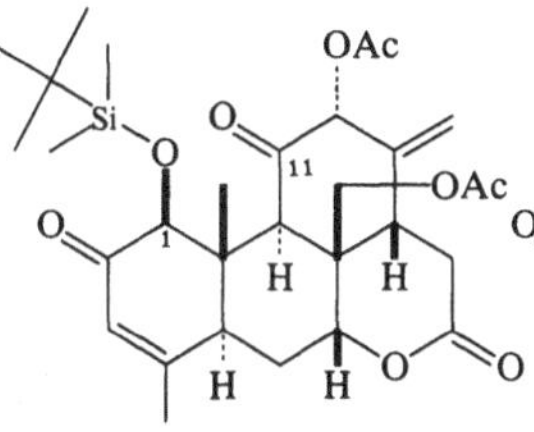 **(21)** 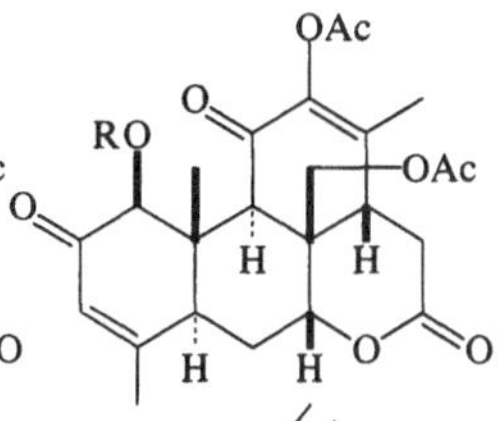 **(22)** R=Si
(23) R=H

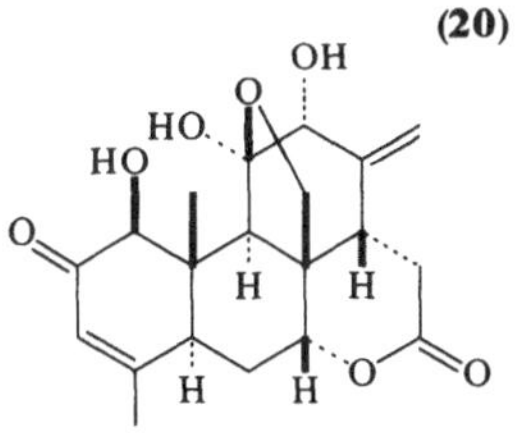

(19)

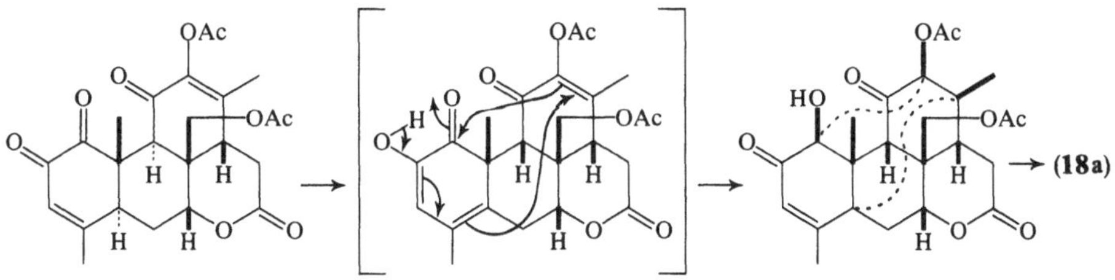

(24) **(25)** → (**18a**)

2-Dihydroailanthone (**26**) has been found as a minor quassinoid in the bark of *Ailanthus glandulosa* (= *A. altissima*) (*8*). The same structure was assigned to shinjulactone A isolated from the Japanese *A. altissima* (*65*).

Another minor quassinoid from the same plant is shinjudilactone (**27**) (*32, 36*) the structure of which was determined by X-ray diffraction.

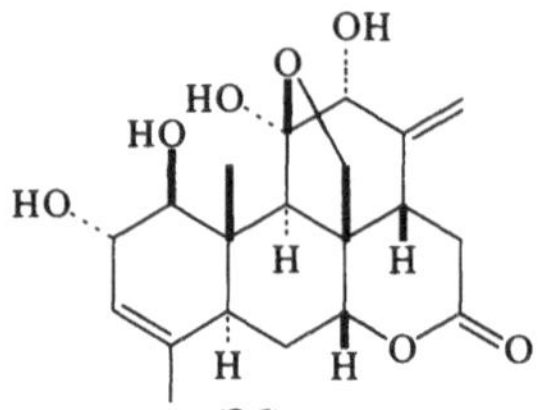

(**26**) 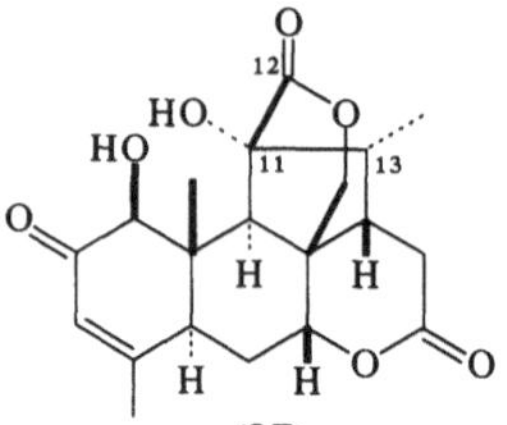   (**27**)

Heating of ailanthone (**19**) in CH_3OH/H_2O (1 : 1) with CO_3HNa afforded shinjudilactone (**27**) and its C-13 epimer (*34*). The conversion which is outlined below probably proceeds through a benzilic acid rearrangement to give a hydroxy carboxylic acid which is subject to lactonisation (*34*).

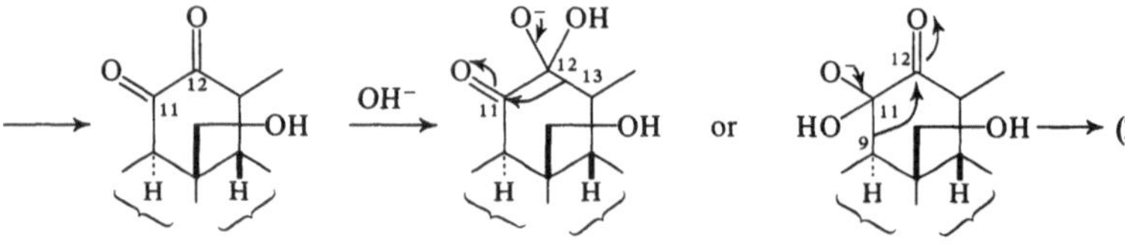

Eurycomanone (**28**) and eurycomanol (**29**) have been isolated from the roots of the Indonesian *Eurycoma longifolia* (*13, 14*). Both quassinoids do not exhibit inhibitory activity against NIH lymphocytic leukemia.

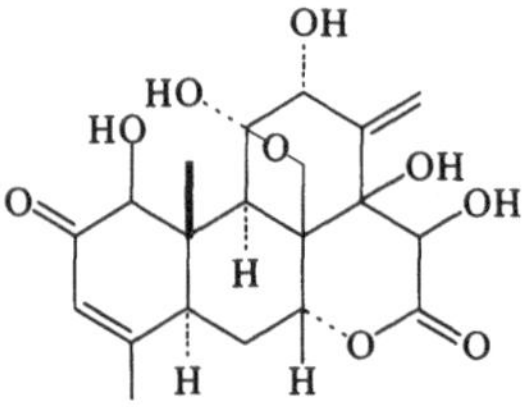

(**28**) 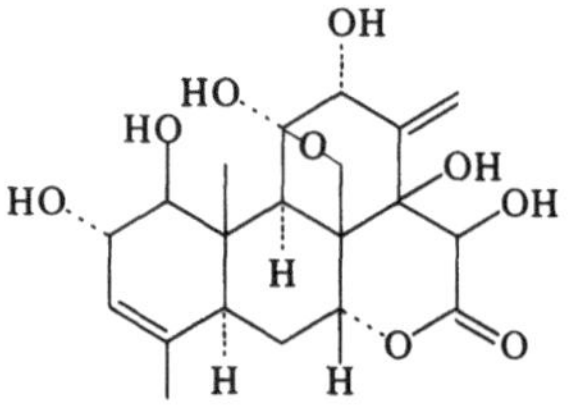   (**29**)

New C-15 esters of glaucarubol (**30**) and glaucarubolone (**31**) have been isolated: 2'-Acetoxyglaucarubine (**32b**) from *Simarouba amara* Aubl. (*91*) and from *Odyendyea gabonensis* Engl. (*112*), excelsin (**33**) from *Ailanthus excelsa* (*40*) and *Odyendyea gabonensis* (*112*), castelanone (**34**) from *Castela tweedii* (*94*) and soularubinone (**35**) from *Soulamea tomentosa* Brongn. et

Gris (*58*). 13,18-Dehydroglaucarubinone (**36b**) was found in *Simarouba amara* (*91*) and in *Ailanthus altissima* (*36*), 13,18-dehydroailanthinone (**37b**) in *Pierrodendron kerstingii* (*46*) and 13,18-dehydroexcelsin (**38**) was found in *Ailanthus excelsa* (*40*) and in *A. malabarica* (*39*). Only those quassinoids which possess an α,β-unsaturated ketol group in ring A display significant antineoplastic activity in the NIH murine lymphocytic leukemia P 388 (PS) system.

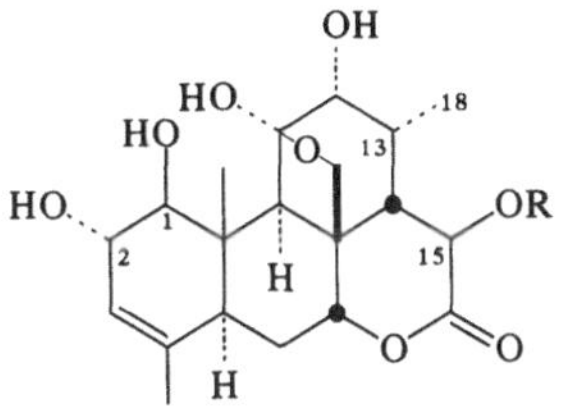

(**30**) R=H: Glaucarubol
(**32a**) R=COC(Me)OHCH$_2$Me
(**32b**) R=COC(Me)OAcCH$_2$Me
(**33**) R=COCH(Me)CH$_2$Me
(**38**) R=COCH(Me)CH$_2$Me; $\varDelta^{13,18}$

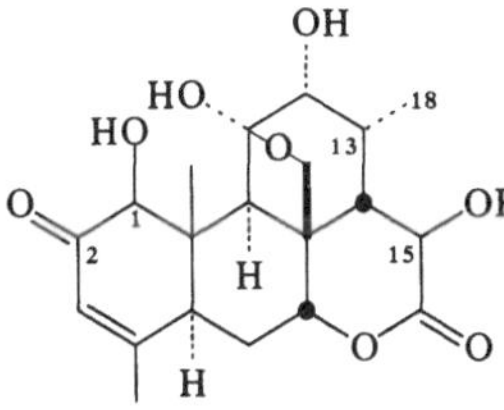

(**31**) R=H: Glaucarubolone
(**34**) R=COCH$_2$CHMe$_2$
(**35**) R=COCH$_2$CMe$_2$OH
(**36a**) R=COC(Me)OHCH$_2$Me
(**36b**) R=COC(Me)OHCH$_2$Me; $\varDelta^{13,18}$
(**37a**) R=COCH(Me)CH$_2$Me
(**37b**) R=COCH(Me)CH$_2$Me; $\varDelta^{13,18}$

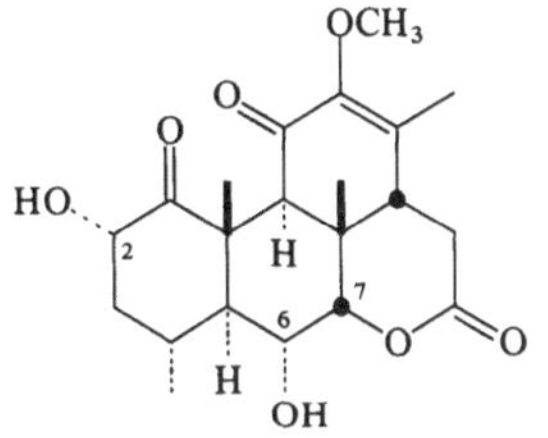

(**39**) 6-Hydroxypicrasin B

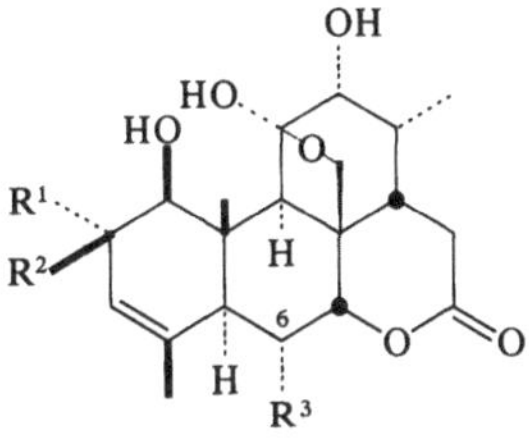

(**40a**) R^1, R^2=O; R^3=H; Chaparrinone
(**40b**) R^1, R^2=O; R^3=OCOCH=C$\begin{smallmatrix}CH_3\\CH_3\end{smallmatrix}$
(**40c**) R^1=OH; R^2=H; R^3=OCOCH=C−CH$_3$ | CH$_3$
(**41a**) R^1, R^2=O; R^3=OCOC=C$\begin{smallmatrix}H\\CH_3\end{smallmatrix}$ | CH$_3$
(**41b**) R^1=OH; R^2=R^3=H; Chaparrin
(**41c**) R^1=OH, R^2=H, R^3=OCOC=C$\begin{smallmatrix}H\\CH_3\end{smallmatrix}$ | CH$_3$

Until 1970 only one quassinoid was known to have a hydroxyl at position 6, namely 6-hydroxypicrasin B (**39**). This structure was later confirmed by X-ray analysis (*75*). Since then quassinoids (**40b**) (*110, 87*) and

(**41a**) (*98, 92*) which have the 6-hydroxyl esterified with senecioic acid and tiglic acid, respectively, have been isolated. Compounds (**40b**) and (**41a**) display significant antileukemic activity. This finding shows that the ester chain required for biological activity can be located either at C-15 or C-6. The quassinoids (**40b**) and (**41a**) are found together with inactive compounds which possess an α-glycol group instead of an α-ketol group in ring A: 6α-Senecioyloxychaparrin (**40c**) (*1, 87*) and 6α-tigloyloxychaparrin (**41c**) (*92*).

Recently another cytotoxic quassinoid, undulatone (**42**) (*109*) was isolated from *Hannoa undulata*. It has ester groups at both C-15 and C-6.

15-Deacetylundulatone (**43a**) and 6α-tigloyloxyglaucarubol (**43b**) were found in *Hannoa klaineana* roots (*56*).

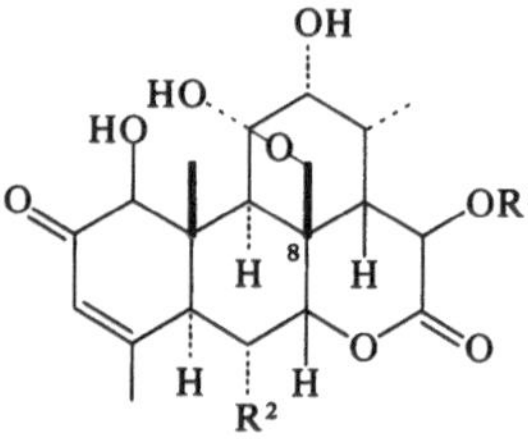

(**42**) Undulatone, R^1=Ac; R^2=OCOC=C$\overset{H}{\underset{CH_3}{}}CH_3$

(**43a**) R^1=H; R^2=OCOC=C$\overset{H}{\underset{CH_3}{}}CH_3$

(**43b**) R^1=H; R^2=OCOC=C$\overset{H}{\underset{CH_3}{}}CH_3$; 2-dihydro

Four new antileukemic quassinoids having an α-ketol group in ring A and an oxide bridge $C_{(30)}$-$C_{(13)}$ in ring C have been described: Isobruceine A (**44**) from *Soulamea tomentosa* Brongn. et Gris (*84*) and *Soulamea soulameoides* Gray (*29*), isobruceine B (**45**) from *Brucea antidysenterica* Mill. (*45*) and *Picrolemma pseudocoffea* Ducke (*60*), quassimarin (**46**) from *Quassia amara* L. (*48*) and samaderine E (**47**) from *Samadera indica* (*111*).

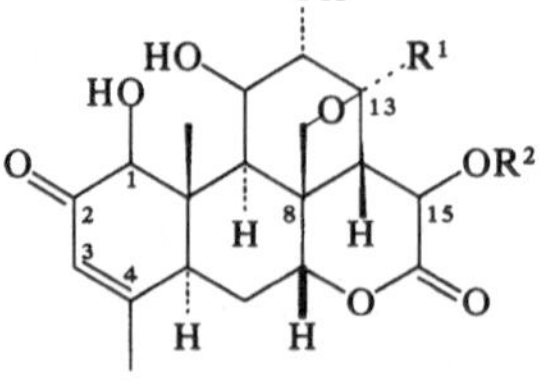

(**44**) R^1=COOCH$_3$
R^2=COCH$_2$CH(CH$_3$)$_2$
(**45**) R^1=COOCH$_3$
R^2=COCH$_3$
(**46**) R^1=CH$_3$
R^2=COC(OAc)(CH$_3$)C$_2$H$_5$

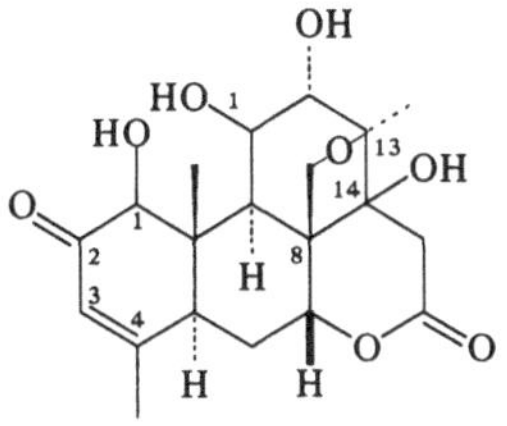

(47)

The bruceolide esters are the only quassinoids which do not have oxygenated functions at positions 1 and 2 but at 2 and 3. They have a diosphenol group at these positions and, in ring C, the hydroxymethyl at C-8 forms an oxide bridge to C-13.

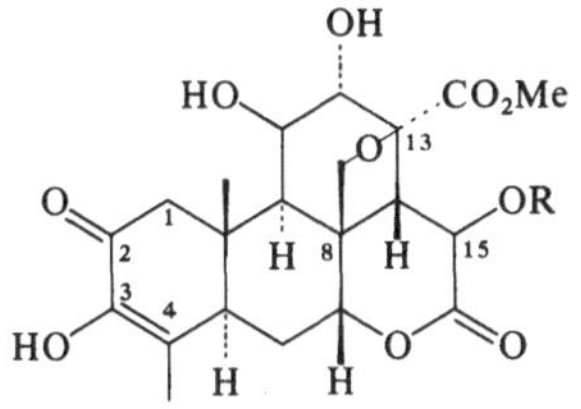

(48) Bruceolide; R=H

(49a) Brucein A; R=COCH$_2$CHMe$_2$

(50) Brucein B; R=Ac

(51) Brucein C; R=COÇ=CÇC—OH (Me, Me, Me, H)

(52) Bruceantarin; R=COC$_6$H$_5$

(53) Bruceantin; R=COÇ=CCH (H, Me, Me)

(54) Bruceantinol; R=COÇ=ÇC—OAc (H, Me, Me)

(55a) Brusatol; R=COÇ=CMe$_2$ (H)

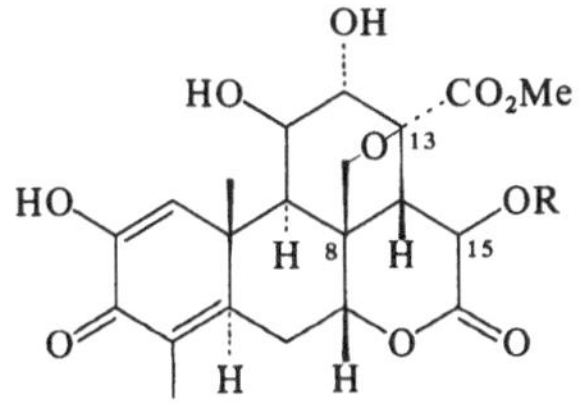

(56) Dehydrobruceantin; R=COÇ=CCH (Me, Me, Me, H)

(57) Dehydrobruceantarin; R=COC$_6$H$_5$

(58) Dehydrobrucein A: R=COCH$_2$CH (Me, Me)

(59) Dehydrobrucein B; R=COCH$_3$

(60) Dehydrobruceantol; R=COÇCH(OH)Me ‖ CMe$_2$

(61) Dehydrobruceantinol; R=COÇ=CÇC—OAc (Me, Me, Me, H)

Brucein A (**49a**), B (**50**) and C (**51**), isolated in 1967 from *Brucea amarissima,* were the first C-15 esters of bruceolide (**48**) to have their structures determined (*82*). These quassinoids differ in the nature of the ester group at C-15. They are esters of isovaleric, acetic and 3,4-dimethyl-4-hydroxy-2-pentenoic acid, respectively.

Later on, KUPCHAN *et al.* (*45*) isolated from *Brucea antidysenterica* along with bruceantarin (**52**) two potent antileukemic principles, bruceantin (**53**) and bruceantinol. Bruceantin is now undergoing Phase II clinical trials as an antineoplastic agent by the U.S. National Cancer Institute. The structure of brucein C (**51**), with a *trans* configuration of the double bond in the ester chain, was originally attributed to bruceantinol (*45*). The antileukemic activity of the bruceolide derivatives varies greatly with the nature of the ester substituent. An unambiguous structure determination of brucein C (**51**) and bruceantinol was therefore necessary. Brucein C proved to have the structure represented in (**51**) and bruceantinol is in fact 4'-O-acetylbrucein C (**54**) (*90*).

The markedly higher antitumor activity of bruceantin and bruceantinol compared with that of brucein C could possibly the attributed to greater lipophilicity of the side chains of (**53**) and (**54**).

The non cytotoxic quassinoids dehydrobruceantin (**56**), dehydrobruceantarin (**57**), dehydrobrucein B (**59**) and dehydrobruceantol (**60**) were also isolated from *B. antidysenterica* (*45*).

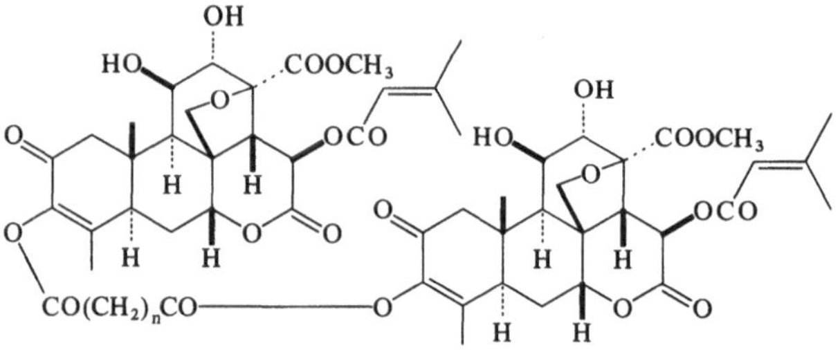

(**55b**) n = 1 to 4 and 8

The roots, fruits and stems of *Brucea javanica* have also been examined and the quassinoids (**49** to **55**) were isolated as well as dehydrobrucein A (**58**), dehydrobrucein B (**59**) and 3,4-dihydrobrucein A (**49b**) (*77, 78*).

Purification of a sample of bruceantinol (**54**) led to the isolation of a small amount of dehydrobruceantinol (**61**) (*90*).

A series of bis-(brusatolyl) esters (**55b**) has been synthetized and tested for *in vivo* antileukemic activity. The bis-(brusatolyl) malonate, succinate, glutarate, adipate and sebacate were as active or more active than brusatol (**55a**) (*51*).

Studies of microbial transformations of bruceantin (**53**) led to the isolation of new analogs of this antitumor quassinoid (*10*). Thus, *Streptomyces griseus* (ATCC 10137) totally metabolized bruceantin (**53**) within 19 h, and a preparative scale incubation using bruceantin provided brucein C (**51**), bruceantin-4′,5′-epoxide and bruceantin-5′-ol. None of the metabolites were as active as bruceantin in the P 388 test system (*10*).

Sergeolide (**62**) (*60*) is the first natural quassinoid to possess a butenolide function. It was isolated along with the known isobruceine B (**45**) (*45*) from the roots of the French Guyanan Simaroubaceae *Picrolemma pseudocoffea*. The structure of sergeolide was established by analysis of the 400 MHz-^{1}H-n.m.r. data and the ^{13}C-n.m.r. spectrum.

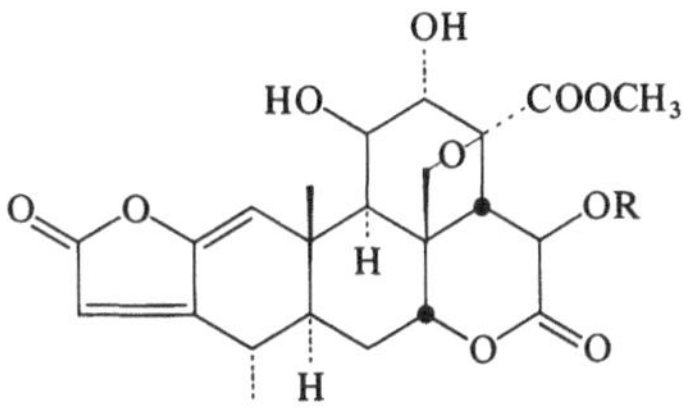

(**62**) R = Ac
(**63**) R = H

The structural similarity of (**45**) and (**62**) suggests that the latter arises from isobruceine B via the intermediacy of its 1-O-acetyl derivative. A Claisen-type rearrangement (*vide infra*) followed by appropriate double bond migration and lactonisation would lead to (**62**).

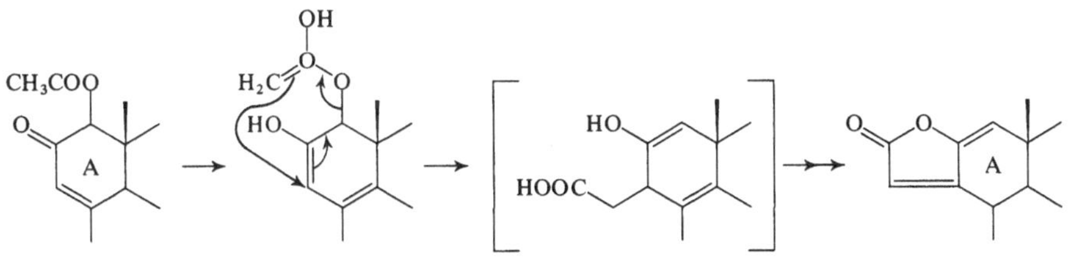

Sergeolide displayed significant antileukemic activity at low dose (0.5 mg/Kg). At slightly higher doses it was highly cytotoxic. It is of interest to note that the butenolide function which might be expected to act as a potent Michael acceptor should confer such a high degree of cytotoxicity to the molecule.

Recently, 15-deacetylsergeolide (**63**) was isolated from the leaf extract of *P. pseudocoffea* (*85*). It seems to be less toxic than sergeolide (**62**) and displays potent antileukemic activity (T/C 169 at 1.25 mg/kg/day level).

Several quassinoid glucosides have been isolated: Bruceoside-A (**64**), bruceoside B (**65**) from *Brucea javanica* (Linn) Merr. seeds (*50*), bruceaninoside-A (**66**) and bruceantinoside-B (**67**) from *Brucea anti-dysenterica* wood (*74*), picrasinoside-A (**68**) from *Picrasma ailanthoides* (Planchon) bark (*72*), 15-O-β-D-glucopyranosylglaucarubolone (**69**) and 15-O-β-D-glucopyranosylglaucarubol (**70**) from *Simarouba glauca* seeds (*6*). Structure (**69**) was established by X-ray analysis. The first four glycosides displayed significant antileukemic activity and the two latter were found to be toxic down to the level of 1 mg/kg).

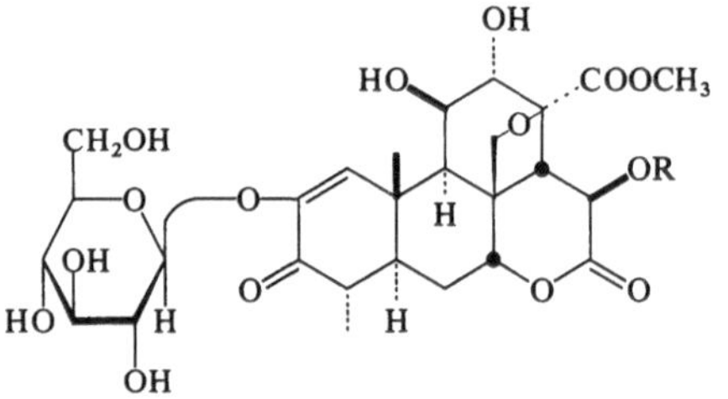

(**64**) Bruceoside A

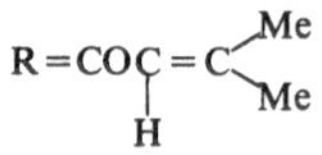

$$R = COC = C \overset{Me}{\underset{H}{\big|}} {}^{Me}$$

(**66**) Bruceantinoside A

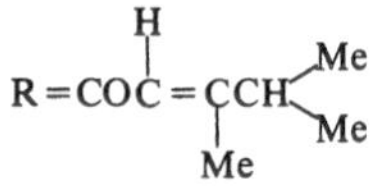

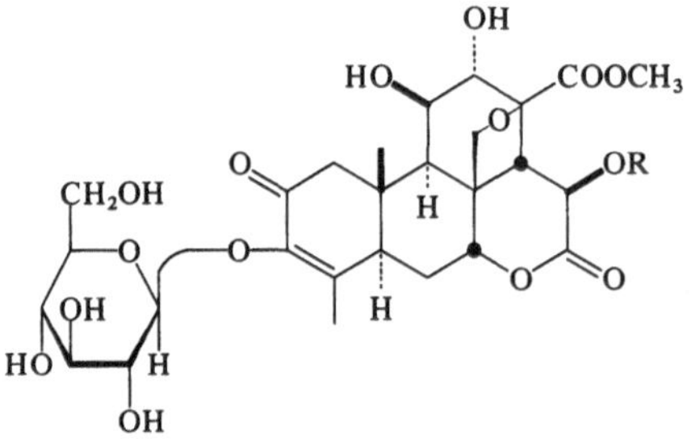

$$R = COC = CCH \overset{H}{\underset{Me}{\big|}} {}^{Me}_{Me}$$

(**65**) Bruceoside B

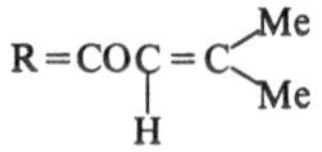

$$R = COC = C \overset{Me}{\underset{H}{\big|}} {}^{Me}$$

(**67**) Bruceantinoside B

$$R = COC = CCH \overset{H}{\underset{Me}{\big|}} {}^{Me}_{Me}$$

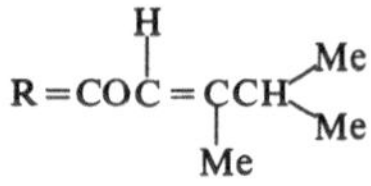

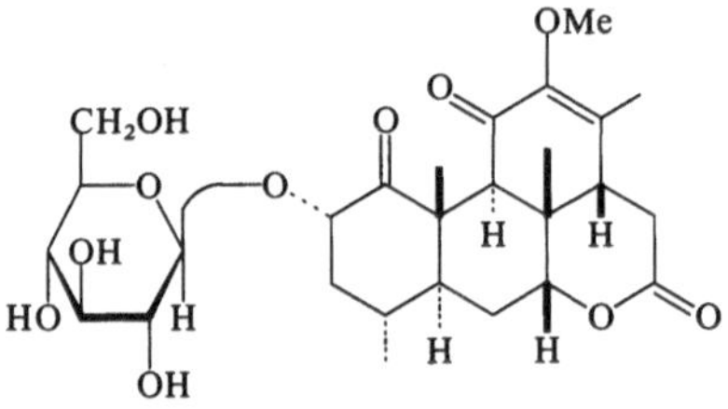

(**68**) Picrasinoside A

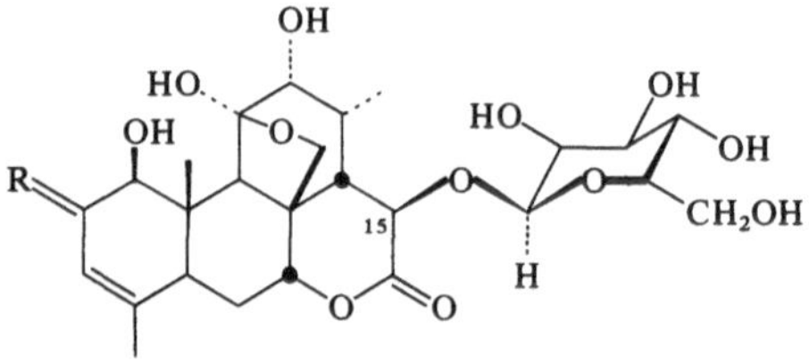

(**69**) R = O   15-*O*-β-D-Glucopyranosyl glaucarubolone

(**70**) R = $\overset{OH}{\underset{H}{<}}$ 15-*O*-β-D-Glucopyranosyl glaucarubol

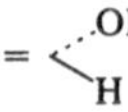

Quite recently the isolation of six additional quassinoid glucosides from the bark of *P. ailanthoides* has been reported: Picrasinoside-B (**71**), -C (**72**), -D (**73**), -E (**74**), -F (**75**) and -G (**76**) (*71*).

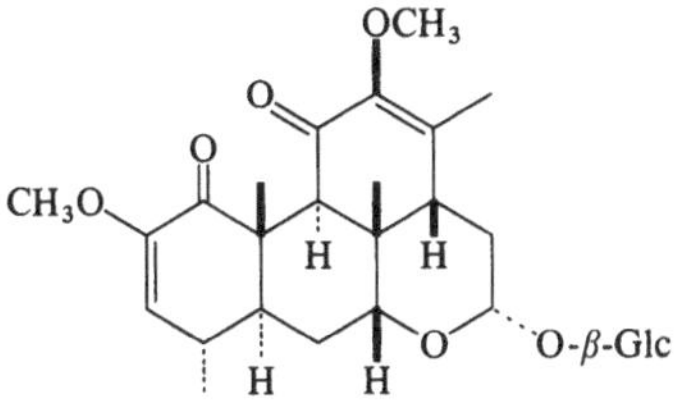

(**71**)

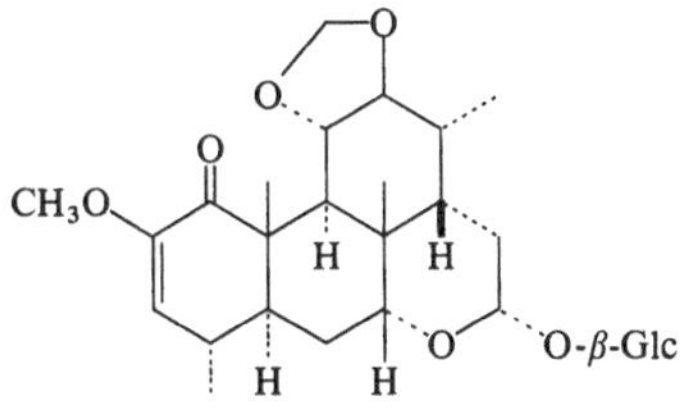

(**72**)

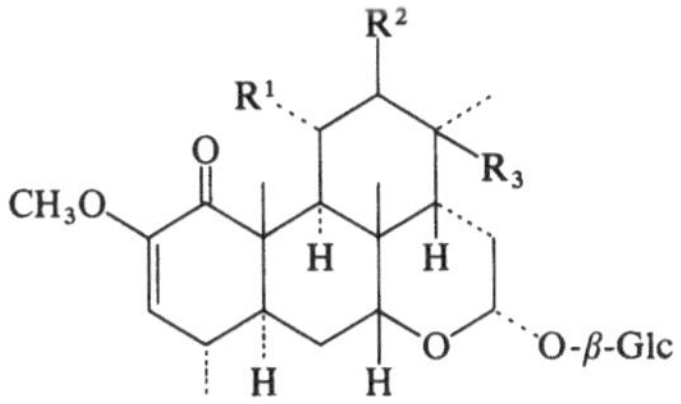

(**73**) R^1=OAc; R^2=OMe; R^3=H
(**74**) R^1=OAc; R^2=OMe; R^3=OH
(**75**) R^1=OH; R^2=OMe; R^3=H
(**76**) R^1=OH; R^2=OMe; R^3=OH

4. C$_{25}$-Quassinoids

Until 1972, only two pentacyclic quassinoids possessing the C$_{25}$ basic skeleton were known: Simarolide (**77**) and picrasin A (**78**) (*80*). Since then four more C$_{25}$ quassinoids have been discovered: Soulameolide (**80**) from the New Caledonian Simaroubaceae *Soulamea tomentosa* Brongn. et Gris (*88*), simarinolide (**81**) and guanepolide (**82**) from the root bark of the French Guianan *Simaba* cf. *orinocensis* (*93*) which has recently been revised to *Simaba moretii* Feuillet (*19*). X-Ray analyses of (**80**) and of the derived primary acetate of (**82**) have been carried out.

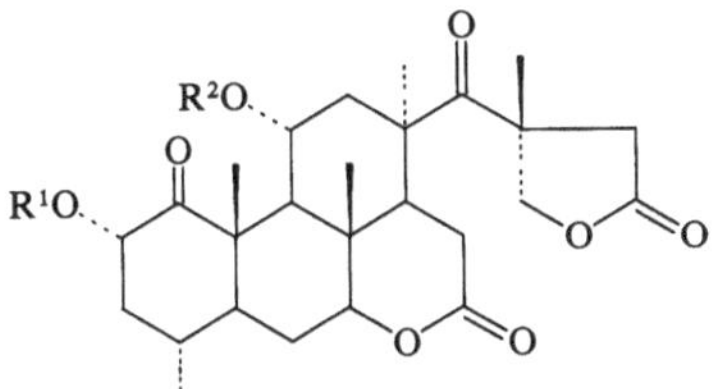

(**77**) Simarolide; R^1=H; R^2=Ac
(**78**) Picrasin A; R^1=Me; R^2=Ac; 2,3-dehydro
(**79**) Deacetylsimarolide; R^1=R^2=H

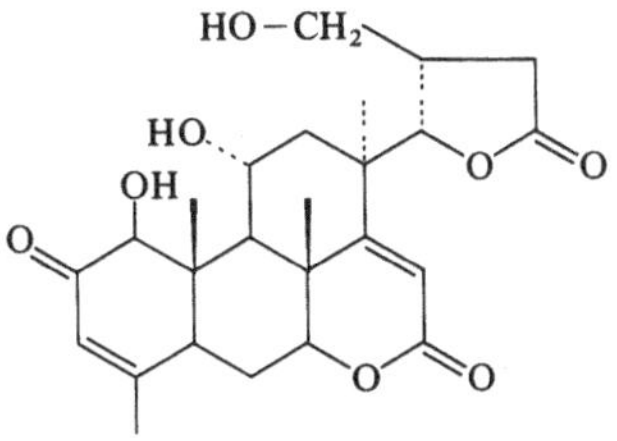

(**80**) Soulameolide

Simarinolide (**81**) and guanepolide (**82**) have A-ring structures not previously encountered among the quassinoids. Compound (**82**) is the first quassinoid to have two substituents at C-4, *i.e.* a hydroxyl and an axially

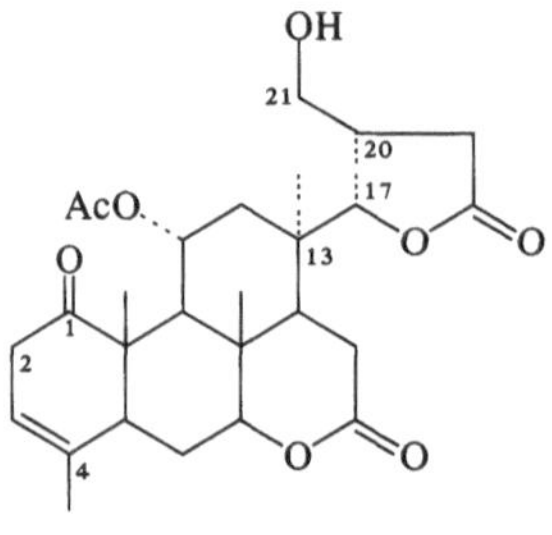

(81) Simarinolide 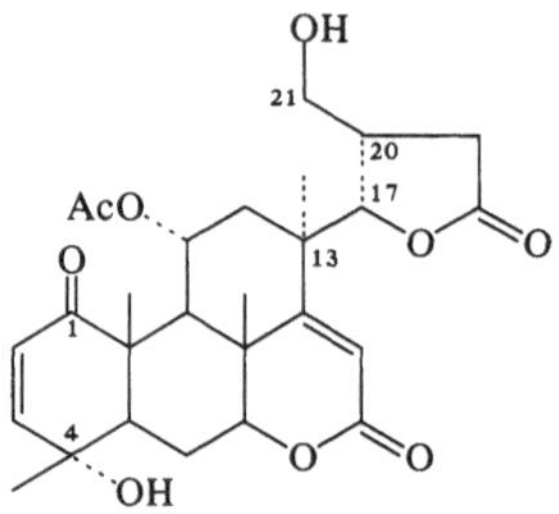 **(82)** Guanepolide

oriented methyl group. The formation of the A-ring in **(81)** and **(82)** from the triterpene precursor can be explained by the following pathway:

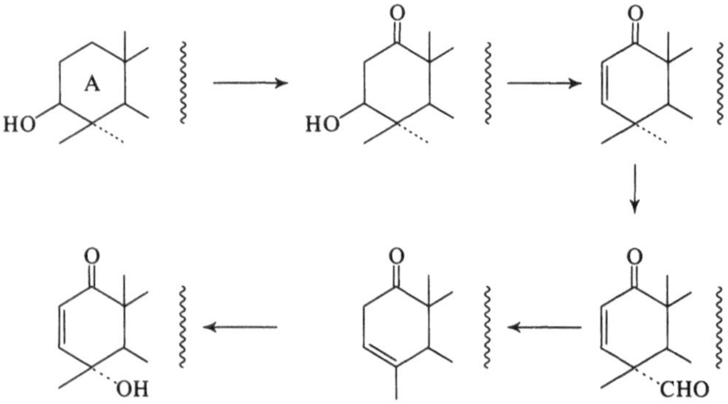

Our previous biogenetic experiments (*61*) support the loss of the equatorial methyl group at C-4 in the sequence leading to **(81)** and hence to **(82)**.

Soulameolide **(81)** and guanepolide **(82)** are rare examples of natural quassinoids which have a 14,15-double bond. This suggests that the oxygenated functions at these carbon atoms, which are present in several quassinoids, are introduced subsequent to the formation of the δ-lactone.

Quite recently deacetylsimarolide **(79)** has been isolated from the fruits of *Simaba moretii* (*86*).

All six C_{25}-quassinoids now known lack on oxygen function at position 12 but possess one at C-17. This fact substantiates the hypothesis (*30, 80*) that cleavage of the C(13)-C(17) bond to give the much more common C_{20} quassinoids is triggered by a C-12 oxygen function.

IV. Physical Methods

During the early structural studies 60 MHz ^{1}H-n.m.r. spectroscopy and electron impact mass spectroscopy were the most important physical tools used. Despite the great amount of information obtained by these physical methods, structural elucidation relied heavily on chemical transformations, degradations and correlations. Prior to 1970, of the 55 quassinoids known only two structures were solved by X-ray analysis.

In recent years higher field ^{1}H-n.m.r. spectroscopy and new mass spectroscopic methods (chemical ionisation, field desorption, FAB and "Mikes" technique) became available. Recently BALDWIN et al. (2) applied some of these new mass spectroscopic techniques to the structure determination of some bruceolides. A detailed ^{13}C-n.m.r. study of a number of quassinoids has been published (83) and ^{13}C-n.n.r. data have since then been recorded in most of the papers dealing with constituents of Simaroubaceae. X-Ray analysis has also become more accessible and structures of many quassinoids are now being established by this method.

HPLC was used for the separation of mixtures of quassinoids (66, 96) and recently a quantitative HPLC analysis of the quassinoids from Simarouba glauca has been described (59). Enzyme-linked immunosorbent assays for quassin (85) and a number of other quassinoids have been carried out (95, 97).

V. Biological Activity

1. Antileukemic Activity

As already mentioned certain quassinoids display in vivo antileukemic activity. In Table 1 are listed some compounds which are most active against the P-388 lymphocytic leukemia (9). T/C is the ratio of test group survival to control group survival in tumored animals, expressed as a percentage. When this ratio is equal or greater than 120, compounds are considered to be active. Bruceantin (53) has been put on clinical trial in the U.S.A. a few years ago (4, 9, 20, 45, 55). It shows activity over a wide dose range and in addition is active against solid tumors. The disposition, excretion and metabolism of bruceantin (53) in the mouse has been investigated (102).

Table 1. *Antileukemic Activity*

Quassinoids	Optimal dose (mg/kg)	T/C
Glaucarubinone (36 a)	0.25	177
Ailanthinone (37 a)	2.0	148
Tigloyloxy-6α-chaparrinone (41 a)	0.6	163
Simalikalactone D (83)	1.0	198
Bruceantin (53)	0.5	220
Bruceantinol (54)	1.0	238
Isobruceine A (44)	2.0	163

As to the mode of action of these compounds, LIAO *et al.* (*54*), using HeLa cells, have shown that the antitumor activity of bruceolide (**48**) esters at the molecular level is due to irreversible inhibition of protein synthesis. The antileukemic activity of a series of esters of brusatol (**55a**), bis-brusatol (**55b**) and bruceantin (**53**) has been correlated with the ability of these compounds to suppress DNA and protein synthesis in P-388 lymphocytic leukemia cells. The active compounds were also shown to inhibit DNA polymerase activity and purine synthesis (*27*).

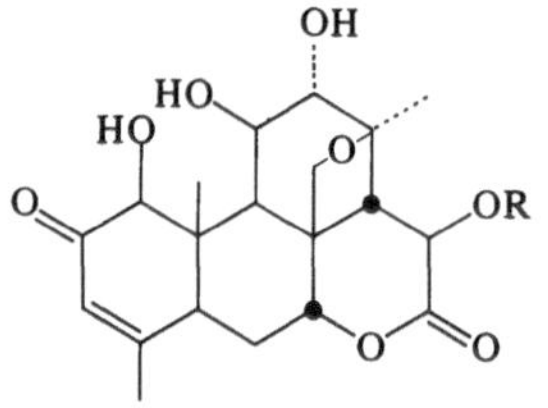

(**83**) Simalikalactone D; R=COCHC$_2$H$_5$
                                                       |
                                                      Me

Studies (*47, 108*) of the structure/activity relationships for several quassinoids have established some of the structural requirements for optimal antineoplastic activity, particularly in P-388 mouse leukemia. These requirements are principally the following (Chart 2): a) Ring A with

Structural Requirements for Antileukemic Activity
[*in vivo*, P-388]

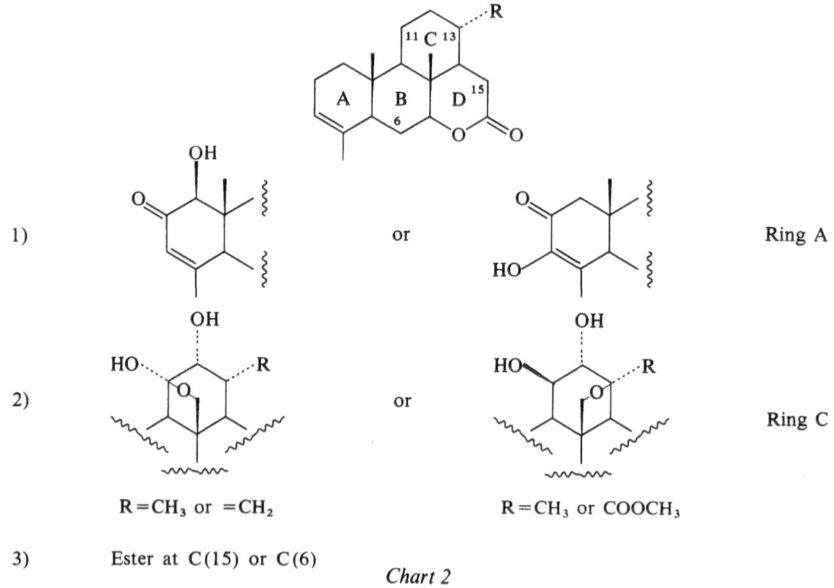

1)   or   Ring A

2)   or   Ring C

      R=CH$_3$ or =CH$_2$ R=CH$_3$ or COOCH$_3$

3)   Ester at C(15) or C(6)

Chart 2

either an α,β-unsaturated ketol group at position 1 and 2 or a diosphenol group at position 2 and 3. b) Ring C with an epoxymethano bridge between C-8 and C-11, or between C-8 and C-13. c) The presence of a free hydroxyl group in ring A and at C-12 in addition to an ester group at C-15 and/or C-6.

2. Antiviral Activity

Certain quassinoids display *in vitro* antiviral activity, namely against the oncogenic Rous sarcoma virus (*79*). This test is performed as follows: Chick-embryo fibroblasts are infected by a known amount of the virus which transforms the morphology of the cells into foci. A number of quassinoids inhibit this transformation at concentration ranging from 0.15 to 1 µg/ml, without having toxic effects on normal cells. Some of the results are shown in Table 2.

Table 2. *Effect of some Quassinoids on RSV-induced Foci Formation*

	I_{50} µg/ml	I^* max µg/ml	% inhibition
Isobruceine A (**44**)	0.15	0.30	96
Simalikalactone D (**83**)	0.10	0.15	76
Chaparrinone (**40a**)	0.35	0.50	92
Glaucarubolone (**31**)	0.16	0.20	83
Castelanone (**34**)	0.78	1.00	88

* Higher concentrations were toxic to normal cells.

While this inhibition does not guarantee antileukemic activity, compounds which do not display this antitransforming behaviour are invariably inactive.

3. Antimalarial Activity

The growth of the chloroquine resistant blood parasite *Plasmodium falciparum* (responsible for malaria) was markedly inhibited *in vitro* by certain quassinoids (*103*). The most active compound simalikalactone D (**83**) gave complete inhibition at 0.002 µg/ml. Glaucarubinone (**36a**) and soularubinone (**35**) were equally effective at 0.006 µg/ml, whereas chaparrinone (**40a**) and simarolide (**77**) had little effect even at 0.01 µg/ml. These relative activities parallel the antineoplastic activities of these compounds. Indeed, the two last mentioned quassinoids (**40a**) and (**77**) do not possess the structural requirements for antileukemic activity.

The 3-, 15-esters and 3,15-diesters of bruceolide (**48**), including bruceantin (**53**) as well as brusatol (**55a**) and bruceolide (**48**), were reported to be effective inhibitors of chloroquine resistant strains of *P. falciparum* (*49*).

4. Antifeedant and Insecticidal Properties

The antifeedant activity of thirteen quassinoids of different structural types has been studied against the Mexican Bean beetle *(Epilachna varivestis)* 4th instar larvae and the Southern armyworm *(Spodoptera eridania)* 5th instar larvae. All quassinoids tested displayed significant activity against the Mexican Been beetle and, thus, do not exhibit a simple structure/activity relationship. Five quassinoids were active against the Southern armyworm. Interestingly, four of these − bruceantin (**53**), glaucarubinone (**36a**), isobruceine A (**44**) and simalikalactone D (**83**) − possess the required structural features for antineoplastic activity. The non-cytotoxic quassin (**85**) is an exception; it is active against both the pests (*52*). Simalikalactone D (**83**) was found to be the most potent antifeedant.

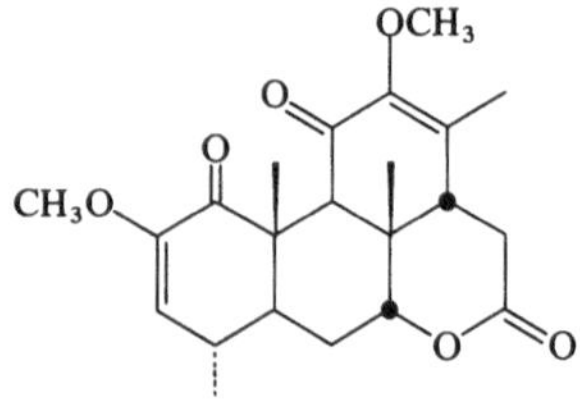

(85) Quassin

The antifeedant and insecticidal behaviour of eight quassinoids − soulameolide (**80**), simarolide (**77**), soulameanone (**14**), chaparrinone (**40a**), glaucarubinone (**36a**), brucein A (**49a**) and brucein B (**50**) − against the third stage cricket larva *(Locusta migratoria)* has also been studied. The first three quassinoids do not display any antifeedant activity, whereas the other five showed moderate antifeedant behaviour. On the contrary, they showed significant insecticidal activity, especially glaucarubinone (**36a**) and brucein B (**50**) (*69*).

5. Amebicidal Activity

Extracts of many Simarouba species including *Castela nicholsoni* or *Chaparro amargoso* (Castamargina) have long been used by local populations in Mexico, China, and elsewhere to treat fevers, dysentery and amebiasis. However, studies of the antiamebic activity of pure quassinoids have been limited. Only two, ailanthone (**19**) (*16*) and glaucarubin (**32a**) (*12*) were shown to be effective amebicides. Quite recently GILLIN and REINER (*23, 24*) have made an extensive study of the *in vitro* activity of 17

References, pp. 259—264

quassinoids against the parasite *Entamoeba histolytica*. Seven of the quassinoids tested were active; they are, in order of decreasing activity: Bruceantin (**53**), simalikalactone D (**83**), ailanthinone (**37a**), glaucarubolone (**31**), glaucarubinone (**36a**), ailanthone (**19**), and glaucarubin (**32a**).

Apparently, there is no simple relationship between structure and activity of the seventeen quassinoids tested.

6. Anti-inflammatory Activity

Several quassinoids related to brusatol (**55a**) were potent inhibitors of induced inflammation and arthritis in rodents. Brusatol (**55a**) was the most potent compound, demonstrating higher activity at 1 mg/kg than indomethacin at 10 mg/kg. The structure-activity relationships have been studied. One of the modes of action of quassinoids as anti-inflammatory agents is to stabilize lysosomal membranes, reducing the release of hydrolytic enzymes that cause damage to surrounding tissues (*28*).

VI. Chemical Modifications

1. Quassin (**85**) was modified, in five steps, by introducing the bruceantin ester side chain at position 15 (*63*). The compound obtained was found inactive against growth of HeLa cells.

2. OKANO and LEE (*73*) converted the quassinoid glycoside bruceoside-A (**64**) to bruceantin (**53**). The methods used involved selective hydrolysis of the ester and glycosyl groups followed by esterification of the hydroxyl at position 15.

3. Chaparrinone (**40a**) which possesses most of the structural requirements for *in vivo* antileukemic activity in the PS test system but lacks an ester side chain at C-15 and/or at C-6, does not display any significant antineoplastic activity. In a search for direct methods of quassinoid functionalization benzeneselininic anhydride (**84**) was considered as a possible reagent for the introduction of a double bond in the lactonic ring of chaparrinone (**40a**) (*41*).

The synthetic pathway is summarized in Scheme 1.

Chaparrinone triacetate (**86**) on reaction with benzeneselininic anhydride afforded the desired $\Delta^{14,15}$-dehydro compound (**87**). In addition, two other products (**88**) and (**89**) arising, interestingly enough, from angular hydroxylation at C-5, were isolated. Deacetylation of these three products gave the free alcohols (**90**), (**91**) and (**92**) which however did not show any significant activity. Further preliminary experiments to introduce an oxygen function at C-15 remained unsuccessful.

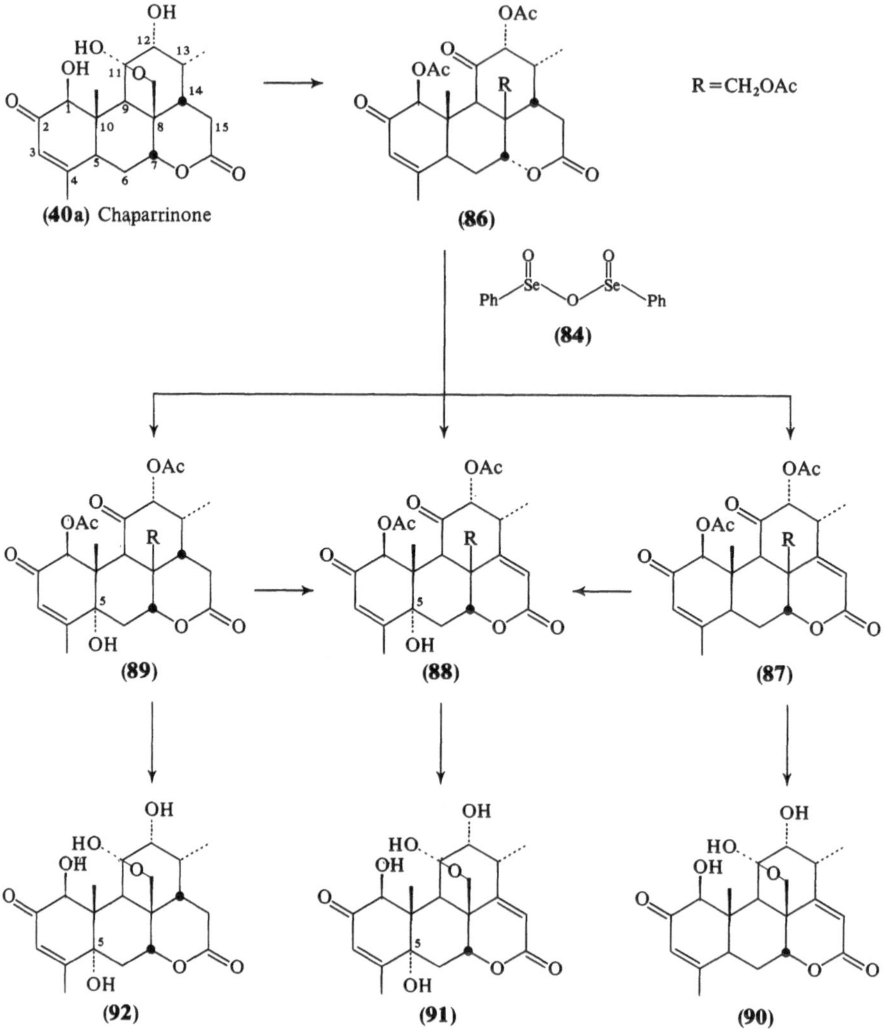

Scheme 1

The mechanism proposed for the observed angular hydroxylation is outlined in Scheme 2. Enolisation of the 2-keto group of chaparrinone triacetate (86) is followed by attack of benzeneselininic anhydride at C-3 of the enol. This mechanism explains both the position and the configuration of the new OH-group.

4. Chaparrin (41b) is inactive and can be isolated in relatively large quantities from Mexican Simaroubaceae. A study aimed at the development of synthetic procedures for the conversion of chaparrin into C-15 glaucarubolone esters and simple analogs in which the C-15 ester side chain has been replaced by an alkyl group was initiated (7).

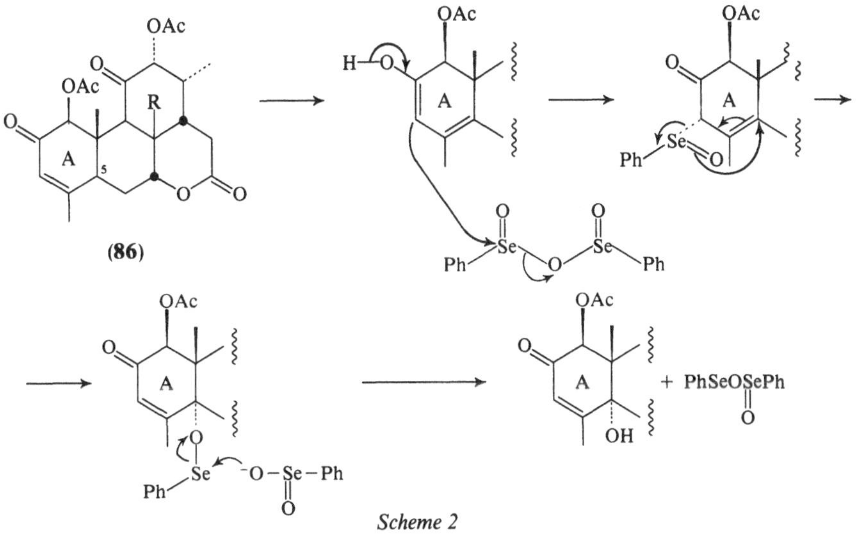

Scheme 2

The synthetic technique is summarized in Scheme 3. Reaction of chaparrin **(41 b)** with *tert*-butyldimethylsilyl chloride *(11)* afforded the crystalline disilyl derivative **(93)**. The latter was obtained in better yield by silylation of **(41 b)** with *tert*-butyldimethylsilyl enol ether of pentane-2,4-dione *(105)*. The hydroxyl function at C-1 of **(93)** was effectively protected using trimethylsilyl triflate to afford the trisilyl lactone **(94)** which upon treatment with lithium diisopropylamide (LDA) and subsequent exposure to MoO$_5$-pyridine-HMPA (MoO$_5$PH) *(104)* gave the required 15-hydroxy lactone **(95)**. Treatment of the latter with isovaleryl chloride afforded the crystalline ester **(96)** which was selectively desilylated to **(97)**. Oxidation of the free allylic hydroxyl and complete desilylation of the resulting disilyl enone with tetrabutylammonium fluoride (Bu$_4$NF) afforded the natural cytotoxic quassinoid castelanone **(34)**.

This reaction sequence could be applied to the synthesis of the quassinoid analogs 15-O-β-octanoylglaucarubolone **(98)** and 15-O-β-3′,4′-dimethyl-2′-pentenoylglaucarubolone **(99)** *(5)*. The latter, which has the bruceantin **(53)** ester side chain, was obtained by esterification of the hydroxylactone **(95)** with 3,4-dimethyl-2-pentenoic anhydride prepared by treatment of the corresponding acid with chlorosulfonyl isocyanate *(38)*.

The alkyl analogs 15-β-butyl- and 15-β-heptylchaparrinone **(102)** and **(103)** have been prepared by treatment of the lithium enolate derived from **(94)** with the corresponding alkyl halides; the alkyl lactones **(100)** and **(101)**, obtained in greater than 90% yield, were subjected to the same series of reactions as **(96)**. Compounds **(98)**, **(102)** and **(103)** cause significant inhibition of cell transformation induced by Rous sarcoma virus *(79)* at the 1 µg/ml level. The analogs **(99)** and **(103)** display significant activity against the murine lymphocytic leukemia.

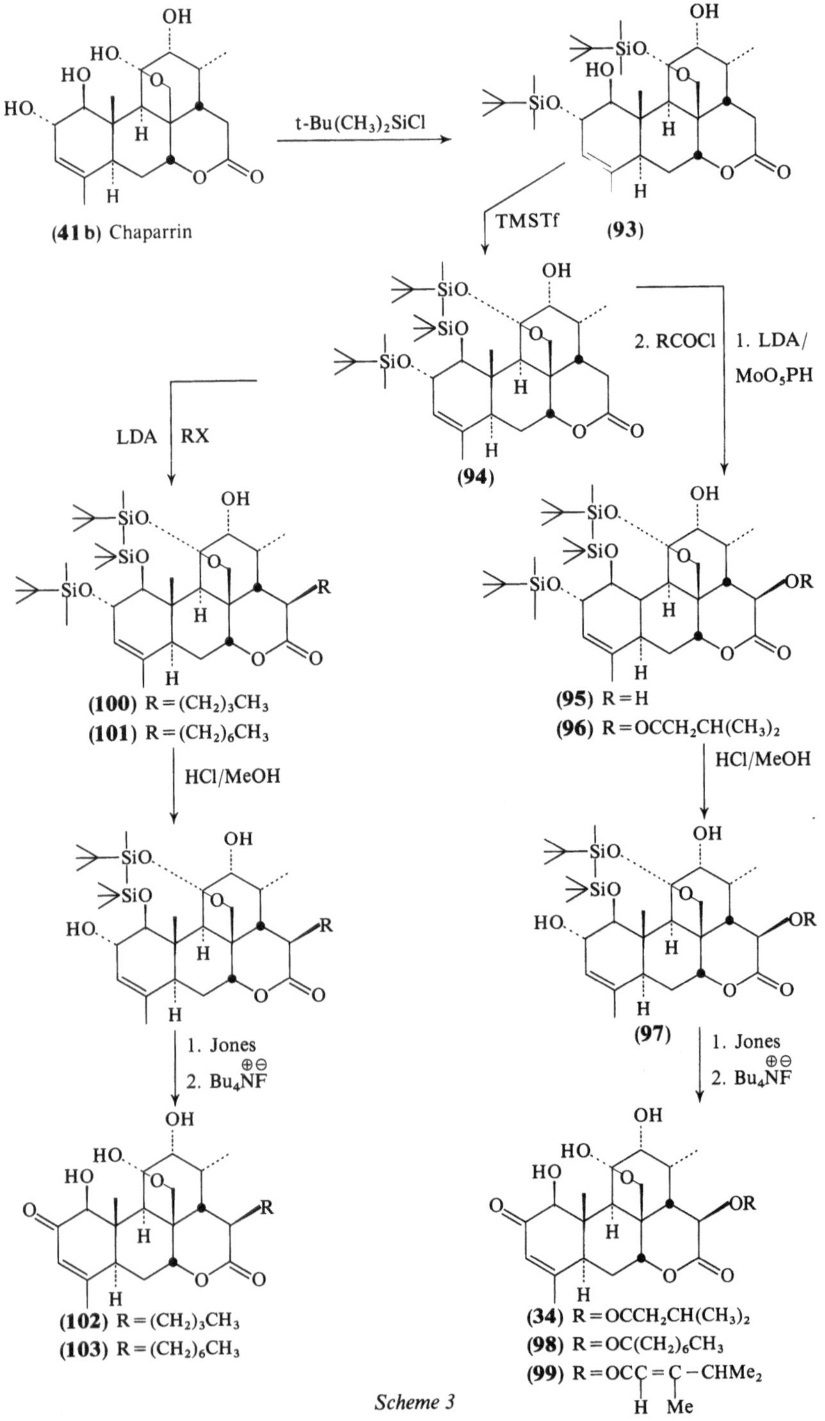

(**41 b**) Chaparrin

t-Bu(CH₃)₂SiCl

(**93**)

TMSTf

(**94**)

2. RCOCl 1. LDA/
MoO₅PH

LDA RX

(**100**) R = (CH₂)₃CH₃
(**101**) R = (CH₂)₆CH₃

(**95**) R = H
(**96**) R = OCCH₂CH(CH₃)₂

HCl/MeOH

HCl/MeOH

(**97**)

1. Jones
2. Bu₄NF
⊕⊖

1. Jones
2. Bu₄NF
⊕⊖

(**102**) R = (CH₂)₃CH₃
(**103**) R = (CH₂)₆CH₃

(**34**) R = OCCH₂CH(CH₃)₂
(**98**) R = OC(CH₂)₆CH₃
(**99**) R = OCC=C—CHMe₂
              H Me

Scheme 3

VII. Synthetic Studies

The broad spectrum of biological properties, especially the antileukemic activity, of the quassinoids as well as their highly oxygenated carbon backbone coupled with the stereochemical features have stimulated a great deal of synthetic activity. A number of approaches towards their synthesis have been described, but success at total synthesis has so far been limited to only two published accounts *(vide infra)*.

The efforts at quassinoid total synthesis can be summarized as follows:

(i) Using a Lewis acid catalyzed Diels-Alder reaction, compound (**104**), a ring A *seco* derivative of quassin (**85**), was synthesized by VALENTA and co-workers (*100, 101*).

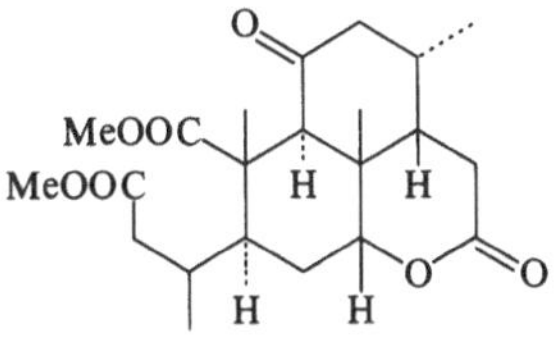

(**104**) 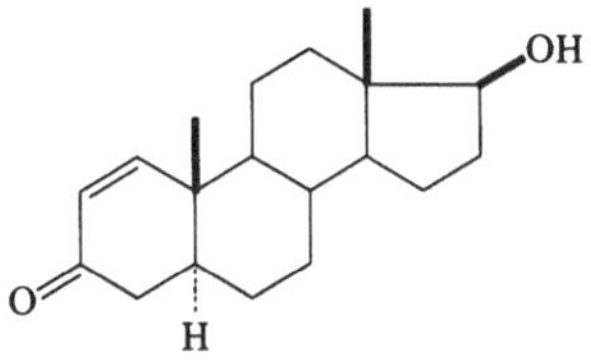   (**105**)

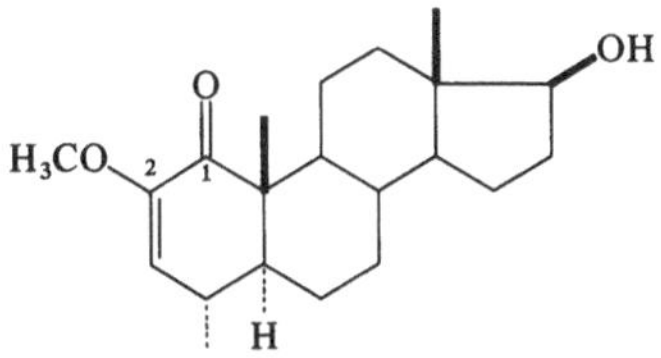

(**106**)

(ii) Starting from the steroid (**105**), the preparation of compound (**106**) with the ring A structure of quassin (**85**) was reported (*42*).

(iii) Compound (**107**), a key intermediate in the partial synthesis of quassin, was synthesized in 28 steps starting from testosterone (**108**). The key features were a) the transformation of (**108**) into (**109**) and the conversion of the latter to the vinylogous α-hydroxyketone (**110**) and b) the photochemically induced [2+2]-cycloaddition of allene to (**110**), affording the derivative (**111**) which was converted into the key compound (**107**) (*76*).

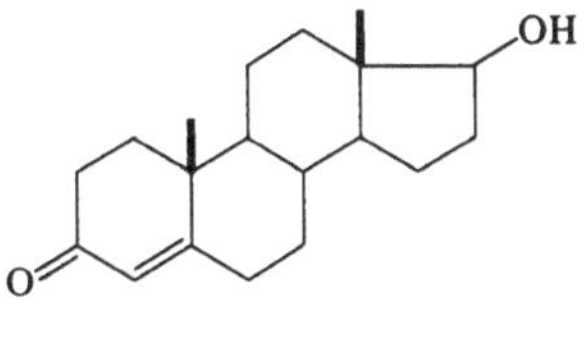

(**108**) 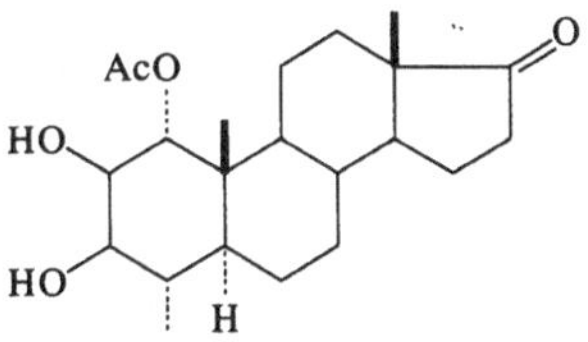   (**109**)

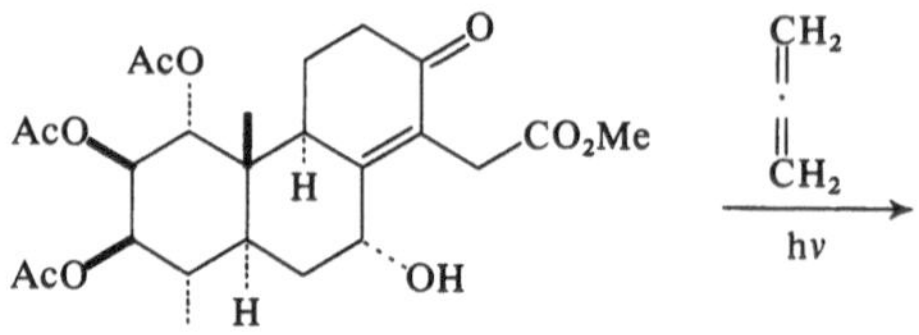

(110)

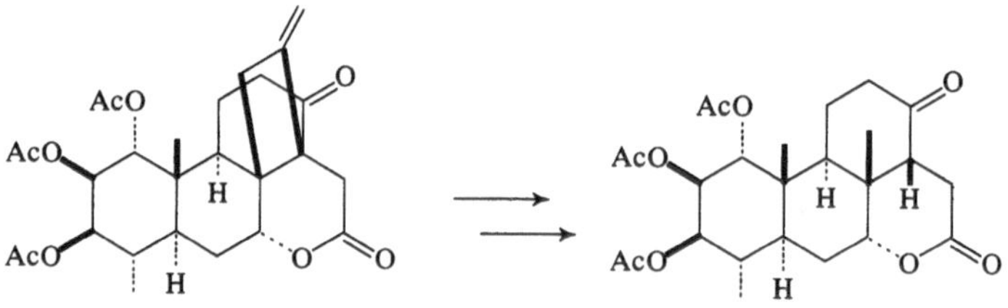

(111)   **(107)**

(iv) D-ring *seco* derivatives of cholic acid **(112)** have been prepared (*17*) and further converted to various δ-lactones (5,14-*epi*-28,30-dinorquassinoids) e.g. **(113)** (*18*).

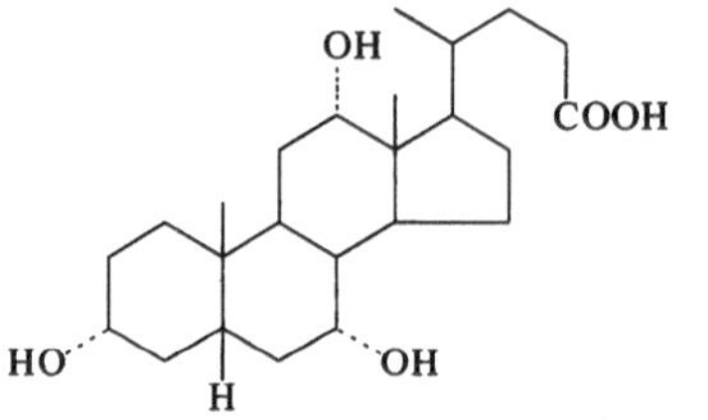

(112)

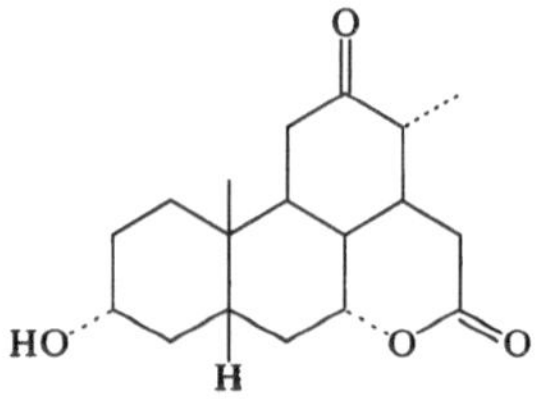

(113)

(v) The lactone **(114)** with a 14αH-picrasane framework has been synthesized and the relative configuration of the chiral centers confirmed by an X-ray analysis. Treatment of the diketone **(115)**, prepared by standard methods, with dilithioacetate afforded after esterification the ester **(116)** which was oxidized to give the unsaturated ketone **(117)**. Reduction of the latter gave the hemiacetal **(118)** which upon oxidation afforded lactone **(114)** (*31*).

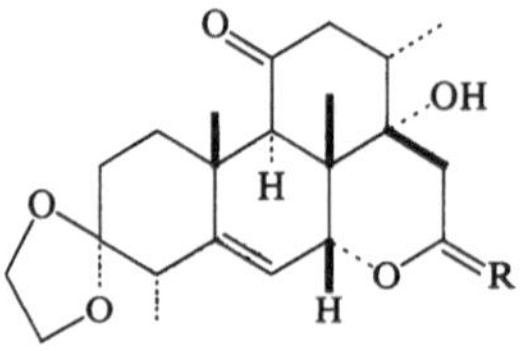

(114) R=O
(118) R=H, OH

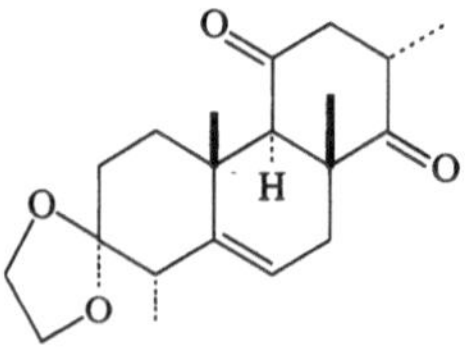

(115)

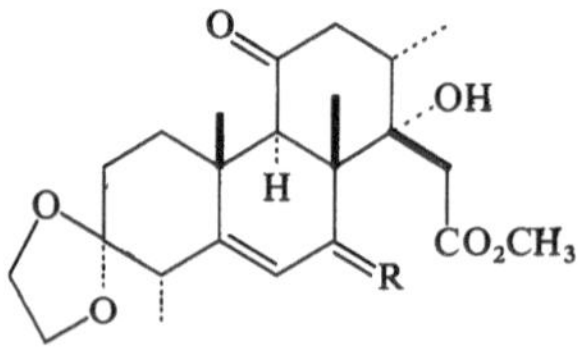

(116) R=H₂
(117) R=O

(vi) A BCE ring model (**119**) of the bruceine type quassinoids has been prepared. The key features of the synthesis include formation of the tetrahydrofuran (**122**) from the tosyloxy enone (**120**) through alcohol tosylate (**121**) via an intramolecular solvolytic ring closure. The olefin → trans diol conversion was achieved by a three-step procedure: a) cis hydroxylation of (**122**) to (**123**); b) regiospecific oxidation of (**123**) to ketol (**124**); c) stereospecific reduction of (**124**) to the trans diol (**119**) (*15*).

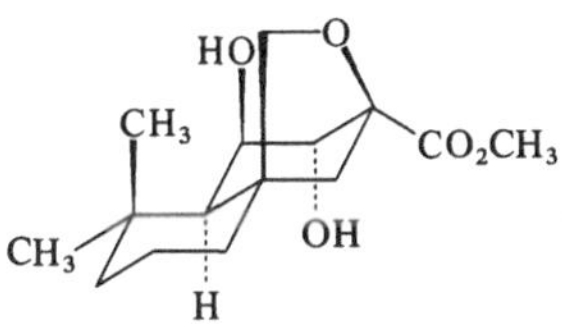

(**119**) 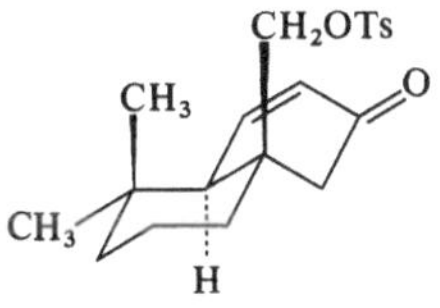　　　　　(**120**) 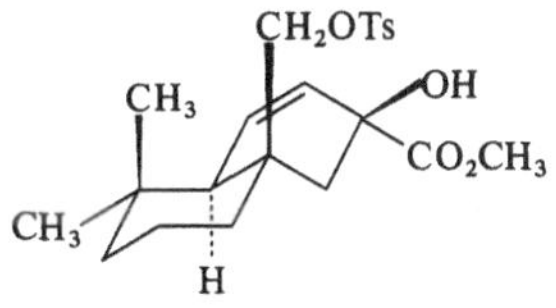　　　　　(**121**)

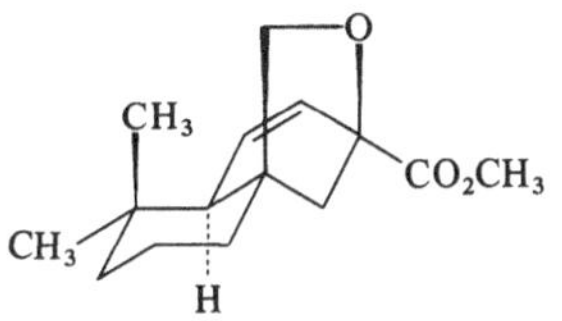

(**122**) 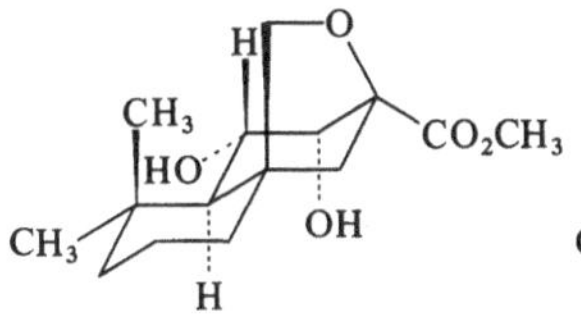　　　　　(**123**) 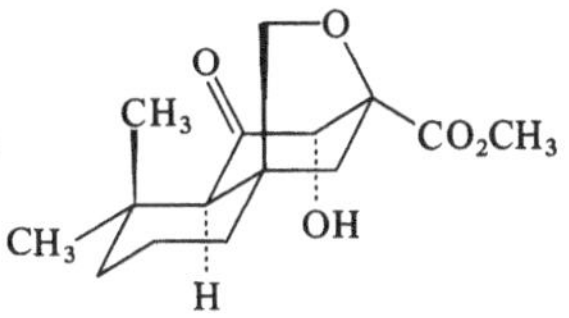　　　　　(**124**)

(vii) A tricyclic intermediate (BCE ring system) (**128**) for the synthesis of quassimarin (**46**) has been synthesized in 9 steps (*43*). Diene (**126**) was reacted with methyl gentisate (**125**) to produce, after reduction and epimerisation, the diketone (**127**). The latter was further transformed to diol (**128**), ring E being introduced by the selenocyclisation method developed by NICOLAOU (*68*). Further investigations allowed transformation of the diketone (**127**) to the BCDE ring system (**129**) (*44*). Compound (**127**) was converted to the epoxide (**130**) which readily afforded the diketone (**131**). Key features of the subsequent reactions include the regioselective protection of diketone (**131**) by use of intramolecular ketal formation, a two-step lactone-to-ether reduction, and a regioselective lactonisation. The tetracyclic system (**129**) was produced in 29% overall yield.

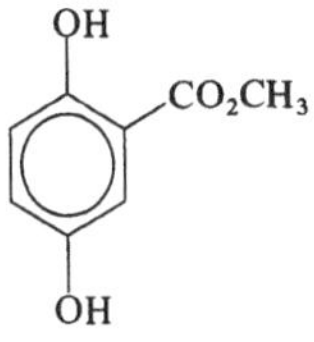

(**125**) 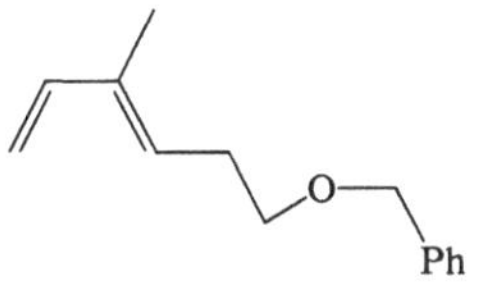　　　　　(**126**) 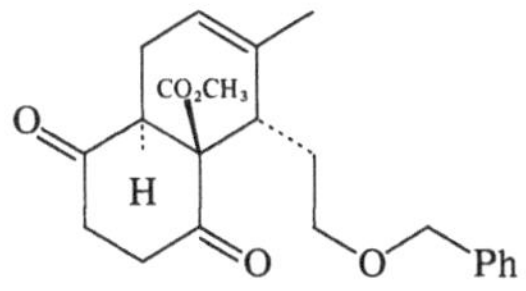　　　　　(**127**)

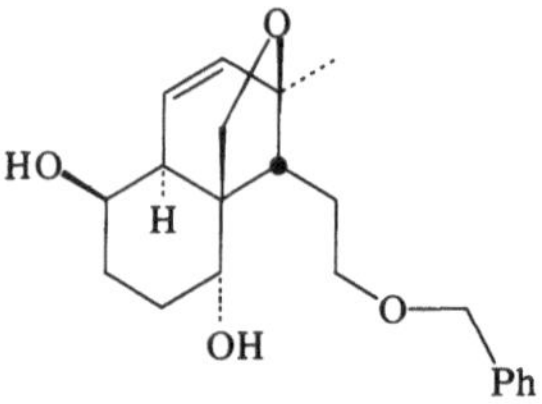

(128) 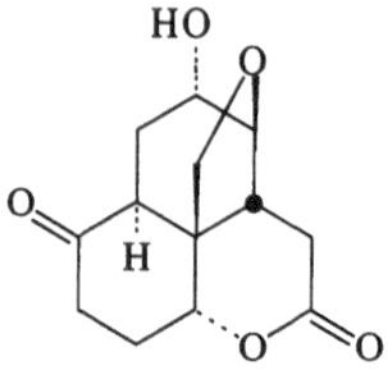   (129) 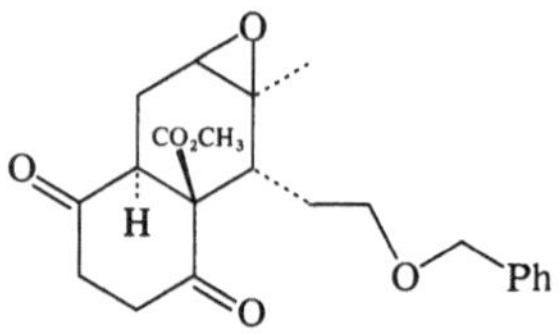   (130)

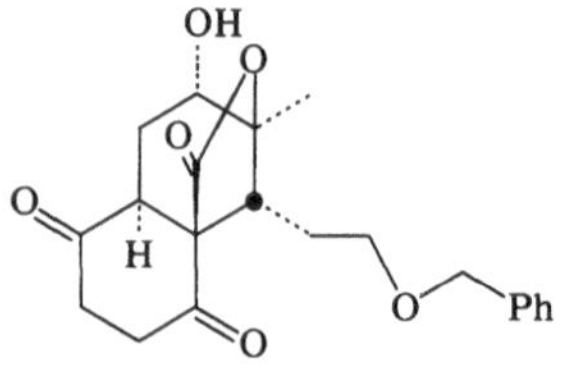

(131)

(viii) The bis-(orthoquinone) (132) has been synthesized as a possible synthetic precursor to quassin (57).

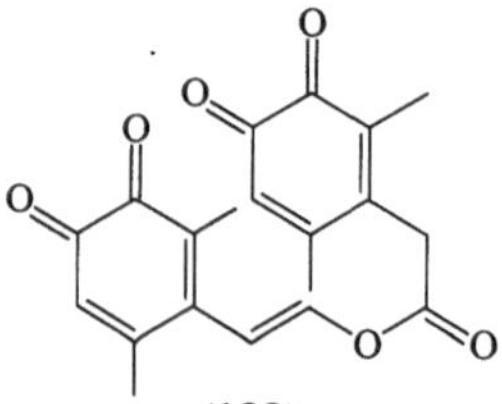

(132)

(ix) Through an intramolecular Diels-Alder approach, thermolysis of the *dl*-benzocyclobutene (133), prepared from norcamphor, gave the tetracyclic compound (134) which was further transformed to aromatic klaineanone (135) (*21a*).

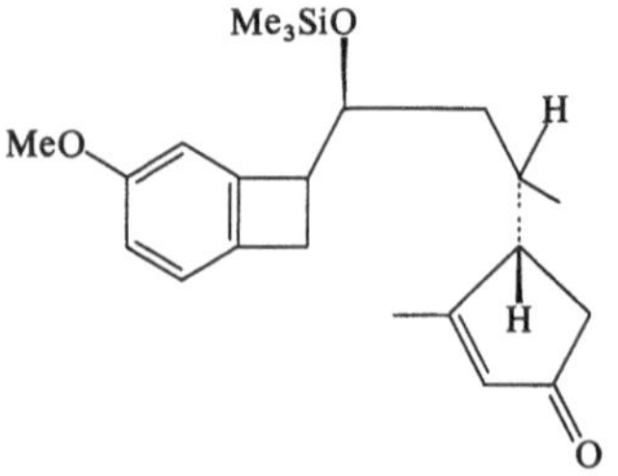

(133) 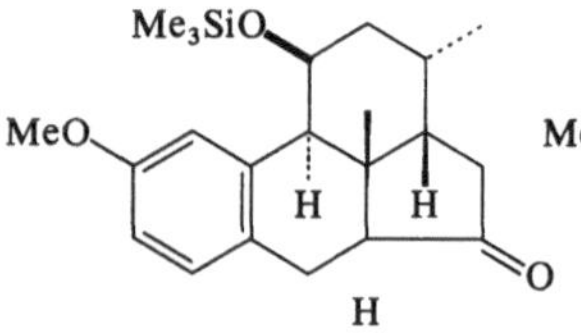   (134) 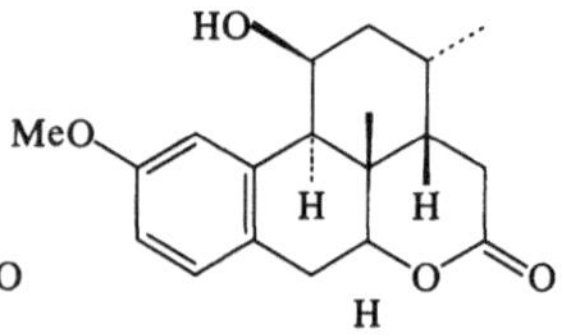   (135)

The tetracyclic compound (134) was also converted into the D-deoxyquassinoid (136) possessing the same ABCD-ring fusion as that of klaineanone (*21b*).

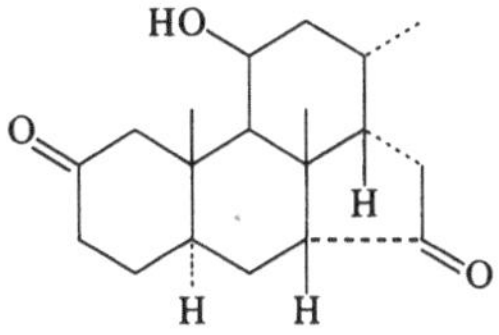

(136) 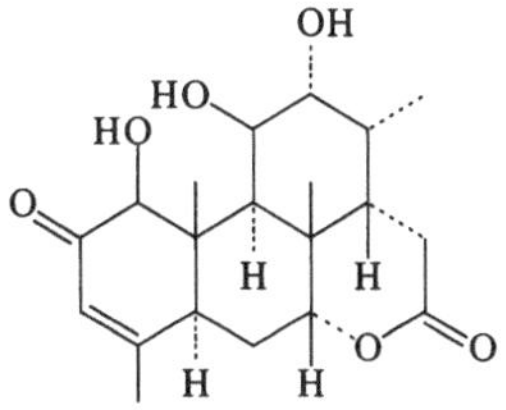   Klaineanone (80)

(x) Recently an efficient synthesis of the pentacyclic system (137) as a model for bruceantin (53) was described (99). The key feature of the synthesis was a stereoselective intramolecular Diels-Alder reaction of the benzocyclobutene derivatives (138a) and (138b) which were conveniently derived from the aldehyde (139) in 7 steps. The tetracyclic cycloadducts (140a), (140b) were converted into alcohol (141) which was then transformed stereoselectively to the pentacyclic lactone (137).

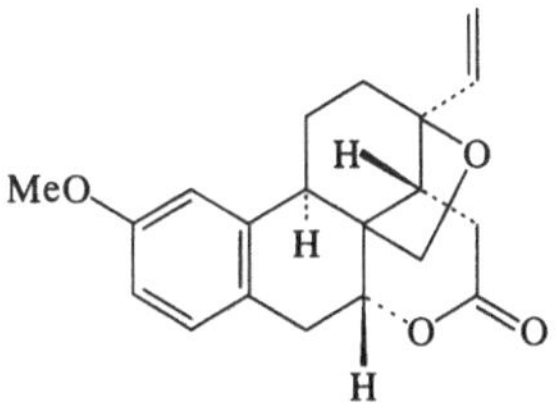

(137) 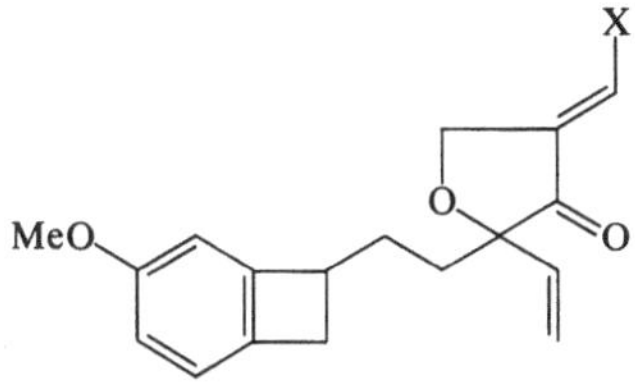   (138) a: X=SPh   (139)
                        b: X=OAc

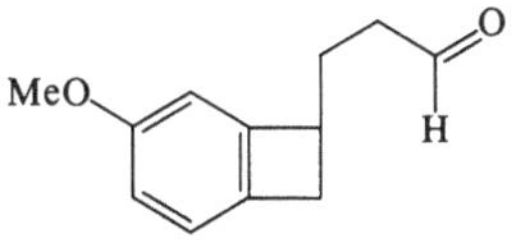

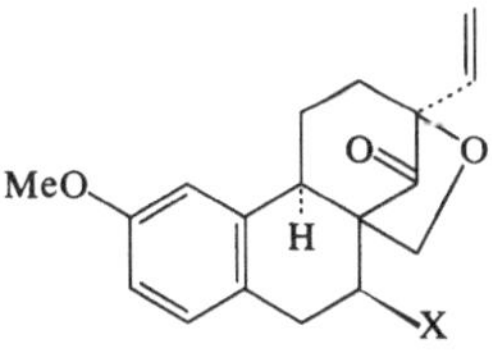

(140) a: X=SPh   (141)
      b: X=OAc

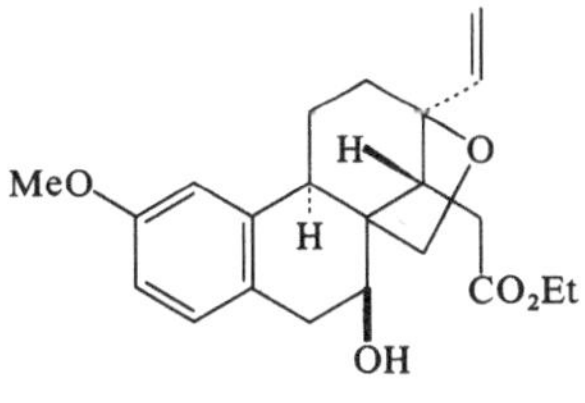

(xi) A synthetic approach to the quassinoids involved an attempted intramolecular Diels-Alder reaction of the linked diene-dienophile (142) (107).

In an alternative approach, the Diels-Alder adduct (144), obtained from the enedione (143) with 1-methoxy-3 [(trimethylsilyl)oxy]-1,3-butadiene, was further transformed into the α-iodo ester (145). Exposure of the latter to iodotrimethylsilane furnished the δ-lactone (146), an attractive ABCD ring precursor to the quassinoids (106).

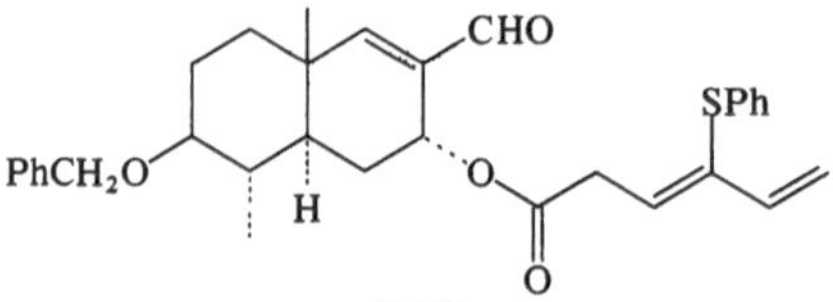

(142) 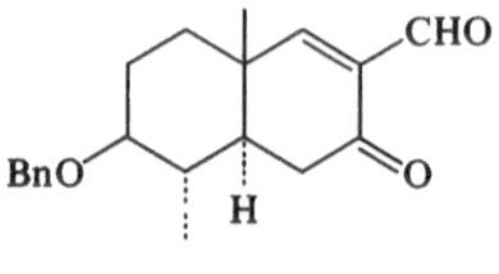   (143)

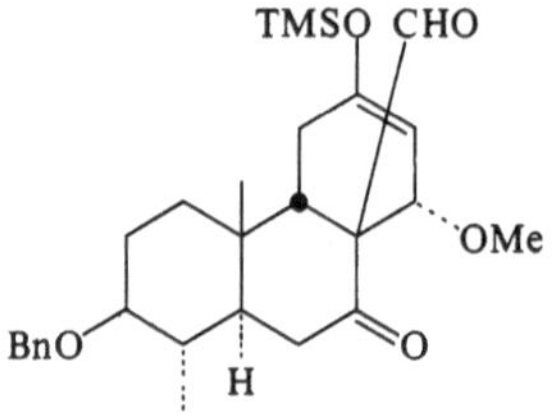

(144) 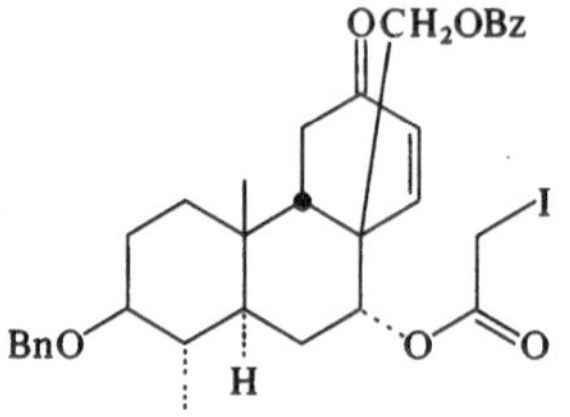   (145) 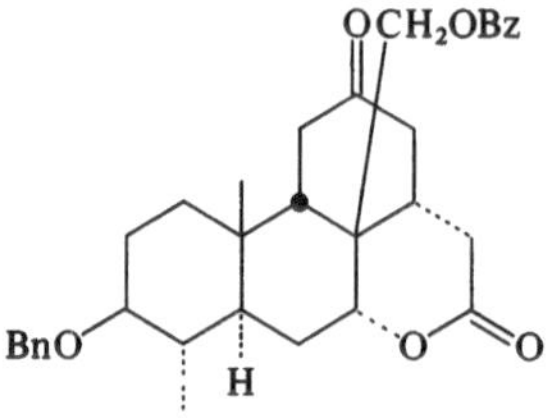   (146)

(xii) Most recently a general synthesis of pentacyclic quassinoids was reported (*3*) (Scheme 4). Much of the carbon skeleton was assembled in a single conjugate addition-enolate trapping reaction using *trans*-1-iodo-3-(benzyloxy)-1-pentene and 4-prenyl-3-methyl-2-cyclohexenone to afford after addition of $(EtO)_2POCl$ the triene (**147**). The latter was transformed to the enone (**148**) to which ethylcyanoacetate was added giving compound (**149**). Ozonolysis and cyclisation of (**149**) led to alcohol (**150**) which was further elaborated to (**151**). Treatment of the latter with 1,1'-carbonyldiimidazole afforded the tetracyclic lactone (**152**) which was converted to the alcohol (**153**) and then by selenocyclisation (*68*) to diene (**154**). Monoepoxidation furnished (**155**) which rearranged to allylic alcohol (**156**).

Upon standing at room temperature or during prolonged exposure to silica gel, (**156**) isomerized completely to the pentacyclic structure (**157**). While diene (**154**) showed no tendancy to rearrange, hydroxy selenide (**158**) also isomerized quantitatively to C_{11}-bridged (**159**) thus implicating some assistance by the C_1 α-hydroxyl group. Both structural classes of ring-C bridging ethers now become synthetically accessible.

Compound (**155**) was transformed by standard methods to enone (**160**).

VIII. Total Synthesis of dl-Quassin

The synthetic strategy applied by GRIECO *et al.* (*25*) is outlined in Scheme 5.

Compound (**165a**) possesses the complete framework of quassin. Furthermore, with the exception of the configuration at C-9, it has six of the

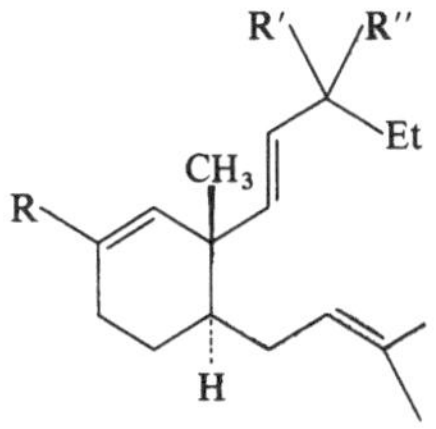

(147) R = OPO(OEt)₂
R' = H; R'' = OBn

(148) R = HR'' = O

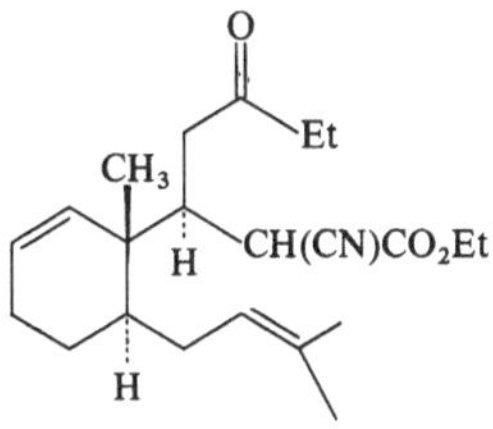

(149)

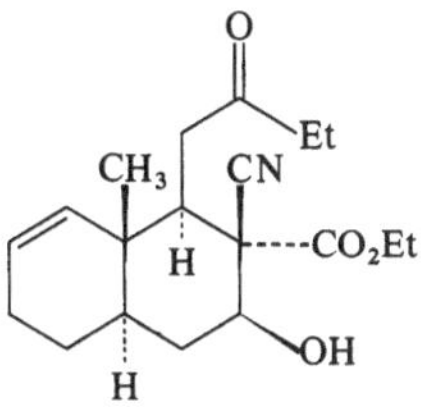

(150)

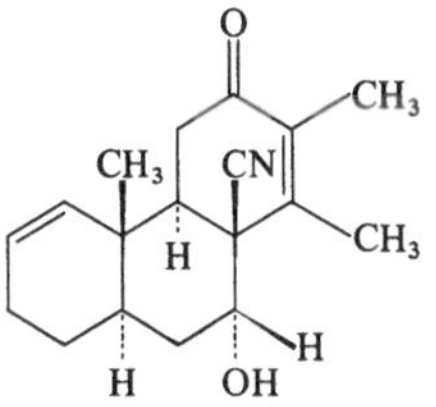

(151)

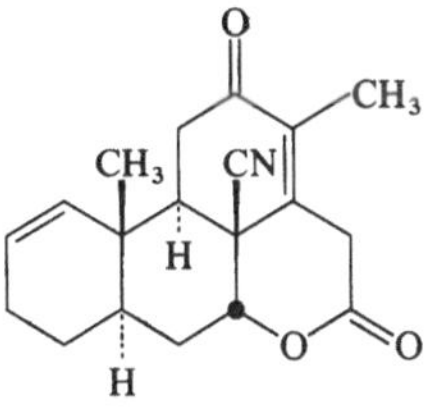

(152)

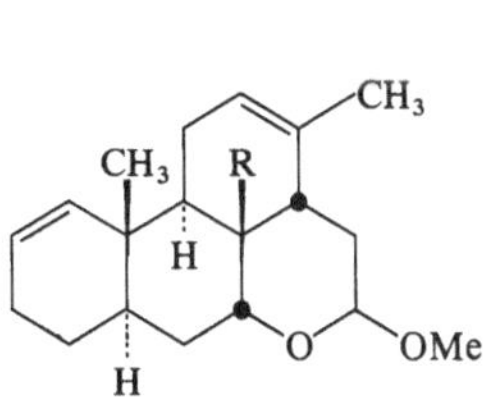

(153) R = CH₂OH

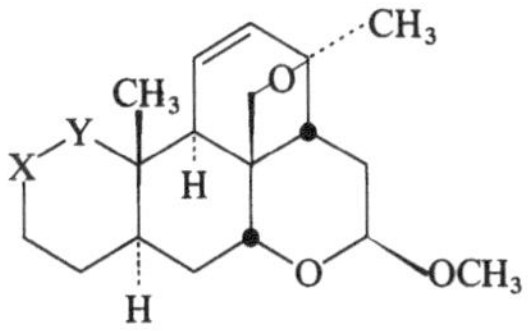

(154) X, Y = CH = CH

(155) X, Y = CH——CH
O

(158) X = β-PhSe——CH
Y = a-OH——CH

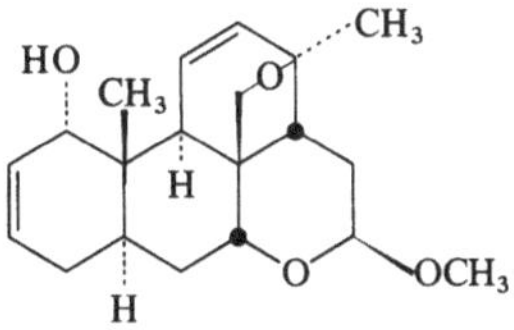

(156)

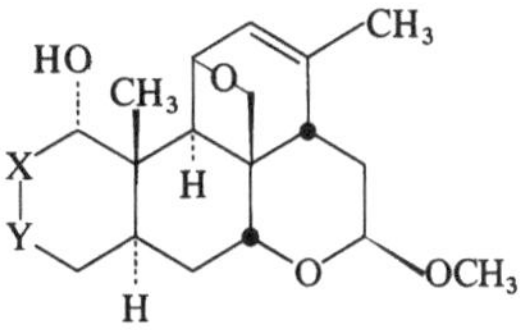

(157) X, Y = CH = CH

(159) X = β-PhSe——CH
Y = CH₂

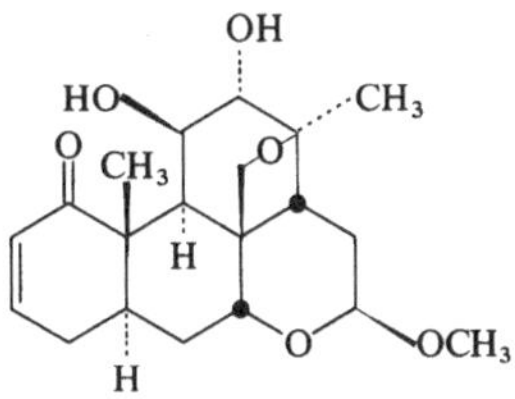

(160)

Scheme 4

seven chiral centers found in quassin. This compound has been prepared *via* a remarkable Diels-Alder reaction. Treatment of a benzene solution of *enone* (**162**) [prepared from (**161**)] containing ethylaluminium chloride with excess of *diene* (**163**) gave as the sole Diels-Alder product the tricyclic keto-

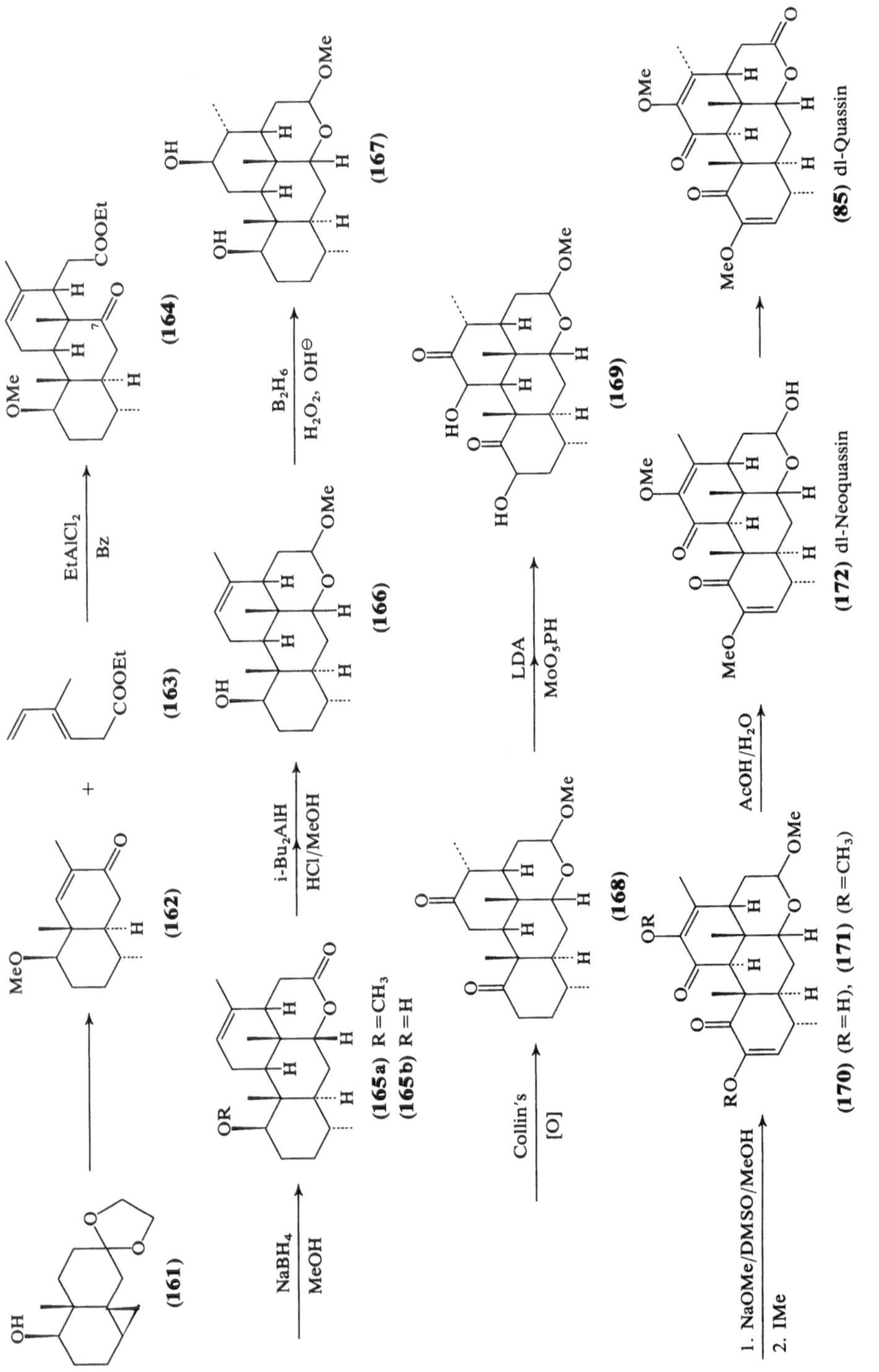
(161)
(162)
(163)
(164)
EtAlCl$_2$
Bz
COOEt
COOEt
NaBH$_4$
MeOH
(165a) R = CH$_3$
(165b) R = H
i-Bu$_2$AlH
HCl/MeOH
(166)
B_2H_6
H_2O_2, $OH^{\ominus}$
(167)
Collin's
[O]
(168)
LDA
MoO$_5$PH
(169)
1. NaOMe/DMSO/MeOH
2. IMe
(170) (R = H), (171) (R = CH$_3$)
AcOH/H$_2$O
(172) dl-Neoquassin
(85) dl-Quassin

ester (**164**) in 64% yield. Reduction of keto ester (**164**) with sodium borohydride in methanol proceeded in a stereospecific fashion giving rise to a single crystalline lactone (**165a**) which was demethylated to the hydroxy lactone (**165b**). This compound represents a versatile intermediate for further elaboration into quassin and other quassinoids. Its stereochemistry has been confirmed by single-crystal X-ray analysis.

Prior to hydroboration of the $\Delta^{12,13}$ olefinic double bond, the lactone carbonyl was reduced by diisobutyl aluminium hydride to the corresponding lactol which was treated with HCl in methanol to give the protected lactol (**166**). Hydroboration afforded the diol (**167**). The latter was oxidized to the diketone (**168**). Elaboration of the disphenol units was achieved in a two-step sequence. The diketone (**168**) was treated with LDA and the resulted di-anion was submitted to the MoO_5PH reagent (*104*) to give in 35% yield the bis-α-hydroxy ketone (**169**), which was smoothly transformed with sodium methoxide in dimethylsulfoxide into diosphenol (**170**). Upon methylation (**170**) gave (**171**) which was selectively hydrolysed to the racemic neoquassin (**172**). Oxidation with Fetizon's reagent yielded racemic quassin. Overall yield from the enone (**162**) was more than 3%.

IX. Total Synthesis of dl-Castelanolide

GRIECO *et al.* (*26*) have synthesized the quassinoid castelanolide (**183**), isolated by GEISSMAN and co-workers over 10 years ago (*80*), starting from the tetracyclic alcohol (**166**) prepared previously (*25*) by a four-step sequence from the known Diels-Alder adduct (**164**) (Scheme 6). Compound (**166**) was subjected to tetrahydropyranylation followed by hydroboration of the C(12)-C(13) olefinic bond thus giving rise to tetracyclic alcohol (**173**). Benzylation of (**173**) and subsequent cleavage of the THP protecting group afforded alcohol (**174**) which was oxidized to ketone (**175**). The latter was directly transformed [LDA (Me$_2$N)$_2$POCl] into (**176a**) and then subjected to reductive cleavage (Li, EtNH$_2$/THF, t-BuOH) to give the tetracyclic olefinic alcohol (**176b**). Construction of the ring C diosphenol moiety was achieved via a three-step sequence. Oxidation of (**176b**) provided ketone (**177**) whose enolate was treated with the MoO_5PH reagent (*104*) to afford the hydroxy ketone (**178**). The latter was smoothly transformed with concomitant inversion of configuration at C(9) into diosphenol (**179**) upon treatment with excess of sodium methoxide in DMSO containing MeOH. This one-pot procedure permits facile elaboration of the *trans, anti, trans* arrangement of the ABC ring system found in castelanolide (**183**). Prior to glycolation of the $\Delta^{1,2}$ double bond, it was necessary to unmask the δ-lactone and protect the sensitive diosphenol unit. Hydrolysis of the protected lactol (**179**) followed by mild oxidation (Ag$_2$CO$_3$) gave rise to tetracyclic lactone (180) which was readily converted into acetate (181). Osmylation of (181) afforded monoacetate (182) which upon mild alkaline hydrolysis led to crystalline *dl*-castelanolide (183).

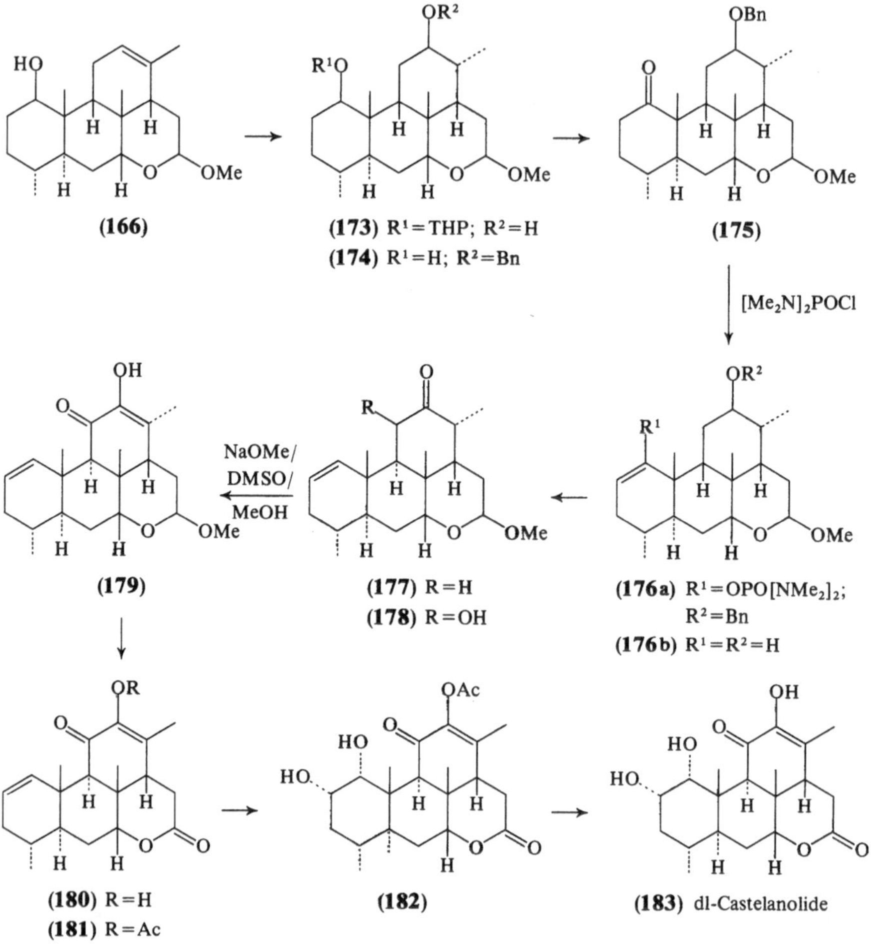

Scheme 6

X. Tables

Table 3. C_{18}-Quassinoids

Compound	Mol. formula	Mol. Wt.	Origin	Mp.	[α]	References
Samaderine A (1)	$C_{18}H_{18}O_6$	330	*Samadera indica*	255°	−31.3°	(111)
Laurycolactone A (2)	$C_{18}H_{22}O_5$	318	*Eurycoma longifolia*	270°	+216°	(67)
Laurycolactone B (3)	$C_{18}H_{20}O_5$	316	*E. longifolia*	230°	+92.6°	(67)

Table 4. C_{19}-Quassinoids

Compound	Mol. formula	Mol. Wt.	Origin	Mp.	[α]	References
Shinjulactone B (4)	$C_{19}H_{22}O_7$	362	*Ailanthus altissima*	268°	+167°	(22)
Eurycomalactone (6)	$C_{19}H_{24}O_6$	348	*E. longifolia*	270°	+100°	(67)

Table 5. C_{20}-Quassinoids

Compound	Mol. formula	Mol. Wt.	Origin	Mp.	[α]	References
Nigakilactone K (8)	$C_{22}H_{30}O_7$	406	*Picrasma ailanthoides*	227°	−26°	(62)
Nigakilactone L (9)	$C_{22}H_{30}O_7$	406	*P. ailanthoides*	296°	+65°	(62)
Nigakilactone M (10)	$C_{21}H_{30}O_7$	394	*P. ailanthoides*	178°	+39°	(62)
Nigakilactone N (11)	$C_{21}H_{30}O_7$	394	*P. ailanthoides*	211°	+36°	(62)
Picrasinol A (12)	$C_{24}H_{36}O_7$	436	*P. ailanthoides*	124°	+49.6°	(62)
Picrasinol B (13)	$C_{22}H_{32}O_6$	392	*P. ailanthoides*	206°	+12.1°	(71)
Soulameanone (14)	$C_{20}H_{28}O_8$	396	*Soulamea muelleri*	265°	+101°	(89)

Table 5 (continued)

Compound	Mol. formula	Mol. Wt.	Origin	Mp.	[α]	References
1,12-Di-O-Acetyl-Soulameanone (15)	$C_{24}H_{32}O_{10}$	480	S. muelleri	268°		(89)
Δ²-Picrasin B (16)	$C_{21}H_{26}O_6$	374	S. muelleri	250°	+31°	(89)
Karinolide (17)	$C_{20}H_{24}O_8$	392	Simaba multiflora	210°		(87)
Shinjulactone C (18a)	$C_{20}H_{22}O_7$	374	Ailanthus altissima	292°	−344°	(36)
Shinjulactone F (18b)	$C_{20}H_{22}O_7$	374	A. altissima	203°	−148°	(37)
Ailanthone (19)	$C_{20}H_{24}O_7$	376	A. altissima	238°	+19°	(64, 65)
2-Dihydroailanthone (26) (= Shinjulactone A)	$C_{20}H_{26}O_7$	378	A. altissima	263°	+2°	(8, 65)
Shinjudilactone (27)	$C_{20}H_{24}O_7$	376	A. altissima	276°	+102°	(32, 36)
Eurycomanone (28)	$C_{20}H_{24}O_9$	408	Eurycoma longifolia	255°	+32°	(13, 14)
Eurycomanol (29)	$C_{20}H_{26}O_9$	410	E. longifolia	275°	+87.7°	(13, 14)
2′-Acetoxyglaucarubine (32b)	$C_{27}H_{38}O_{11}$	538	Simarouba amara	246°	+29.5°	(91, 112)
Excelsin (33)	$C_{25}H_{36}O_9$	480	Ailanthus excelsa, Odyendyea gabonensis			(40, 112)
Castelanone (34)	$C_{25}H_{34}O_9$	478	Castela tweedii	239°	+46.6°	(94)
Soularubinone (35)	$C_{25}H_{34}O_{10}$	494	Soulamea tomentosa	238°	+47.6°	(58)
13,18-Dehydroglaucarubinone (36b)	$C_{25}H_{32}O_{10}$	492	Simarouba amara, A. altissima	218°	+34.7°	(36, 91)
13,18-Dehydroailanthinone (37b)	$C_{25}H_{32}O_9$	476	Pierrodendron kerstingii	260°	+39.6°	(46)
13,18-Dehydroexcelsin (38)	$C_{25}H_{34}O_9$	478	A. excelsa, A. malabarica			(39, 40)
6α-Senecioyloxychaparrinone (40b)	$C_{25}H_{32}O_9$	476	Simaba multiflora	257°	+213°	(87, 110)
6α-Tigloyloxychaparrinone (41a)	$C_{25}H_{32}O_9$	476	Ailanthus integrifolia, Simaba cuspidata, Ailanthus grandis	229°	+195.7°	(92, 98)
6α-Senecioyloxychaparrin (40c)	$C_{25}H_{34}O_9$	478	Simaba multiflora	282°	+240°	(1, 87)
6α-Tigloyloxychaparrin (41c)	$C_{25}H_{34}O_9$	478	Simaba cuspidata, Ailanthus grandis	275°	+130°	(92)
Undulatone (42)	$C_{27}H_{34}O_{11}$	534	Hannoa undulata, H. klaineana	235°	+205.9°	(56, 109)
15-Desacetylundulatone (42a)		402				(56)

Compound	Source	Formula	MW	mp	[α]	Ref.
6α-Tigloyloxyglaucarubol (43b)	H. klaineana	$C_{25}H_{34}O_{10}$	494			(56)
Isobruceine A (44)	Soulamea tomentosa / S. soulameoides	$C_{26}H_{34}O_{11}$	522	200°	+43°	(29, 84)
Isobruceine B (45)	Bruca antidysenterica / Picrolemma pseudocoffea	$C_{23}H_{28}O_{11}$	480	258°	+17°	(45, 60)
Quassimarin (46)	Quassia amara	$C_{27}H_{36}O_{11}$	536	238°	+22.4°	(48)
Samaderine E (47)	Samadera indica	$C_{20}H_{26}O_{8}$	394	207°	−11.7°	(111)
Bruceantarin (52)	B. antidysenterica	$C_{28}H_{30}O_{11}$	542	185°	−20.7°	(45)
Bruceantin (53)	B. antidysenterica	$C_{28}H_{36}O_{11}$	548	226°	−43°	(45)
Bruceantinol (54)	B. antidysenterica	$C_{30}H_{38}O_{13}$	606		−14.5° / +12.8°	(45, 90)
Dehydrobruceantin (56)	B. antidysenterica	$C_{28}H_{34}O_{11}$	546		+79°	(45)
Dehydrobruceantarin (57)	B. antidysenderica	$C_{28}H_{28}O_{11}$	540		+68°	(45)
Dehydrobrucein A (58)	B. javanica	$C_{26}H_{32}O_{11}$	520			(77, 78)
Dehydrobrucein B (59)	B. antidysenterica / B. javanica	$C_{23}H_{26}O_{11}$	478			(45, 77, 78)
Dehydrobruceantol (60)	B. antidysenterica	$C_{28}H_{34}O_{12}$	562		+30°	(45)
Dehydrobruceantinol (61)	B. antidysenterica	$C_{30}H_{36}O_{13}$	604			(90)
3,4-Dihydrobrucein A (49b)	B. javanica	$C_{26}H_{36}O_{11}$	524			(77, 78)
Sergeolide (62)	Picrolemma pseudocoffea	$C_{25}H_{28}O_{11}$	504	206°	−103.3°	(60)
15-Deacetylsergeolide (63)	P. pseudocoffea	$C_{23}H_{26}O_{10}$	462	300°	−145°	(85)
Bruceoside-A (64)	B. javanica	$C_{32}H_{42}O_{16}$	682	∼180°	+9.2°	(50)
Bruceoside-B (65)	B. javanica	$C_{32}H_{42}O_{16}$	682	224°	+3.7°	(50)
Bruceantinoside-A (66)	B. antidysenterica	$C_{34}H_{40}O_{16}$	704	∼150°	+7.8°	(74)
Bruceantinoside-B (67)	B. antidysenterica	$C_{34}H_{40}O_{16}$	704	∼200°	−3.6°	(74)
Picrasinoside-A (68)	Picrasma ailanthoides	$C_{27}H_{38}O_{12}$	538	166°	−54°	(72)
15-Glaucarubolone glucoside (69)	Simarouba glauca	$C_{26}H_{36}O_{13}$	556	254°	−25.7°	(6)
15-Glaucarubol glucoside (70)	S. glauca	$C_{26}H_{38}O_{13}$	558	264°	+12°	(6)
Picrasinoside-B (71)	Picrasma ailanthoides	$C_{28}H_{40}O_{11}$	552	153°	−15.1°	(71)
Picrasinoside-C (72)	P. ailanthoides	$C_{26}H_{42}O_{11}$	554	164°	−41.1°	(71)
Picrasinoside-D (73)	P. ailanthoides	$C_{30}H_{46}O_{12}$	598	144°	+3.3°	(71)
Picrasinoside-E (74)	P. ailanthoides	$C_{30}H_{46}O_{13}$	614	164°	−14.7°	(71)
Picrasinoside-F (75)	P. ailanthoides	$C_{28}H_{44}O_{11}$	556	154°	−8.2°	(71)
Picranoside-G (76)	P. ailanthoides	$C_{28}H_{44}O_{12}$	572	163°	+24.2°	(71)

Table 6. C_{25}-Quassinoids

Compound	Mol. formula	Mol. Wt.	Origin	Mp.	[α]	References
Deacetylsimarolide (79)	$C_{25}H_{34}O_8$	462	*Simaba moretii (19)*	194°		(86)
Soulameolide (80)	$C_{25}H_{32}O_8$	460	*Soulamea tomentosa*	263°	−72.6°	(88)
Simarinolide (81)	$C_{27}H_{36}O_8$	488	*Simaba moretii*	210°	+23.7°	(93)
Guanepolide (82)	$C_{27}H_{34}O_9$	502	*S. moretii*	284°	−71°	(93)

References

1. Arisawa, M., A. D. Kinghorn, G. A. Cordell, and N. R. Farnsworth: Plant Anticancer Agents. XXIII. 6α-Senecioyloxy Chaparrin, a New Antileukemic Quassinoid from *Simaba multiflora*. J. Nat. Prod. **46**, 218 (1983).
2. Baldwin, M. A., D. M. Carter, F. A. Darwish, and J. D. Phillipson: The Mass Spectral Behaviour of Bruceolides. Biomed. Mass Spectrom. **8**, 362 (1981).
3. Batt, D. G., N. Takamura, and B. Ganem: General Synthesis of Pentacyclic Quassinoids. J. Amer. Chem. Soc. **106**, 3353 (1984).
4. Bedikian, A. Y., M. Valdivieso, G. P. Bodey, W. K. Murphy, and E. J. Freireich: Initial Clinical Studies with Bruceantin. Cancer Treat. Rep. **63**, 1843 (1979).
5. Bhatnagar, S., and J. Polonsky: Conversion of Chaparrin into Quassinoid Analogs. Unpublished results.
6. Bhatnagar, S., J. Polonsky, T. Prangé, and C. Pascard: New Toxic Quassinoid Glucosides from *Simarouba Glauca* (X-Ray Analysis). Tetrahedron Lett. **25**, 299 (1984).
7. Caruso, A. J., J. Polonsky, and B. Soto Rodriguez: Synthetic Studies in the Quassinoid Series. Conversion of Chaparrin into Castelanone and Quassinoid Analogs. Tetrahedron Lett. **23**, 2567 (1982).
8. Casinovi, C. G., P. Ceccherelli, G. Fardella, and G. Grandiloni: Isolation and Structure of a Quassinoid from *Ailanthus glandulosa*. Phytochemistry **22**, 2871 (1983).
9. Cassady, J. M., and M. Suffness: Terpenoid Antitumor Agents. In: Anticancer Agents Based on Natural Product Models, p. 254. New York: Academic Press, Inc. 1980.
10. Chien, M. M., and J. P. Rosazza: Microbial Transformation of Natural Antitumour Agents. Part 15. Metabolism of Bruceantin by *Streptomyces griseus*. J. C. S. Perkin I, 1352 (1981).
11. Corey, E. J., and A. Venkateswarlu: Protection of Hydroxyl Groups as *tert*-Butyldimethylsilyl Derivatives. J. Am. Chem. Soc. **94**, 6190 (1972).
12. Cuckler, A. C., S. Kuna, C. W. Mushett, R. H. Silber, R. B. Stebbins, H. C. Stoerk, R. N. Arison, F. Cuchie, and C. M. Malanga: Chemotherapeutic and pharmacological studies on glaucarubin, a specific amebacide. Arch. Int. Pharmacodyn. **114**, 307 (1958).
13. Darise, M., H. Kohda, K. Mizutani, and O. Tanaka: Eurycomanone and Eurycomanol, Quassinoids from the Roots of *Eurycoma Longifolia*. Phytochemistry **21**, 2091 (1982).
14. — — Revision of Configuration of the 12-Hydroxyl Group of Eurycomanone and Eurycomanol, Quassinoids from *Eurycoma Longifolia*. Phytochemistry **22**, 1514 (1983).
15. Dailey, O. D., and P. L. Fuchs: Synthesis of a Model for the BCE Ring System of Bruceantin. A Caveat on the Cyclohexene→Trans Diaxial Diol Conversion. J. Org. Chem. **45**, 216 (1980).
16. De Carneri, I., and C. G. Casinovi: Un potente antiamebico d'origine vegetale: l'ailantone, principio attivo di *Ailanthus glandulosa*. Parasitologia **10**, 215 (1968).
17. Dias, J. R., and R. Ramachandra: Studies Directed toward Synthesis of Quassinoids. 2. D-Ring cleavage of Cholic Acid Derivatives. J. Org. Chem. **42**, 1613 (1977).
18. Dias, J. R., and R. Ramachandra: Studies Directed toward Synthesis of Quassinoids. 5. Conversion of D-Ring Seco Derivatives of Cholic Acid to δ-Lactones. J. Org. Chem. **42**, 3584 (1977).
19. Feuillet, C.: Etudes sur les Simaroubaceae. II. Un Simaba nouveau de Guyane française dans la section Floribundae Engl.: S. morettii. Candollea **38**, 745 (1983).
20. Fong, K. L. L., D. H. W. Ho, R. S. Benjamin, N. S. Brown, A. Y. Bedikian, B. S. Yap, C. L. Wiseman, W. Kramer, and G. P. Bodey: Clinical Pharmacology of Bruceantin by Radioimmunoassay. Cancer Chemother. Pharmacol. **9**, 169 (1982).

21a. FUKUMOTO, K., M. CHIHIRO, Y. SHIRATORI, M. IHARA, T. KAMETANI, and T. HONDA: An Intramolecular Diels-Alder Approach to Quassinoids — A Stereoselective Construction of A Aromatic Klaineanone. Tetrahedron Lett. **23**, 2973 (1982).

21b. FUKUMOTO, K., M. CHIHIRO, M. IHARA, T. KAMETANI, and T. HONDA: A New Synthetic Approach to Quassinoids *via* an Intramolecular Diels-Alder Reaction: A Stereoselective Construction of the Klaineanone Ring System. J. Chem. Soc. (London). Perkin Trans. I, 2569 (1983).

22. FURUNO, T., H. NAORA, T. MURAE, H. HIROTA, T. TSUYUKI, T. TAKAHASHI, A. ITAI, Y. IITAKA, and K. MATSUSHITA: Structure of Shinjulactone B, A New Bitter Principle from *Ailanthus altissima*. Chemistry Lett. 1797 (1981).

23. GILLIN, F. D., and D. S. REINER: In vitro Activity of Certain Quassinoid Antitumor Agents against *Entamoeba histolytica*. Arch. Invest. Med. **13** (Suppl. 3), 43 (1982).

24. GILLIN, F. D., D. S. REINER, and M. SUFFNESS: Bruceantin, a Potent Amebicide from a Plant, *Brucea antidysenterica*. Antimicrob. Agents Chemother. **22** (2), 342 (1982).

25. GRIECO, P. A., S. FERRIÑO, and G. VIDARI: Total Synthesis of *dl*-Quassin. J. Amer. Chem. Soc. **102**, 7586 (1980).

26. GRIECO, P. A., R. LIS, S. FERRIÑO, and J. Y. JAW: Synthetic Studies on Quassinoids: Total Synthesis of *dl*-Castelanolide. J. Org. Chem. **47**, 601 (1982).

27. HALL, I. H., Y. F. LIOU, M. OKANO, and K. H. LEE: Antitumor Agents XLVI: *In vitro* Effects of Esters of Brusatol, Bisbrusatol, and Related Compounds on Nucleic Acid and Protein Synthesis of P-388 Lymphocytic Leukemia Cells. J. Pharm. Sci. **71**, 345 (1982).

28. HALL, I. H., K. H. LEE, Y. IMAKURA, M. OKANO, and A. JOHNSON: Anti-Inflammatory Agents III: Structure-Activity Relationships of Brusatol and Related Quassinoids. J. Pharm. Sci. **72**, 1282 (1983).

29. HANDA, S. S., A. D. KINGHORN, G. A. CORDELL, and N. R. FARNSWORTH: Plant Anticancer Agents XXV. Constituents of *Soulamea Soulameoides*. J. Nat. Prod. **46**, 359 (1983).

30. HIKINO, H., T. OHTA, and T. TAKEMOTO: Picrasins, Simaroubolides of Japanese Quassia Tree *Picrasma Quassinoides*. Phythochemistry **14**, 2473 (1975).

31. HONDA, T., T. MURAE, S. OHTA, Y. KURATA, H. KAWAI, T. TAKAHASHI, A. ITAI, and Y. IITAKA: Synthesis of ($\pm$)-3,3-Ethylenedioxy-14α-Hydroxy-5-Picrasene-11,16-Dione, A 14α-H-Picrasane Derivative. Chemistry Lett. 299 (1981).

32. ISHIBASHI, M., T. MURAE, H. HIROTA, H. NAORA, T. TSUYUKI, T. TAKAHASHI, A. ITAI, and Y. IITAKA: Shinjudilactone, a New Bitter Principle from *Ailanthus Altissima* Swingle. Chemistry Lett. 1597 (1981).

33. ISHIBASHI, M., T. MURAE, H. HIROTA, T. TSUYUKI, T. TAKAHASHI, A. ITAI, and Y. IITAKA: Shinjulactone C, a New Quassinoid with a 1α,12α:5α,13α-Dicyclo-9β H-Picrasane Skeleton from *Ailanthus Altissima* Swingle. Tetrahedron Lett. 1205 (1982).

34. ISHIBASHI, M., T. TSUYUKI, T. MURAE, and T. TAKAHASHI: Conversion of Ailanthone into Shinjudilactone, a Backbone-Rearranged Picrasane. Chem. Pharm. Bull. **30** (5), 1917 (1982).

35. ISHIBASHI, M., T. TSUYUKI, and T. TAKAHASHI: Conversion of Ailanthone into Shinjulactone C through an Ionic (4+2) Cycloaddition Reaction. Tetrahedron Lett. 4843 (1983).

36. ISHIBASHI, M., T. TSUYUKI, T. MURAE, H. HIROTA, T. TAKAHASHI, A. ITAI, and Y. IITAKA: Constituents of the Root Bark of *Ailanthus altissima* Swingle. Isolation and X-Ray Crystal Structures of Shinjudilactone and Shinjulactone C and Conversion of Ailanthone into Shinjudilactone. Bull. Chem. Soc. Jpn. **56**, 3683 (1983).

37. ISHIBASHI, M., S. YOSHIMURA, T. TSUYUKI, T. TAKAHASHI, A. ITAI, Y. IITAKA, and K. MATSUSHITA: Shinjulactone F, A New Bitter Principle with a 5βH-Picrasane Skeleton from *Ailanthus Altissima* Swingle. Chemistry Lett. 555 (1984).

38. KESHAVAMURTHY, K. S., Y. D. VANKAR, and D. N. DHAR: Preparation of Acid

Anhydrides, Amides and Esters Using Chlorosulfonyl Isocyanate as a Dehydrating Agent. Synthesis 506 (1982).

39. KHAN, S. A., B. I. FOZDAR, and K. M. SHAMSUDDIN: Quassinoids from *Ailanthus malabarica*. Indian J. Chem. Sect. B **21 B** (12), 1133 (1982).

40. KHAN, S. A., S. S. ZUBERI, and K. M. SHAMSUDDIN: Isolation and Structure of Excelsin, a New Quassinoid from *Ailanthus excelsa*. Indian J. Chem. **19 B**, 183 (1980).

41. KHÔI, N., and J. POLONSKY: Structural Modifications of Chaparrinone Using Benezeneseleninic Anhydride. Helv. Chim. Acta **64**, 1540 (1981).

42. KOCH, H. J., H. PFENNINGER, and W. GRAF: Quassinoid Bitterstoffe I 1-Oxo-2-methoxy-4α-methyl-17β-hydroxy-Δ²-5α-androsten als Modell für den Ring A des Quassins. Helv. Chim. Acta **58**, 1727 (1975).

43. KRAUS, G. A., and M. J. TASCHNER: Model Studies for the Synthesis of Quassinoids. 1. Construction of the BCE Ring System. J. Org. Chem. **45**, 1175 (1980).

44. KRAUS, G. A., M. TACHNER, and M. SHIMAGAKI: An Approach to the BCDE Ring of Quasimarin. J. Org. Chem. **47**, 4271 (1982).

45. KUPCHAN, S. M., R. W. BRITTON, J. A. LACADIE, M. F. ZIEGLER, and C. W. SIGEL: The Isolation and Structural Elucidation of Bruceantin and Bruceantinol, New Potent Antileukemic Quassinoids from *Brucea antidysenterica*. J. Org. Chem. **40**, 648 (1975).

46. KUPCHAN, S. M., and J. A. LACADIE: Dehydroailanthinone, a New Antileukemic Quassinoid from *Pierreodendron kerstingii*. J. Org. Chem. **40**, 654 (1975).

47. KUPCHAN, S. M., J. A. LACADIE, G. A. HOWIE, and B. R. SICKLES: Structural Requirements for Biological Activity among Antileukemic Glaucarubolone Ester Quassinoids. J. Med. Chem. **19**, 1130 (1976).

48. KUPCHAN, S. M., and D. R. STREELMAN: Quassimarin, a New Antileukemic Quassinoid from *Quassia amara*. J. Org. Chem. **41**, 5481 (1976).

49. LEE, K. H.: Bruceolides. U.S. Patent **4**, 350, 638 (1982).

50. LEE, K. H., Y. IMAKURA, Y. SUMIDA, R. Y. WU, J. H. HALL, and H. CH. HUANG: Antitumor Agents. 33. Isolation and Structural Elucidation of Bruceoside-A and -B, Novel Antileukemic Quassinoid Glycosides, and Brucein-D and -E from *Brucea javanica*. J. Org. Chem. **44**, 2180 (1979).

51. LEE, K. H., M. OKANO, I. H. HALL, D. A. BRENT, and B. SOLTMANN: Antitumor Agents XLV: Bisbrusatolyl and Brusatolyl Esters and Related Compounds as Novel Potent Antileukemic Agents. J. Pharm. Sci. **71**, 338 (1982).

52. LESKINEN, V., J. POLONSKY, and S. BHATNAGAR: Antifeedant Activity of Quassinoids. J. Chem. Ecol. **10**, 1497 (1984).

53. LE-VAN-THOI, and NGUYÊN-NGOC-SUONG: Constituents of *Eurycoma longifolia* Jack. J. Org. Chem. **35**, 1104 (1970).

54. LIAO, L. L., S. M. KUPCHAN, and S. B. HORWITZ: Mode of Action of the Antitumor Compound Bruceantin, an Inhibitor of Protein Synthesis. Mol. Pharmacol. **12**, 167 (1976).

55. LIESMANN, J., R. J. BELT, C. D. HAAS, and B. HOOGSTRATEN: Phase I Study on Bruceantin Administered on a Weekly Schedule. Cancer Treat. Rep. **65**, 883 (1981).

56. LUMONADIO, L., and M. VANHAELEN: Canthin-6-one, Undulatone and two New Quassinoids from *Hannoa klaineana* Roots. Phytochemistry **23**, 2121 (1984).

57. MANDEL, L., D. E. LEE, and L. F. COURTNEY: Toward the Total Synthesis of Quassin. J. Org. Chem. **47**, 610 (1982).

58. MAI VAN TRI, J. POLONSKY, C. MERIENNE, and T. SEVENET: Soularubinone, a New Antileukemic Quassinoid from *Soulamea tomentosa*. J. Nat. Products **44**, 279 (1981).

59. MONSEUR, X., and J. C. MOTTE: Quantitative high performance liquid chromatographic analysis of the bitter quassinoid compounds from *Simaruba glauca* seeds. J. Chromat. **264**, 469 (1983).

60. MORETTI, C., J. POLONSKY, M. VUILHORGNE, and T. PRANGÉ: Isolation and Structure of Sergeolide, a Potent Cytotoxic Quassinoid from *Picrolemma pseudocoffea*. Tetrahedron Lett. **23**, 647 (1982).

61. MORON, J., and J. POLONSKY: Sur l'origine triterpénique des constituants amers des Simarubacées. Tetrahedron Lett. 385 (1968).

62. MURAE, T., A. SUGIE, T. TSUYUKI, S. MASUDA, and T. TAKAHASHI: Bitter Principles of *Picrasma ailanthoides* Planchon Nigakilactones J, K, L, M, and N. Tetrahedron **29**, 1515 (1973).

63. MURAE, T., and T. TAKAHASHI: Conversion of Quassin into 15β-[(E)-3,4-Dimethyl-2-pentenoyloxy]quassin. A D-Ring Analog of Bruceantin. Bull. Chem. Soc. Jpn. **54**, 941 (1981).

64. NAORA, H., T. FURUNO, M. ISHIBASHI, T. TSUYUKI, T. TAKAHASHI, A. ITAI, Y. IITAKA, and J. POLONSKY: On the Structure of Ailanthone, a Bitter Principle from *Ailanthus Altissima*. Chemistry Letters 661 (1982).

65. NAORA, H., M. ISHIBASHI, T. FURUNO, T. TSUYUKI, T. MURAE, H. HIROTA, T. TAKAHASHI, A. ITAI, and Y. IITAKA: Structure Determination of Bitter principles in *Ailanthus altissima*. Structure of Shinjulactone A and Revised Structure of Ailanthone. Bull. Chem. Soc. Jpn. **56**, 3694 (1983).

66. NESTLER, T., G. TITTEL, and H. WAGNER: Quantitative Bestimmung der Bitter Quassinoide von *Quassia amara* und *Picrasma excelsa*. Planta Medica **38**, 204 (1980).

67. NGUYÊN-NGOC-SUONG, S. BHATNAGAR, J. POLONSKY, M. VUILHORGNE, T. PRANGÉ, and C. PASCARD: Structure of Laurycolactone A and B, New C_{18}-Quassinoids from *Eurycoma Longifolia* and revised Structure of Eurycomalactone (X-Ray analysis). Tetrahedron Lett. 5159 (1982).

68. NICOLAOU, K. C., and Z. LYSENKO: Phenylselenolactonization. An Extremely Mild and Synthetically Useful Cyclization Process. J. Am. Chem. Soc. **99**, 3185 (1977).

69. ODJO, A, J. PIART, J. POLONSKY, and M. ROTH: Etude de l'effet insecticide de deux quassinoïdes sur des larves de *Locusta migratoria migratorioides* R et F *(Orthoptera, Acrididae)*. C.R. Acad. Sci. Paris **293**, Série III, 241 (1981).

70. OEI-KOCH, A., and LJ. KRAUS: Inhaltsstoffe von Eurycoma Longifolia JACK III. Bitterstoff (Eurycomalacton). Sci. Pharm. **48**, 110 (1980).

71. OKANO, M., T. FUJITA, N. FUKAMIYA, and T. ARATANI: New Quassinoid Glycosides and Hemiacetals from *Picrasma Ailanthoides* Planchon. Picrasinoside-B, -C, -D, -E, -F, and -G, and Picrasinol-A and -B. Chemistry Lett. 221 (1984).

72. OKANO, M., N. FUKAMIYA, K. KONDO, T. FUJITA, and T. ARATANI: Picrasinoside-A, A Novel Quassinoid Glucoside from *Picrasma Ailanthoides* Planchon. Chemistry Lett. 1425 (1982).

73. OKANO, M., and K. H. LEE: Antitumor Agents. 43. Conversion of Bruceoside-A into Bruceantin. J. Org. Chem. **46**, 1138 (1981).

74. OKANO, M., K. H. LEE, I. H. HALL, and F. E. BOETTNER: Antitumor Agents. 39. Bruceantinoside-A and -B, Novel Antileukemic Quassinoid Glucosides from *Brucea Antidysenterica*. J. Nat. Prod. **44**, 470 (1981).

75. PASCARD, C., T. PRANGÉ, and J. POLONSKY: Crystal and Molecular Structure of the Quassinoid 6-Hydroxypicrasin B. J. Chem. Res. (S) 324 (1977).

76. PFENNINGER, J., and W. GRAF: Quassinoide Bitterstoffe II Partialsynthetischer Zugang zu Quassin: Überführung von Testosteron in eine Schlüsselverbindung mit angularer 8β-Methylgruppe. Helv. Chim. Acta **63**, 1562 (1980).

77. PHILLIPSON, J. D., and F. A. DARWISH: TLX-5 Lymphoma Cells in Rapid Screening for Cytotoxicity in Brucea Extracts. Planta Med. **35**, 308 (1979).

78. PHILLIPSON, J. D., and F. A. DARWISH: Bruceolides from Fijian Brucea Javanica. Planta Med. **41**, 209 (1981).

79. PIERRÉ, A., M. ROBERT-GERO, C. TEMPETE, and J. POLONSKY: Structural Requirements of Quassinoids for the Inhibition of Cell Transformation. Biochem. Biophys. Res. Comm. **93,** 675 (1980).

80. POLONSKY, J.: Quassinoid Bitter Principles. In: Fortschr. Chem. Organ. Naturstoffe Vol. **30** (L. Zechmeister, ed.), p. 101. Wien-New York: Springer. 1973.

81. — Chemistry and Biological Activity of the Quassinoids. In: The Chemistry and Chemical Taxonomy of the Rutales (P. G. WATERMAN and M. F. GRUNDON, ed.), p. 247. New York: Academic Press. 1983.

82. POLONSKY, J., Z. BASKEVITCH, A. GAUDEMER, and B. C. DAS: Constituants amers de *Brucea amarissima:* structures des brucéines A, B, et C. Experientia **23,** 427 (1967).

83. POLONSKY, J., Z. BASKEVITCH, H. E. GOTTLIEB, E. W. HAGAMAN, and E. WENKERT: Carbon-13 Nuclear Magnetic Resonance Spectral Analysis of Quassinoid Bitter Principles. J. Org. Chem. **40,** 2499 (1975).

84. POLONSKY, J., Z. BASKEVITCH, Z. VARON, and T. SEVENET: Constituants amers de *Soulamea tomentosa* (Simaroubaceae). Structure d'un nouveau quassinoide, l'iso-brucéine A. Experientia **31,** 1113 (1975).

85. POLONSKY, J., S. BHATNAGAR, and C. MORETTI: 15-Deacetylsergeolide, a Potent Antileukemic Quassinoid. J. Nat. Prod. **47,** 994 (1984).

86. POLONSKY, J., S. BHATNAGAR, and C. MORETTI: Deacetylsimarolide, a new C_{25} quassinoid from *Simaba morettii.* Unpublished results.

87. POLONSKY, J., J. GALLAS, J. VARENNE, T. PRANGÉ, C. PASCARD, H. JACQUEMIN, and C. MORETTI: Isolation and Structure (X-Ray Analysis) of Karinolide, A New Quassinoid from *Simaba multiflora.* Tetrahedron Lett. 869 (1982).

88. POLONSKY, J., MAI VAN TRI, T. PRANGÉ, and C. PASCARD: Isolation and Structure (X-Ray Analysis) of a New C_{25} Quassinoid Soulameolide from *Soulamea tomentosa.* J. C. S. Chem. Comm. 641 (1979).

89. POLONSKY, J., MAI VAN TRI, Z. VARON, T. PRANGÉ, C. PASCARD, T. SEVENET, and J. PUSSET: Quassinoids. Isolation from *Soulamea muelleri* and Structures of 1,12-Di-O-Acetyl Soulameanone and Δ^2-Picrasin B, X-Ray Analysis of Soulameanone. Tetrahedron **36,** 2983 (1980).

90. POLONSKY, J., J. VARENNE, T. PRANGÉ, and C. PASCARD: Antileukaemic Quassinoids: Structure (X-Ray Analysis) of Bruceine C and Revised Structure of Bruceantinol. Tetrahedron Lett. **21,** 1853 (1980).

91. POLONSKY, J., Z. VARON, H. JACQUEMIN, and G. R. PETTIT: The Isolation and Structure of 13,18-Dehydroglaucarubinone, a new antineoplastic quassinoid from Simarouba amara. Experientia **34,** 1122 (1978).

92. POLONSKY, J., Z. VARON, C. MORETTI, G. R. PETTIT, C. L. HERALD, J. RIDEOUT, S. B. SAHA, and H. N. KHASTGIR: The Antineoplastic Quassinoids of *Simaba Cuspidata* Spruce and *Ailanthus Grandis* Prain. J. Nat. Prod. **43,** 503 (1980).

93. POLONSKY, J., Z. VARON, T. PRANGÉ, C. PASCARD, and C. MORETTI: Structures of Simarinolide and Guanepolide (X-Ray Analysis). New Quassinoids from *Simaba* cf *Orinocensis.* Tetrahedron Lett. **22,** 3605 (1981).

94. POLONSKY, J., Z. VARON, and E. SOLER: Constituants amers de *Castela tweedii* (Simaroubaceae). Structure d'un nouveau quassinoïde, la Castelanone. Compt. Rend. Acad. Sc., Série C **288,** 269 (1979).

95. ROBINS, R. J., M. R. A. MORGAN, M. J. C. RHODES, and J. M. FURZE: An Enzyme-Linked Immunosorbent Assay for Quassin and Closely Related Metabolites. Anal. Biochem. **136,** 145 (1984).

96. ROBINS, R. J., and M. J. C. RHODES: High-performance liquid chromatographic methods for the analysis and purification of quassinoids from *Quassia amara* L. J. Chromat. **283,** 436 (1984).

97. ROBINS, R. J., M. R. A. MORGAN, M. J. C. RHODES, and J. M. FURZE: Determination of

Quassin in Picogram Quantities by an Enzyme-Linked Immunosorbent Assay. Phytochemistry **23**, 1119 (1984).

98. SEIDA, A., A. D. KINGHORN, G. A. CORDELL, and N. FARNSWORTH: Potential Anticancer agents. IX. Isolation of a New Simaroubolide, 6α-Tigloyloxychaparrinone from *Ailanthus integrifolia* ssp. calcyna (Simaroubaceae). J. Nat. Prod. **41**, 584 (1978).

99. SHISHIDO, K., T. SAITOH, and R. FUKUMOTO: A New Synthetic Approach to Bruceantin *via* an Intramolecular Diels-Alder Reaction: Stereoselective Construction of the Pentacyclic Model System. J. Chem. Soc. Perkin Trans I, 2139 (1984).

100. STOJANAC, N., A. SOOD, Ž. STOJANAC, and Z. VALENTA: A Synthetic Approach to Quassin. Introduction of Functionality and Stereochemistry by a Diels-Alder Reaction. Can. J. Chem. **53**, 619 (1975).

101. STOJANAC, N., Ž. STOJANAC, P. S. WHITE, and Z. VALENTA: A Synthetic Approach to Quassin. Synthesis of a Ring A *seco* Derivative. Can. J. Chem. **57**, 3346 (1979).

102. SULING, W. J., C. W. WOOLLEY and W. M. SHANNON: Disposition and Metabolism of Bruceantin in the Mouse. Cancer Chemother. Pharmacol. **3**, 171 (1979).

103. TRAGER, W., and J. POLONSKY: Antimalarial Activity of Quassinoids against Chloroquine-Resistant *Plasmodium falciparum in vitro*. Am. J. Trop. Med. Hyg. **30** (3), 531 (1981).

104. VEDEJS, E., D. A. ENGLER, and J. E. TELSCHOW: Transition-Metal Peroxide Reactions. Synthesis of α-Hydroxycarbonyl Compounds from Enolates. J. Org. Chem. **43**, 188 (1978).

105. VEYSOGLU, T., and L. A. MITCHER: A Class of New Silylating Reagents. I. A Mild Method for Introduction of the Tert-Butyldimethylsilyl Group. Tetrahedron Lett. **22**, 1299 (1981).

106. VOYLE, M., N. K. DUNLAP, D. S. WATT, and O. P. ANDERSON: Quassinoids. 2. Synthesis of an ABCD Ring Precursor Involving a Lactonization Induced by Iodotrimethylsilane. J. Org. Chem. **48**, 3242 (1983).

107. VOYLE, M., K. S. KYLER, S. ARSENIYADIS, N. D. DUNLAP, and D. S. WATT: Quassinoids. 1. Attempted Intramolecular Diels-Alder Approach for Assembling the ABCD Rings. J. Org. Chem. **48**, 470 (1983).

108. WALL, M. E., and M. C. WANI: Plant Antitumor Agents. 17. Structural Requirements for Antineoplastic Activity in Quassinoids. J. Med. Chem. **21**, 1186 (1978).

109. WANI, M. C., H. L. TAYLOR, J. B. THOMPSON, M. E. WALL, A. T. MCPHAIL, and K. D. ONAN: Plant Antitumour Agents. XV. Isolation and X-Ray Crystal Structure of a New Antileukaemic Quassinoid Undulatone from *Hannoa Undulata*. Tetrahedron **35**, 17 (1979).

110. WANI, M. C., H. L. TAYLOR, J. B. THOMPSON, and M. E. WALL: Plant Antitumor Agents. XVI. 6α-Senecioyloxychaparrinone, a New Antileukemic Quassinoid from *Simaba multiflora*. J. Nat. Prod. **41**, 578 (1978).

111. WANI, M. C., H. L. TAYLOR, M. E. WALL, A. T. MCPHAIL, and K. D. ONAN: Plant Antitumour Agents: Isolation and Structure of Samaderine A (X-Ray Analysis) and a New Antileukaemic Quassinoid Samaderine E from *Samadera indica*. J.C.S. Chem. Comm. 295 (1977).

112. WATERMAN, P. G., S. A. AMPOHO: Cytotoxic Quassinoids from *Odyendyea gabonensis* Stem Bark; Isolation and High-Field NMR. Planta Medica **50**, 261 (1984).

(Received October 29, 1984)

Author Index

Page numbers printed in *italics* refer to References

Subject Index

By

A. SIEGEL, Wien

Satz: Austro-Filmsatz Richard Gerin, A-1020 Wien

Fortschritte der Chemie organischer Naturstoffe

Progress in the Chemistry of Organic Natural Products

Volume 46:

1984. 7 figures. IX, 253 pages.
Cloth DM 178,–. ISBN 3-211-81804-9

Contents: O. TANAKA and R. KASAI: Saponins of Ginseng and Related Plants. – E. FUJITA, M. NODE: Diterpenoids of *Rabdosia* Species. – S. JOHNE: The Quinazoline Alkaloids.

Volume 45:

1984. 2 figures. VIII, 288 pages.
Cloth DM 194,–. ISBN 3-211-81755-7

Contents: D. A. H. TAYLOR: The Chemistry of the Limonoids from Meliaceae. – J. A. ELIX, A. A. WHITTON, and M. V. SARGENT: Recent Progress in the Chemistry of Lichen Substances. – Y. SHIMIZU: Paralytic Shellfish Poisons.

Volume 44:

1983. 72 partly coloured figures. IX, 326 pages.
Cloth DM 208,–. ISBN 3-211-81754-9

Contents: F. J. EVANS and S. E. TAYLOR: Pro-Inflammatory, Tumour-Promoting and Anti-Tumour Diterpenes of the Plant Families Euphorbiaceae and Thymelaeaceae. – A. MONDON and B. EPE: Bitter Principles of Cneoraceae. – S. NAYLOR, F. J. HANKE, L. V. MANES, and P. CREWS: Chemical and Biological Aspects of Marine Monoterpenes. – J. G. BUCHANAN: The C-Nucleoside Antibiotics.

Volume 43:

1983. VIII, 383 pages.
Cloth DM 208,–. ISBN 3-211-81741-7

Contents: J. L. INGHAM: Naturally Occurring Isoflavonoids (1855–1981). – A. KOSKINEN and M. LOUNASMAA: The Sarpagine-Ajmaline Group of Indole Alkaloids.

Volume 42:

1982. VII, 323 pages.
Cloth DM 164,–. ISBN 3-211-81706-9

Contents: Y. ASAKAWA: Chemical Constituents of the Hepaticae. – M. HEIDELBERGER: Cross-Reactions of Plant Polysaccharides in Antipneumococcal and Other Antisera, an Update.

Volume 41:

1982. 37 figures. VIII, 373 pages.
Cloth DM 196,–. ISBN 3-211-81690-9

Contents: E. HASLAM: The Metabolism of Gallic Acid and Hexahydroxydiphenic Acid in Higher Plants. – D. G. ROUX and D. FERREIRA: The Direct Biomimetic Synthesis, Structure and Absolute Configuration of Angular and Linear Condensed Tannins. – ST. J. GOULD and ST. M. WEINREB: Streptonigrin. – D. J. ROBINS: The Pyrrolizidine Alkaloids. – J. W. DALY: Alkaloids of Neotropical Poison Frogs (Dendrobatidae).

All Volumes and Cumulative Index 1–20 available

Price reduction for subscribers: 10%

Special reduced price (20% reduction) for the complete Series Vols. 1–47 incl. the Cumulative Index to Vols. 1–20

Springer-Verlag Wien New York